LIPID BIOCHEMISTRY

of Fungi and Other Organisms

LIPID BIOCHEMISTRY

of Fungi and Other Organisms

John D. Weete

Department of Botany, Plant Pathology, and Microbiology
Auburn University, Auburn, Alabama

With Contributions by

Darrell J. Weber
Department of Botany and Range Science
Brigham Young University, Provo, Utah

PLENUM PRESS · NEW YORK AND LONDON

Library of Congress Cataloging in Publication Data

Weete, John D 1942-
   Lipid biochemistry of fungi and other organisms.

   Edition for 1974 published under title: Fungal lipid biochemistry.
   Includes bibliographical references and index.
   1. Fungi—Physiology.   2. Lipids.   I. Weber, Darrell J., 1933-   joint author.
II. Title.
QK601.W44 1980   589.2'0419246   80-22158
ISBN 0-306-40570-9

© 1980 Plenum Press, New York
A Division of Plenum Publishing Corporation
227 West 17th Street, New York, N.Y. 10011

Printed in the United States of America

To Anna Blaire

PREFACE

Only six years have passed since the precursor to this
book, "Fungal Lipid Biochemistry," was published. It seemed
to satisfy the need of a central comprehensive reference to
which students and researchers could turn for information
on the lipid composition and metabolism in fungi. This book
was concerned with the distribution and biochemistry of
lipids in fungi, and in many instances lipid metabolism was
presented in a comparative context. The principal lipids
covered were the aliphatic hydrocarbons, fatty acids, sterols,
acylglycerols, phospholipids, and sphingolipids. The final
two chapters of the book, contributed by Drs. William Hess
and Darrell J. Weber, summarized fungal metabolism and
ultrastructure during fungal spore germination and sporulation. The information in that book has been completely
re-written, re-organized, expanded, updated, and is now presented under the altered title of LIPID BIOCHEMISTRY of Fungi
and Other Organisms. Some of the noteworthy additions include (1) an expanded presentation of lipid classification,
(2) brief description of the historical development of research on fungal lipids, (3) expanded presentation of lipid
production during vegetative growth, and in relation to
nutrient utilization, (4) the relatively new interpretation
of the yeast fatty acid synthetase as a multifunctional
enzyme rather than multienzyme complex, (5) the chemistry,
distribution, and biosynthesis of polyprenols and carotenoids, and (6) condensation of the information on spore
germination and sporulation into one chapter with greater
emphasis on the involvement and role of lipids in these
processes. While all possible publications on fungal lipids
are not cited, I feel that the topics covered are thoroughly
reviewed. Information from some published work previously
overlooked along with work published since the first book
appeared has been incorporated into this volume. My objec-
tive is to present the current state of our knowledge in the
topics of lipid biochemistry covered with emphasis on fungi.
In many cases, comparisons of bacterial, plant and animal

systems are made. Any mistakes that may occur in this book
are the responsibility of this author.

I would like to express my appreciation to Mrs. Sharon
Harper for typing this book, and to Ms. Terry Rodriguez for
the art work. Finally, I would like to express my gratitude
to Dr. Darrell J. Weber of the Department of Botany and
Range Science, Brigham Young University for his contribution
of the final chapter of this book.

JOHN D. WEETE
Auburn

CONTENTS

CHAPTER 1   INTRODUCTION TO LIPIDS

DEFINITION AND CLASSIFICATION

Historically, biological materials have been classified
as fats, proteins, carbohydrates, and minerals. One of the
main characteristics that distinguishes fats from the other
natural products is their solubility in non-polar solvents
such as ether, chloroform, benzene, etc., and their insolu-
bility in water. These organic solvent-soluble substances
have been grouped according to their physical state at room
temperature; hard solids, soft solids, and liquids have been
called waxes, fats, and oils, respectively. The heterogeneity of 'fat extracts' has become evident with the development of more efficient extraction procedures, and particularly with the development and application of chromatographic
techniques to the separation of the extract components. 'Fat
extracts' may be composed of a complex mixture of substances
with widely different chemical structures, but with similar
solubilities in organic solvents. The term presently used
to include this diverse group of substances is *Lipid*. Lipids
are still most often defined according to their solubility
properties. For example, according to Davenport and Johnson
(1971), lipids are substances that have as part of their
structure a substituted or unsubstituted aliphatic hydrocarbon chain which confers hydrophobic properties to at least
part of the molecule. According to Burton (1974), lipids are
broadly defined as materials which are soluble in organic
solvents and essentially insoluble in water. Kate's (1972)
definition generally agrees with those given above. It
should be pointed out that certain lipids possess some but
limited solubility in water depending on polar groups linked

1

to the hydrophobic portion of the molecule. Some of these
lipids include lysophosphatides (see Chapter 6) and certain
glycolipids such as gangliosides (see Chapter 7). Most
lipids contain carbon, hydrogen, and oxygen but some may
contain only the first two elements. Lipids may also contain phosphorus, nitrogen, and/or sulfur.

Most systems of lipid classification are based on chemical structure. However, because of the physicochemical
heterogeneity of substances defined as lipids there is no
entirely unambiguous system of classification of these substances. Stoffel (1974) suggests that a simple classification system that separates acyl lipids from those of isoprenoid origin may be useful (Table 1.1). According to this
system, acyl lipids are designated as simple or complex, the
former including glycerides, cholesterol esters, and waxes.
Complex lipids include glycerophospholipids and sphingolipids.
The term 'simple lipid' is often used to refer to non-acyl
lipids classified according to the functional group present,
i.e. hydrocarbons, aliphatic alcohols, acids, ketones and
aldehydes. Burton (1974) separates more complex lipids into
three groups, the glycerol-containing lipids, sphingosine-
containing lipids, and steroids (Table 1.1). A third system
of classification separates lipids on the basis of polarity,
neutral and amphiphilic lipids (Table 1.1) (Davenport and
Johnson, 1971). Classes of neutral lipids include hydrocarbons, glycerides, fatty acids, waxes, estolides, isoprenoids, and other esters. Amphiphilic lipids include
glycerolipids (phosphoglycerides and glycosylglycerides) and
sphingolipids.

The most comprehensive and detailed lipid classification system has been developed by Kates (1972). Lipids are
classified on the basis of the presence of functional groups,
types of linkages between molecules, and the presence of non-
lipid substances and elements other than carbon, hydrogen,
and oxygen (Table 1.1).

There is no attempt in this text to add another system
of lipid classification to those already presented. Five
major classes of non-isoprenoid lipids (fatty acids, acylglycerols, phosphoacylglycerols, sphingolipids, and hydrocarbons) and three classes of isoprenoids (sterols, carotenoids, and polyprenols) are included in this book.

TABLE 1.1   Systems of Lipid Classification

Stoffel (1974)	Burton (1974)
a) Fatty acid as a component	A. Hydrocarbons
1) Simple lipids	B. Aliphatic alcohols
Glycerides	C. Aliphatic acids
Cholesterol esters	D. Wax
Waxes	E. Glycerol-containing lipids
2) Complex lipids	F. Sphingosine-containing
Glycerophospholipids	lipids
Sphingolipids	G. Steroids
b) Isoprene derived	H. Miscellaneous lipids
Steroids	
Carotenoids	
Vitamins A, D, E, K	
Ubiquinones	

Davenport and Johnson (1971)[a]

I.   Neutral Lipids
   A.   Hydrocarbons--long chain, branched and normal,
       saturated and unsaturated aliphatic hydrocarbons.
   B.   Glycerides--compounds containing ester, vinyl
       ether, or saturated ether linkages with the
       hydroxyl functions of glycerol.
   C.   Fatty Acids--long-chain monocarboxylic acids.
   D.   Waxes--esters of long-chain fatty acids and alco-
       hols.
   E.   Estolides--intermolecular lactones of hydroxy fatty
       acids.
   F.   Isoprenoids
      1.   Carotenoids--polyisoprenoid hydrocarbons, alco-
         hols, epoxides, and carboxylic acids containing
         40 carbon atoms.
      2.   Terpenoids--polyisoprenoid compounds varying in
         carbon number, including vitamin A, squalene,
         etc.
      3.   Steroids--alicyclic compounds having the cyclo-
         pentanoperhydrophenanthrene carbon skeleton
         (includes the sterols).
   G.   Other esters (excluding those containing phosphate
       or sphingosine bases)--these include naturally occur-
       ring esters of long-chain fatty acids and short-chain

TABLE 1.1   Continued

---

        alcohols, such as methyl palmitate, and sterol
        esters, such as cholesteryl palmitate.
II. Amphiphilic Lipids
    A. Glycerolipids
        1. Phosphoglycerides--derivatives of *sn*-glycero-
            3-phosphoric acid that contains at least one
            0-acyl, 0-alkyl, or 0-alk-1-enyl group.
        2. Glycosylglycerides--glycosides of diacylglyc-
            erol.
    B. Sphingolipids
        1. Phosphosphingolipids--phosphate esters of N-acyl
            (ceramides) sphinganines.
        2. Glycosphingolipids--glycosides of ceramides.
            a. Ceramide monoglycosides
                (1) Cerebrosides--glucosides or galacto-
                    sides.
                (2) Sulfatides--contain sulfate ester of
                    galactose.
            b. Ceramide of oligoglycosides--contain poly-
                saccharide residues.
        3. Glycophosphosphingolipids--ceramides containing
            both sugars (polysaccharides) and phosphate
            esters, phytoglycolipids in plants and myco-
            glycolipids in fungi.

Kates (1972)

1.1 Hydrocarbons
    1.1.1 Normal, saturated paraffins
    1.1.2 Monobranched-chain paraffins
    1.1.3 Multibranched paraffins: saturated isoprenoids
    1.1.4 Normal, mono-unsaturated paraffins
    1.1.5 Monobranched, unsaturated paraffins
    1.1.6 Isoprenoid polyenes
    1.1.7 Carotenoids
1.2 Alcohols
    1.2.1 Normal, saturated alcohols
    1.2.2 Monobranched-chain alcohols
    1.2.3 Multibranched alcohols: saturated isoprenoids
    1.2.4 Unsaturated alcohols
    1.2.5 Isoprenoid alcohols (terpenols or polyprenols)
    1.2.6 Sterols

TABLE 1.1   Continued

TABLE 1.1   Continued

      1.11.2 Sphingosyl phosphatides
1.12 Glycolipids
      1.12.1 Glycosyl diglycerides
      1.12.2 Glycosides of hydroxy fatty acids
      1.12.3 Fatty acid esters of sugars
      1.12.4 Sugar-phosphatide derivatives
      1.12.5 Phytoglycolipid (plants)
      1.12.6 Lipopolysaccharides (bacteria)
      1.12.7 Steryl glycosides
      1.12.8 Cerebrosides (ceramide hexosides)
      1.12.9 Gangliosides
1.13 Sulfur-containing lipids
      1.13.1 Long-chain alcohol sulfates (alkyl sulfates)
      1.13.2 Cerebroside sulfates (sulfatides)
      1.13.3 Glycolipid sulfates
      1.13.4 Sulfoglycosyl diglyceride
1.14 Amino acid-containing lipids
      1.14.1 N-acyl amino acids and ester derivatives
      1.14.2 O-acyl carnitines
      1.14.3 Peptidolipids

---

[a]Several lipid classes not discussed in this text are also
omitted from this classification. Davenport and Johnson
(1971) should be consulted for these lipids (cytosides,
globosides, gangliosides, lipoamino acids, lipopeptides,
and redox lipids).

                    NOMENCLATURE OF LIPIDS

     A wide variety of names for various types of lipids is
found in the literature. Various systematic and trivial
names are in common use. The nomenclature of each class of
lipid discussed in this book is described at the beginning
of the chapter introducing the particular lipid. Generally,
recommendations of the IUPAC-IUB Commission on Biochemical
Nomenclature on the naming of non-isoprenoid lipids are
followed in this book [Lipids 12:455 (1977)]. However, the
repeated use of systematic names becomes cumbersome, particularly for complex lipids, and trivial names or symbols are
often used after the lipid is first introduced by its systematic name.

HISTORICAL ASPECTS OF FUNGAL LIPID RESEARCH

Some of the earliest interest in fungal lipids occurred in the early 1870's when it was reported that the ergot fungus *Claviceps purpurea* contained 30% fat. During the following 50 years most work of this nature was devoted to screening fungi for potential high fat producers. During this time it was recognized that fungi vary considerably in their capacity to produce fat, but perhaps more importantly, it was recognized that the degree of fat production varies according to the culture conditions and media composition. Research in the following half century emphasized defining culture conditions and nutritional factors favorable for fat production. Several species became known as "fat-producing" or oleaginous fungi. These included some species of the filamentous fungi *Aspergillus*, *Penicillium*, *Mucor*, *Fusarium*, and the yeast-like fungi *Lipomyces*, *Rhodotorula*, and *Candida* (*Torulopsis*). Some of the objectives of this type of research was to determine if fungal fat can be a suitable dietary supplement, and if it is feasible to produce fungal fat on a commercial basis. Therefore much of the research was focused on carbon substrate utilization efficiency, and the utilization of relatively inexpensive waste products as substrates for growth.

In the early 1950's, after about a century of work in this area, there was a transition from an age of fat production to a period of lipid research. The development of chromatographic techniques, particularly gas chromatography, was mainly responsible for the transition. This was also accompanied by improvements in extraction techniques. While it was recognized that triacylglycerides were the predominant fats that accumulated in the fungi studied, the complex nature of fat extracts became more apparent. Initially, most emphasis was on the fatty acid composition of fungi, but as analytical techniques became more refined individual sterols, phospholipids, and other lipids were identified.

The availability of radioisotopes suitable for tracer studies in the late 1940's, combined with the chromatographic techniques, made possible the next major thrust in lipid research. This was the elucidation of biochemical pathways and reaction mechanisms, and studies on enzyme chemistry. Of course research on the biochemistry of lipids was not restricted to fungi, but much of our fundamental knowledge in this area has been gained through work with this group of

organisms, particularly baker's yeast *Saccharomyces cerevisiae*
and *Neurospora crassa*.

CHAPTER 2   FUNGAL LIPIDS

INTRODUCTION

For many years there has been considerable concern over
the nutritional problems accompanying the rapid growth of
the world population. The need to find sources of fat and
protein to supplement agricultural production has developed
with this concern, and microbial sources of these essential
nutrients have been explored. Many microorganisms, particularly fungi, have been screened for their potential to accumulate relatively large quantities of fat. Under appropriate
conditions, there are several fungi that accumulate high
(>20%) amounts of fat (oleaginous fungi). We have gained
considerable knowledge from these studies about the environmental and nutritional conditions that influence fat production. Woodbine (1959) has comprehensively reviewed these
studies, many of which were conducted in the 1940's and 1950's.
Similar studies have been reported since, but they have added
little more to our understanding of the relationships between
environmental and nutrient conditions and fat production.
The most important recent advances have been in the development of continuous culture techniques and methods for inducing
synchronous division of yeast cells in culture. The application of these techniques in studying lipid production and
metabolism in fungi is limited. Most of the early studies
on fat production have been conducted using batch culture
techniques.

This chapter is concerned with fungal lipid content and
production. The lipid content of mycelium and yeast cells,
reproductive structures, subcellular components such as cell

9

walls and membranes are described, as well as a brief summary
of the functions of major lipids. The relationships between
environmental and nutritional factors and lipid production
are discussed. Also, lipid production by fungi grown under
batch and continuous culture conditions are compared; and
lipid turnover in synchronously dividing yeast cells is
described.

                        TOTAL LIPID CONTENT

     The lipid content of numerous fungal species has been
reported, and is highly variable depending on the growth
conditions and species. Certainly, lipid production can be
manipulated by varying culture conditions; and studies to
determine the potential of fungi as commercial sources of
fat have lead to an understanding of the relationships between fungal growth and lipid production as they relate to
environmental conditions and nutritional requirements. As
we shall see, the optimum growth conditions are not necessarily best for maximum lipid production. Other studies, on
the other hand, have been concerned simply with fungal lipid
content under suitable, but not necessarily optimum, growth
conditions rather than conditions for maximum lipid production. While some species appear to have greater potential
for lipid production than others, there is no apparent
taxonomic value of fungal lipid content. However, this cannot be fairly evaluated from the available data since it
would require comparison of lipid content under optimum growth
conditions for each species, and relatively few species have
been studied with taxonomic objectives in mind.

     Vegetative Hyphae and Yeast Cells. The total lipid con-
tent of vegetative hyphae ranges from about 1 to 56% of the
dry weight depending on the species, developmental stage of
growth, and culture conditions. Although several species
have a high capacity for lipid production, mycelium of most
species generally contains between 3 and 32% lipid when the
fungus is grown under favorable conditions for growth. Fungi
generally average about 17% lipid (Table 2.1). There may be considerable variation in lipid content between different species
of the same genus and strains (isolates) of the same species cultured under identical conditions. For example, two isolates
of *Fusarium lini* grown on the same medium may differ as much
as 100% in lipid content (Table 2.2). Variation in lipid
content can also occur in the same fungus grown on different

TABLE 2.1   Total Lipid Content of Certain Filamentous Fungi

Fungus	Lipid Content (% Dry Weight)	Reference
PHYCOMYCETES		
Absidia blakesleeana	7.2-30.4	Satina & Blakeslee, 1928
A. glauca	19.5	Satina & Blakeslee, 1928
Absidia ("whorled")	17.3-21.8	Satina & Blakeslee, 1928
A. dauci	2.3	Gunasekaran & Weber, 1972
Blakeslea trispora	15.1-22.1	Satina & Blakeslee, 1928
Choanophora curcubitarum	6.2-24.6	Satina & Blakeslee, 1928; White & Werkman, 1948
Circinella umbellata	7.4	Satina & Blakeslee, 1928
C. spinosa	12.8-21.9	Satina & Blakeslee, 1928
Conidiobolus species (12 species)	8.0-23.7	Tyrrell, 1971
Cunninghamella bertholletiae	12.7-22.0	Satina & Blakeslee, 1928
Mucor albo-ater	6.5-41.8	Woodbine et al., 1951; Murray et al., 1953
M. plumbeus	5.3-17.9	Woodbine et al., 1951; Murray et al., 1953
M. circinelloides	12.9-45.4	Woodbine et al., 1951; Murray et al., 1953
M. spinosus	5.6-46.2	Woodbine et al., 1951; Murray et al., 1953
M. mucedo	2.0-33.3	Satina & Blakeslee, 1928; Sumner & Morgan, 1969; Woodbine et al., 1951; Sumner et al., 1969
M. dispersus	10.6	Satina & Blakeslee, 1928
M. griseo-cyanus	6.5	Satina & Blakeslee, 1928
Mucor isolates	2.1-36.4	Satina & Blakeslee, 1928; Sumner et al., 1969
M. racemosus	8.2-19.0	Sumner & Morgan, 1969; Sumner et al., 1969
M. hiemalis	15.1-19.3	Sumner & Morgan, 1969; Sumner et al., 1969
M. miehei	7.8-25.1	Sumner & Morgan, 1969; Sumner et al., 1969
M. pusillus	18.3-26.2	Mumma et al., 1970; Sumner et al., 1969

TABLE 2.1   Continued

Fungus	Lipid Content (% Dry Weight)	Reference
M. ramannianus	15.2–55.5	Sumner et al., 1969; Sumner & Morgan, 1969; Woodbine et al., 1951
M. strictus	7.8–24.4	Sumner et al., 1969
M. oblongisporus	2.5–20.0	Sumner et al., 1969
M. globosus	16.7	Mumma et al., 1970
Parasitella simplex	7.2	Satina & Blakeslee, 1928
Phycomyces blakesleeanus	15.7–22.1	Satina & Blakeslee, 1928
Pythium ultimum	3.0–48.0	Borrow et al., 1961
P. irregulare	7.9–17.1	Cantrell & Dowler, 1971
P. vexans	5.9–13.3	Cantrell & Dowler, 1971
Rhizopus arrhizus	2.2–19.9	Weete et al., 1973; Shaw, 1966; Gunasekaran & Weber, 1972
R. nigricans	5.3–18.3	Woodbine et al., 1951; Satina & Blakeslee, 1928; Murray et al., 1953
R. oryzea	4.9–35.8	Woodbine et al., 1951; Murray et al., 1953
Rhizopus species	8.8–45.3	Sumner & Morgan, 1969; Sumner et al., 1969
Syncephalastrum sp.	12.9	Satina & Blakeslee, 1928
Zygorhynchus moelleri	7.7–18.4	Woodbine et al., 1951; Murray 3t al., 1953

ASCOMYCETES AND FUNGI IMPERFECTI

Aspergillus clavabus	14.2–20.7	Woodbine et al., 1951; Murray et al., 1953
A. fischeri	10.5–37.0	Prill et al., 1935
A. flavipes	39.7	Woodbine et al., 1951
A. flavus	4.0–35.5	Woodbine, et al., 1951; Uen & Ling, 1969; Murray et al., 1953; Ward et al., 1935
A. insuetus	16.6	Woodbine et al., 1951
A. minutus	13.5	Woodbine et al., 1951; Murray et al., 1953

TABLE 2.1   Continued

Fungus	Lipid Content (% Dry Weight)	Reference
A. nidulans	5.2–39.8	Singh & Walker, 1956; Murray et al., 1953; Gregory & Woodbine, 1953; Singh et al., 1955; Woodbine et al., 1951
A. niger	0.9–2.2	Woodbine et al., 1951
Chaetomium thermophile[d]	9.4	Mumma et al., 1970
C. globosum	54.1	Mumma et al., 1970
Cladosporium herbarum	0.7	Woodbine et al., 1951
Claviceps purpurea[c]	1.9–31.3	Leegwater et al., 1962; Mantle et al., 1969
Cylindrocarpon radicicola	7.5	Hartman et al., 1962
Fusarium bulbigenum	6.9–20.0	Murray et al., 1953; Woodbine et al., 1951
F. graminearum	10.3–24.4	Murray et al., 1953; Woodbine et al., 1951
F. lini	5.5–34.6	Murray et al., 1953; Woodbine et al., 1951
F. lycopersici	7.1–16.1	Murray et al., 1953
F. oxysporum	7.7–33.9	Murray et al., 1953; Woodbine et al., 1951
F. solani f. phaseoli	2.0	Gunasekaran & Weber, 1972
Helicostylum piriforme	9.1	Satina & Blakeslee, 1928
Humicola brevis	14.6	Mumma et al., 1970
H. grisea	10.8	Mumma et al., 1970
H. grisea var. thermoidea	13.0	Mumma et al., 1970
H. insolens[d]	14.2	Mumma et al., 1970
H. linuginosa	17.2	Mumma et al., 1970
H. nigrescens	8.0	Mumma et al., 1970
Malbranchea pulchella	26.5	Mumma et al., 1970
Mortierella vinacea	3.2–51.4	Chesters and Peberdy, 1965
Neurospora crassa	6.4–11.9	Bianchi & Turian, 1967
Paecilomyces variati	6.4	Murray et al., 1953
P. aurantiolbrunneum	6.1	Murray et al., 1953
Penicillium chrysogenum	1.2–9.8	Gaby et al., 1957; Mumma et al., 1970

TABLE 2.1   Continued

Fungus	Lipid Content (% Dry Weight)	Reference
P. flavo-cinereum	5.3-28.5	Woodbine et al., 1951; Murray et al., 1953; Ward et al., 1935
P. purpurogenum	1.3	Woodbine et al., 1951
P. variata	6.8	Woodbine et al., 1951
P. spinulosum	5.0-19.0	Woodbine et al., 1951; Murray et al., 1953; Gregory & Woodbine, 1953
P. oxalicum	15.7-24.4	Woodbine et al., 1951; Ward et al., 1935
P. lilacinum (NRRL898)	6.0-47.3	Osman et al., 1969
P. cyaneum	9.6	Koman et al., 1969
P. soppi	20.2-34.8	Murray et al., 1953; Ward et al., 1935
P. luteum	16.0	Murray et al., 1953
P. javanicum	3.5-19.4	Gregory & Woodbine, 1953; Murray et al., 1953; Woodbine et al., 1951; Ward et al., 1935
P. piscarum	28.0	Ward et al., 1935
P. roquefortii	22.9	Ward et al., 1935
P. hirsutum	18.4	Ward et al., 1935
P. citrinum	18.1	Ward et al., 1935
P. bialowiezense	17.0	Ward et al., 1935
P. duponti[d]	14.8	Mumma et al., 1970
Pithomyces chartarum	4.6	Hartman et al., 1962
Sclerotinia sclerotiorum	1.1-12.4	Sumner & Colotelo, 1970
Sclerotium rolfsii	2.8	Gunasekaran & Weber, 1972
Sporotrichum thermophile[d]	15.5	Mumma et al., 1970
S. exile	9.5	Mumma et al., 1970
Stemphylium dendriticum	2.4	Hartman et al., 1962
Stilbella thermophila	38.1	Mumma et al., 1970
Trichoderma viride	4.4-24.0	Ballance and Crombie, 1961; Murray et al., 1953; Jones, 1969
Trichosporan cutaneum	45.0-56.0	Ilina et al., 1970

TABLE 2.1   Continued

Fungus	Lipid Content (% Dry Weight)	Reference
Trichothecium roseum	8.1-17.0	Montant & Sancholle, 1969
BASIDIOMYCETES		
Clitocybe illudens	9.1	Bentley et al., 1964
Tilletia controversa	5.8	Trione & Ching, 1971
T. nudum	0.2-47.2	Leegwater et al., 1962
Ustilago zeae	7.0-36.6	Woodbine et al., 1951; Murray et al., 1953

[a]Ten races.
[b]Twenty-four races.
[c]Many isolates.
[d]Thermophilic.

media considered good for growth. In the examples given,
mycelial lipid content varies by a factor of 2 to 4 when the
fungi are grown on different media (Table 2.2). There are
several fungal (mycelial) genera with species capable of
producing high (>20%) quantities of fat; such as *Claviceps*,
Penicillium, *Aspergillus*, *Mucor*, *Fusarium*, and *Phycomyces*
(Woodbine, 1959). These fungi are readily grown in culture
and perhaps have potential for economic production of fat.
There are probably other mycelial fungi that produce high
percentages of fat, but culture conditions have not been
developed for producing large quantities of biomass.

    Yeasts and yeast-like fungi also vary considerably in
their ability to produce lipid (Table 2.3). Lipid of most
yeasts comprise 5 to 15% of the cell dry weight, but there
are several species known as "fat yeasts" that produce 30
to 70% lipid. Some of these include *Rhodotorula* and *Lipomyces*
species. However, *Saccharomyces cerevisiae* (ATCC 7754) may
produce up to 87% lipid (Castelli et al., 1969). Lipids of
yeasts and yeast-like fungi have been reviewed by Hunter and
Rose (1971) and Rattray et al. (1975).

TABLE 2.2   Total Lipid Content of Fungi Grown on Different
Media[a]

Fungus	Media[b,c]			
	A	B	C	D
Ustilago zeae	7.0	18.3	27.8	16.7
Fusarium lini[d]	5.5	28.4	26.4	15.0
F. lini[d]	6.8	–	–	32.2
F. oxysporum	7.7	13.7	13.9	24.4
F. graminearum	10.3	–	24.0	24.4

[a] From Hunter & Rose, 1971.
[b] See Woodbine et al., 1951 for media composition.
[c] Total lipid is expressed as percent of the mycelial dry
weight.
[d] Different isolates of *F. lini*.

TABLE 2.3   Total Lipid Content of Certain Yeasts

Fungus	Total Lipid (% Dry Weight)	Reference
Blastomyces dermatitidis	5.0	Domer & Hamilton, 1971
Candida albicans (ATCC 10231)	0.3–6.3	Combs et al., 1968
C. albicans 1 Ha 582	13.9	Jansons & Nickerson, 1970
C. lipolytica	8.5	Kates & Baxter, 1962
C. scottii	8.2–10.7	Kates & Baxter, 1962
C. utilis	0.5	Babij et al., 1969
C. stellatoidea	0.6–19.9	Guarneri, 1966
Histoplasma capsulatum	18.0	Domer & Hamilton, 1971
Lipomyces starkeyi	7.7–31.4	Cullimore & Woodbine, 1961
Pullularia pullulans	11.0	Merdinger et al., 1968
Rhodotorula glutinis	12.0	Kates & Baxter, 1962
R. gracilis	20.3–63.2	Castelli et al., 1969a; 1969b
Rhodotorula sp.	49.8	Holmberg, 1948
Saccharomyces cerevisiae (ATCC 7754)	68.5–87.1	Castelli et al., 1969a; 1969b; 1969c

TABLE 2.3   Continued

Fungus	Total Lipid (% Dry Weight)	Reference
S. cerevisiae (ATCC 7755)	7.0-10.2	Castelli et al., 1969a; 1969b; 1969c
S. cerevisiae	3.3-10.2	Jollow et al., 1968
Saccharomyces sp.	17.0	Paltauf & Schatz, 1969
"Soil yeast"	5.5-65.3	Starkey, 1946
Torulopsis utilis	3.0	Enebo et al., 1946
Torulopsis sp.	6.4	Reichert, 1945

Spores and Sclerotia. Fungal spore lipid content varies
considerably depending on the species and conditions of spore
formation, but ranges between 1 and 35% of the spore dry
weight (Table 2.4). There is considerable variation in spore
lipid content among members of the same genus. For example,
the lipid content of *Ustilago* species ranges from 4 to 22%
of the spore dry weight. Similar differences are found among
Puccinia and *Mucor* species. In the case of rusts and smuts,
spores have been collected from their natural hosts and the
lipid content may have been subject to host influence, and
conditions such as the type and amount of carbon source
available during spore development. However, the total lipid
content of *Cronartium fusiforme* aeciospores (average 20%)
from pine is similar to basidiospores (28%) of the same species from oak (Table 2.4).

Spores of thermophilic and thermotolerant *Mucor* species
have a higher lipid content (10.4 to 19.4%) than spores of
mesophilic species (3.7 to 8.4%) (Sumner and Morgan, 1969).
Spores of these fungi generally contain less lipid than the
mycelia on which they are formed. Like seeds of some higher
plants that have a high oil content, endogenous lipid is
used as a source of energy and carbon skeletons during spore
germination in some species. This seems to be restricted to
rust fungi, but has not been studied in a large number of
species. Not all spore lipids are present as globules in
the cytoplasm and available for degradation during germination. For example, *Neurospora crassa* conidia contain 19%
lipid, 95% of which is phospholipid and carotenoids (Bianchi
and Turian, 1967). Little change in lipid content occurs

TABLE 2.4   Total Lipid Content of Spores from Certain Fungi

Fungus	Total Lipid (% Dry Weight)	Reference
PHYCOMYCETES		
Rhizopus arrhizus	2.7	Weete et al., 1973
Rhizopus sp	10.4-16.1	Sumner & Morgan, 1969
Mucor mucedo	3.7	Sumner & Morgan, 1969
M. ramannianus	7.6	Sumner & Morgan, 1969
M. racemosus	4.1	Sumner & Morgan, 1969
M. hiemalis	8.4	Sumner & Morgan, 1969
M. miehei	11.3-19.4	Sumner & Morgan, 1969
M. pusillus	16.1-19.3	Sumner & Morgan, 1969
M. rousii	9.8	Bartnicki-Garcia, 1968
ASCOMYCETES AND FUNGI IMPERFECTI		
Pithomyces chartarum	1.4	Hartman et al., 1962
Trichoderma viride	9.6	Ballance & Crombie, 1961
Candida albicans	20.0	Peck, 1947
Neurospora crassa	19.0	Bianchi & Turian, 1967
Sphaerotheca humili var. *fuliginia*	10.0	Tulloch & Ledingham, 1960
Erysiphe graminis	12.0	Tulloch & Ledingham, 1960
BASIDIOMYCETES		
Melampsora medusae[d]	22.0	Tulloch & Ledingham, 1960
M. lini[d]	14.6-22.0	Jackson & Frear, 1967; Tulloch & Ledingham, 1960
Tilletia controversa[c]	35.0	Trione & Ching, 1971
T. foetens[a]	20.0	Tulloch & Ledingham, 1960
Cronartium fusiforme[e]	20.0	Carmack et al., 1976
C. fusiforme[g]	28.0	Weete & Kelley, 1977
Uromyces psoraleae	10.2	Tulloch & Ledingham, 1960
Phragmidium speceosum[c]	9.5	Tulloch & Ledingham, 1960
Puccinia hieracii[d]	8.4	Tulloch & Ledingham, 1960
P. hieracii[c]	7.3	Tulloch & Ledingham, 1960
P. helianthi[c]	13.3	Tulloch & Ledingham, 1960
P. carthami	9.0	Tulloch & Ledingham, 1960
P. graminis tritici 56[d]	18.0	Tulloch & Ledingham, 1960

TABLE 2.4   Continued

Fungus	Total Lipid (% Dry Weight)	Reference
P. graminis tritici (mixed races)	19.7	Shu et al., 1954
P. graminis avenae 7A	16.0	Tulloch & Ledingham, 1960
P. triticina[d]	17.0	Tulloch & Ledingham, 1960
P. coronata[d]	17.0	Tulloch & Ledingham, 1960
Ustilago zeae[a]	5.0	Tulloch & Ledingham, 1960
U. nigra[a]	4.0	Tulloch & Ledingham, 1960
U. levis[c]	14.5	Tulloch & Ledingham, 1960
U. bullata[c]	0.6	Gunasekaran et al., 1972
U. maydis[c]	0.4	Gunasekaran et al., 1972
Ravinelia hobsoni[c]	18.8	Tulloch & Ledingham, 1960
Gymnosporangium juvenescens[f]	6.3	Tulloch & Ledingham, 1960

[a] Chlamydospores.   [e] Aeciospores.
[b] Conidia.   [f] Dried gelatinous horns.
[c] Teliospores.   [g] Basidiospores.
[d] Uredospores.

during germination of these conidia, and the principal source of energy appears to be sugar trehalose.

Sclerotia are spherical, tightly packed masses of hyphae that are formed by some fungi with the onset of unfavorable growth conditions. Generally, they have a low lipid content of 2.5 to 6% of the dry weight. However, sclerotia of several isolates of the ergot fungus *Claviceps purpurea* contain 20 to 30% lipid (Morris, 1967).

LIPID COMPOSITION

As described in Chapter 1, organic solvent extracts of biological materials are usually complex mixtures of different types of lipophilic substances. These extracts can be easily separated into neutral and polar lipid fractions. Common neutral lipids are triacylglycerides and sterol esters, which tend to accumulate with increasing age of the fungus.

Predominant polar lipids are the phospholipids, but also may include glycolipids. The relative proportions of these and other lipids may vary according to the stage of fungal development, age, and conditions under which the fungus is cultured. The lipid composition of some fungi is given in Table 2.5. Each type of lipid is discussed in more detail in subsequent chapters.

LIPID FUNCTIONS

The roles of lipids in fungi are generally similar to those of other organisms, yet there are some lipid functions that may be unique to fungi. The precise functions of all lipids are not well understood. Triacylglycerides are generally considered reserve lipid that may be used for energy and carbon skeletons during growth and development. Lipid (acylglycerides) is the primary carbon substrate for spore germination of some rusts. Triacylglycerides represent the major lipid storage material of fungi and may be observed as globules (sphaerosomes, liposomes) when the fungus is viewed microscopically. Triacylglycerides are about two times more efficient for metabolic energy than either proteins or carbohydrates. In addition to the importance of fatty acid moieties of acyl lipids as sources of energy, fatty acids with specific structures are required for growth. For example, under anaerobic conditions, unsaturated acids (mainly oleic acid) are required for mitochondrial function and growth of certain yeasts. The degree of lipid unsaturation influences membrane function, tolerance to extreme temperatures, and perhaps adaptation to aquatic habitats.

Phospholipids are important structural components of biological membranes, forming the lipid bilayer in which proteins are embedded (Figure 2.1). Some phospholipids have been implicated in the active transport of ions across membranes and are also essential for the activity of some membrane-bound enzymes.

Sterols are also major components of membranes and are required for the anaerobic growth of some yeasts. There is evidence that sterols exhibit a condensing or liquifying effect on acyl lipids depending on the physical state of the lipid. Sterols may regulate permeability by affecting internal viscosity and molecular motion of lipids in the membrane

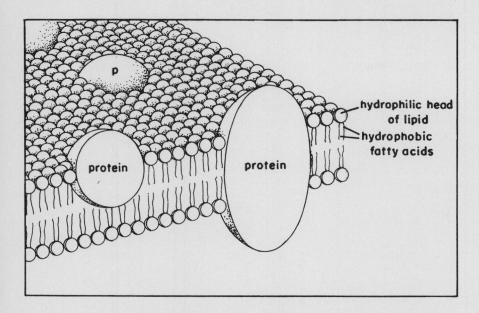

Figure 2.1   Model of a biological membrane showing the
  globular protein embedded in the lipid bilayer.

(Demel and Dekruyff, 1976). Sensitivity to polyene antibiotics is related to membrane sterol content; resistance to
these antibiotics occurs in the absence of sterols. Sterols
may also serve as precursors of steroid hormones involved in
the sexual reproduction of some fungi. Other isoprenoid
lipids such as carotenoid pigments may serve as light recep-
tors for phototrophic responses; and polyprenols complexed
with sugars serve as intermediates or carriers in carbohydrate and glycopeptide biosynthesis. The function of sphingolipids in fungi is not known.

SUBCELLULAR DISTRIBUTION OF LIPIDS

   Cell Wall. The structure and chemical composition of
fungal cell walls has been studied extensively, and the carbohydrate composition has been reviewed by Bartnicki-Garcia
(1968). Although both qualitative and quantitative changes
occur during morphogenesis, the fungal cell wall is composed
of 80 to 90% polysaccharides with the remainder consisting
mainly of lipid and protein. Fungal cell wall preparations

TABLE 2.5   Lipid Composition of Some Fungi[a]

Fungi	Hydro-carbons	Sterol Esters	Triacyl-glycerols	Fatty Acids	Sterols	Phospho-lipids
Cephalosporium falciforme[b]	–	9.0	25.0	4.0	13.0	49.0
C. kiliense[b]	–	16.0	18.0	3.0	14.0	49.0
Saccharomyces cerevisiae[c]	–	16.0	25.0	6.0	3.0	49.0
Pithomyces chartarum[d,e]	–	–	64.5	21.1	–	14.0
Stemphylium dendriticum[d,e]	–	–	65.0	15.0	–	19.7
Cylindrocarpon radicicola[d,e]	–	–	65.0	11.0	–	24.0
Neurospora crassa[f,g]	<0.5	0.6	62.0	1.5	4.0	21.0
N. crassa[h]	–	1.5	0.2	1.5	3.5	93.3
Smittium culisetae[i]	–	7.2	26.6	11.4	12.9	33.9
Rhizopus arrhizus[j]	1.0	–	22.1	11.7	16.7	44.4
Ustilago bullata[k,l]	–	–	19.4	5.2	9.3	52.9
U. maydis[k,m]	–	–	3.5	0.7	2.2	84.3
Blastocladiella emersonii[n]	3.8	3.2	8.1	2.2	4.1	60.4
Mucor rouxii[o]	–	–	22.6	2.5	13.4	60.5
Pythium ultimum[p]	–	–	57.3	31.5	–	9.2
Saccharomyces cerevisiae (NCYC 714)[q,r]	–	45.9	8.4	1.3	7.1	32.9
S. carlsbergensis (NCYC 74)[q,r]	–	53.3	8.2	1.9	6.1	27.7
Candida utilis (NCYC 321)[q,r]	–	52.8	11.2	–	5.6	30.4
Schwanniomyces occidentalis (NCYC 133)[q,r]	–	11.5	27.6	3.6	20.6	24.7

[a]All values expressed as a percentage of the total lipid; [b]Sawick & Pisano, 1977; [c]Hunter & Rose, 1972; [d]iacyglycerols, 2%; [d]Hartman et al., 1962; [e]values for triacylglycerols were

reported as "neutral glycerides"; [f]Kushwaha et al., 1976; [g]squalene; carotenoids, 0.2%; geranylgeraniol, 2%; [h]Bianchi & Turian, 1967; conidia, 6 h after germination; value for phospholipids includes carotenoids; [i]Shearer, 1973; value for sterol esters includes hydrocarbons; [j]Weete et al., 1970; fatty acid methyl esters, 4.1%; value for phospholipids reported as polar lipids; hydrocarbon is squalene; [k]Gunasekaran et al., 1972; sterol esters, hydrocarbons, and methyl esters reported as single value, *U. bullala*, 5.5%; *U. maydis*, 0.7%; [l]monoacylglycerols, 1.0%; diacylglycerols, 6.6%; [m]monoacylglycerols, 1.3%; diacylglycerols, 7.3%; [n]Mills & Cantino, 1974; zoospores; value for phospholipids reported as polar lipid; glycolipid, 12.4%; monoacylglycerols, 2.4%; diacylglycerols, 3.2%; [o]Safe & Duncan, 1974; phospholipids reported as polar lipids; 0.6% diacylglycerols; [p]Bowman & Mumma, 1967; 2.1% monoacylglycerols; [q]Johnson et al., 1972; organisms grown on 0.2% glucose in chemostat culture; [r]mono- and diacylglycerides, 0 to 7%.

contain both freely extractable and bound lipids (Kessler
and Nickerson, 1959). Care should be taken that lipids isolated from cell walls are not contaminants from the cytoplasm, particularly fragments of the plasma membrane adher-
ing to the wall.

There is usually 2 to 3 times more bound lipid present
in fungal cell walls than unbound lipid. The lipid content
of most fungal cell walls ranges from 0.6 to 18.9% of the
dry wall material (Table 2.6). Although the cell wall lipid
content of relatively few species has been reported, there
appears to be no taxonomic significance to the amount of
lipid present in these structures. The cell wall lipid content of *Saccharomyces* species ranges from <2 to 13.5%, averaging about 5.8%. Cell wall lipids from *S. cerevisiae* are composed of neutral lipids (including sterols) and phosphoacyl-
glycerides with $C_{16:1}$ and $C_{18:1}$ as the principal fatty acids
(Suomalainen, 1969; Northcote and Horne, 1952). The cell
wall lipid content of different strains and morphological
forms of *Candida albicans* ranges from 0.6 to 10.6%, but
averages about 3.1%. There is over 2 times more lipid in
the cell wall of the mycelial forms of this fungus than
either the juvenile or yeast-like forms (Bianchi, 1967).
Sterol esters, triacylglycerides, sterols, fatty acids, and
phospholipids are components of the cell wall lipid from
each of the three morphological forms, but phosphatidylglycerol and sphingolipids are present in only the juvenile
and mature forms. Unlike *Saccharomyces* species studied,
sterols and sterol esters comprise 40 to 60% of the lipid in
C. albicans cell walls. Phospholipids are the next most
abundant type of lipid. It has been noted that the lipid
content of the yeast and mycelial forms is variable depending
on the growth temperature and medium composition.

There is no difference in the amount of lipid in the
cell wall of yeast and mycelial forms of *Paracoccidioides
brasileensis*, but the cell wall lipid content of the yeast
form of *Blastomyces dermatitidis* is about half that of the
corresponding mycelial form (Kanetsuna et al., 1969). Although the amount of cell wall lipid of *Histoplasma capsulatum*
varies according to the report, there appears to be no significant difference between the yeast and mycelial forms (Domer
and Hamilton, 1971; Domer et al., 1967). Triacylglyceride is
the principal type of cell wall lipid of this fungus, and is
accompanied by phospholipids, sterol esters, sterols, fatty
acids, and diacylglycerides. There are also no qualitative

TABLE 2.6   Cell Wall Lipid Content of Certain Fungi

Fungus	Lipid Content (% Dry Weight)	Reference

PHYCOMYCETES

Apodachlya brachynema	3.1	Sietsma et al., 1969
Dictyuchus sterile	4.7	Sietsma et al., 1969
Mucor rouxii (yeast)	5.7	Bartnicki-Garcia, 1968
M. rouxii (hyphae)	7.8	Bartnicki-Garcia, 1968
M. rouxii (sporangiosphore)	4.8	Bartnicki-Garcia & Reyes, 1964; 1968
Pythium butleri	12.0	Mitchell & Sabar, 1966
P. myriotylum	8.0	Mitchell & Sabar, 1966
Pythium sp. PRL 2142	8.2	Sietsma et al., 1969
Saprolegnia ferax	5.0	Sietsma et al., 1969

ASCOMYCETES AND FUNGI IMPERFECTI

Aspergillus nidulans	4.6-10.5	Bull, 1970; Zonneveld, 1971
Aspergillus sp.	18.9	Ruiz-Herrera, 1967
Blastomyces dermatitidis (yeast)	4.6	Kanetsuna et al., 1969
B. dermatitidis (mycelium)	9.6	Kanetsuna et al., 1969
Candida albicans	1.2	Kessler & Nickerson, 1959
C. albicans (yeast, blastospore)	0.6-10.6	Chattaway et al., 1968
C. albicans (mycelium	4.5-5.5	Chattaway et al., 1968; Bianchi, 1967
C. albicans (juvenile)	2.1	Bianchi, 1967
C. albicans (yeast)	1.8	Bianchi, 1967
Chaetomium globosum	ca. 3.0	Manet, 1972
Cordyceps militaris	15.0	Marks et al., 1969
Epidermophyton floccosum	3.3	Shah & Knights, 1968
Geotrichum candidum	8.0	Sietsma & Woutern, 1971
Helminthosporium sativum	11.4	Applegarth & Bozoian, 1968
Histoplasma capsulatum (yeast)	2.0	Domer & Hamilton, 1971
H. capsulatum (yeast)	6.8	Domer et al., 1967

TABLE 2.6   Continued

Fungus	Lipid Content (% Dry Weight)	Reference
H. capsulatum (mycelium)	1.0	Domer & Hamilton, 1971
H. capsulatum (mycelium)	6.2	Domer et al., 1967
Microsporum canis	4.0	Shah & Knights, 1968
M. gypseum	3.6	Shah & Knights, 1968
Paracoccidioides brasiliensis (yeast)	8.3–10.7	Kanetsuna et al., 1969
P. brasiliensis (mycelium)	4.8–10.4	Kanetsuna et al., 1969
Penicillium roquefortii	2.0	Applegarth & Bozoian, 1968
Saccharomyces cerevisiae	2.0–13.5	Kessler & Nickerson, 1959; Eddy, 1958; Suomalainen & Nurminen, 1970; Masschelein, 1959; Northcote & Horne, 1952
S. fragilis CECT 1207	3.3	Reuvers et al., 1969
S. carlsbergensis	<2.0	Eddy, 1958
S. oviformis	<2.0	Eddy, 1958
Trichophyton mentagrophytes	3.1	Shah & Knights, 1968
Trigonopsis variabilis	8.7	SentheShanmuganathan & Nicherson, 1962

differences between morphological forms of *Blastomyces* and *Histoplasma* species.

The role of cell wall lipids is unknown. It has been suggested that they confer rigidity or protection against drying since morphological distortion is observed after the walls of dermatophytic fungi are extracted with fat solvents (Shah and Knights, 1968). It has also been proposed that lipids play a role in the pathogenicity of dermatophytes (Peck, 1947), but unhydrolyzed lipids do not possess antigenic properties (Elinor and Zaikina, 1963). Polyprenols are known to be involved in cell wall synthesis (see Chapter 9).

Protoplast. Depending on the stage of growth, lipids
may accumulate as globules (liposomes) which are believed to
be composed mainly of triacylglycerides. However, a considerable portion of cellular lipid occurs in membranes. The
principal membranes include the vacuolar membrane (tonoplast),
endoplasmic reticulum, mitochondrial membranes, and the plasmalemma. It is well-known that the principal components of
biological membranes are lipid and protein, but the arrangement of these materials has been the subject of considerable
conjecture (Bretscher, 1973). Membrane structure is cur-
rently viewed as globular protein embedded in a lipid bilayer
matrix. The protein may be randomly distributed on either
side of the lipid belayer, or it may extend through the
lipid. The bilayer has hydrophilic surfaces with a hydrophobic center. A typical model of a biological membrane is
shown in Figure 2.1.

Most studies on the composition of fungal membranes
have been concerned with the plasmalemma of yeast. The development of mechanical and enzyme digestion methods for
removal of the cell wall from yeast cells has facilitated
isolation of the cell envelope. The lipid content of fungal
membranes ranges from about 30 to 50%, and the protein:lipid
ratio ranges from 0.95 to 2.1. Other substances often present in yeast plasma membranes are carbohydrates (glucans and
mannans), sterols, and nucleic acids. (When considering membrane composition, sterols are considered separately and are
not included with the value for lipid.) Protoplast membranes
represent 13 to 20% of the dry cell weight of *S. cerevisiae*,
and contain 46 to 49.3% protein, 37.8 to 45.6% lipid, 0.97%
DNA, 6.7 to 7% RNA, 3 to 6% carbohydrate, and 5.6 to 6%
sterol (Longley et al., 1968; Boulton, 1965). Phospholipids
comprise 15% to 25% of the membrane, with the remainder of
the lipid being mono-, di-, and triacylglycerides, fatty
acids, sterol esters, and sterols (Longley et al., 1968;
Boulton, 1965; Suomalainen and Nurminen, 1970; Kramer et al.,
1978). Plasmalemma and tonoplast lipids of *S. cerevisiae*
are quite different quantitatively (Table 2.7). The tonoplast contains significantly more lipid and less protein than
the plasmalemma. The neutral lipid fraction of both membranes contain unusually high amounts of fatty acids (nonesterified), but the plasmalemma contains sterol esters which
are absent in the tonoplast, and about 4 times more free
sterols (Figure 2.2). The tonoplast contains considerably
more triacylglyceride and hydrocarbon than the plasmalemma.
Phospholipid represents 40% of the tonoplast lipid, but only

TABLE 2.7   Composition of the *Saccharomyces cerevisiae*
Plasmalemma and Tonoplast[a]

	Plasmalemma	Tonoplast
Protein: total lipid (TL)	2.1	0.66
Phospholipid (% of TL)	6.4	40.0
Neutral lipid (% of TL)	93.5	60.0
Ergosterol (% of neutral lipid)	26.2	6.05
Sterol: phospholipid		
weight	4.65	0.16
molar	8.97	2.9
phosphatidylcholine[b]	34	33
lysophosphatidylcholine	0	5
phosphatidylethanolamine	20	15
phosphatidylserine }		
phosphatidylinositol	28	43
phosphatidic acid	15	0
other phospholipids	3	4

[a] From Kramer et al., 1978.
[b] Phospholipids expressed as % lipid phosphorus.

6% of the plasmalemma lipid (Table 2.7). The main difference
in individual phospholipids in the two membranes is the
higher level of phosphatidylserine and phosphatidylinositol
in the tonoplast, and absence of phosphatidic acid. A low
phospholipid (3 to 5%) content of a "heavy" fraction containing 10% lipid from *S. cerevisiae* has also been reported
(Klein et al., 1967).

    Reports of membrane lipids of other fungi are limited.
The protoplast membrane of *Candida utilis* comprises about
10% of the cell dry weight and is composed of 38.5% protein,
40.4% lipid, 5.2% carbohydrate, 1.1% RNA, and no DNA (Garcia
et al., 1967). The protein:lipid ratio of membranes of the
sporophore cap of the mushroom *Agaricus bisporus* is 1.1 and
a molar ratio of sterol to phospholipid of 0.77 (Holtz et al.,
1972). This ratio for the plasmalemma and tonoplast of *S.
cerevisiae* is reported to be 8.97 and 2.9, respectively
(Kramer et al., 1978), which seem high, since a sterol:phos-
pholipid ratio of 2:1 is generally considered high when based

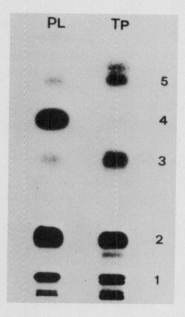

Figure 2.2   Neutral lipids of the plasmalemma (PL) and tono-
  plasts (TP) of *S. cerevisiae*. 1, sterols, diacylglycerides
  2, free fatty acids; 3, triacylglycerides; 4, sterol esters;
  5, hydrocarbons. (Redrawn from Kramer et al., 1978.)

on stoichiometric relationship between the two substances
measured with NMR, and a 1:1 ratio is reported as their maximum association in ultrasonicated solutions (Demel and
DeKruyff, 1976).

   The principal difference between promitochondrial and
mitochondrial membranes of *S. cerevisiae* is that the latter
contains about 70% more ergosterol (Paltauf and Schatz,
1969). Respiratory particles from *Claviceps purpurea* contain 25.7% lipid, 57.8% protein, 2.2% ergosterol, and no
plasmalogens (Anderson et al., 1964).

   Protein components of yeast membranes have been studied
by electrophoresis. Twenty to 30 proteins with a molecular
weight range of 10,000 to 300,000 have been detected in the
plasmalemma of *S. cerevisiae* (Santos et al., 1978; Kramer
et al., 1978). Up to 12 glycoproteins with a molecular
weight range of 28,000 to 240,000 have also been reported.
It is suspected that the main structural component of the

plasmalemma has a molecular weight of 30,000 and the main glycoprotein 50,000. Under the same analytical conditions, 20 proteins from the yeast plasmalemma could be distinguished, but only 17 from the tonoplast could be resolved (Kramer et al., 1978). Based on the analytical conditions employed, the two membranes differ by only three proteins.

RELATION OF CULTURE CONDITIONS TO FUNGAL GROWTH AND LIPID PRODUCTION

Optimum fungal growth occurs under optimal nutritional and environmental conditions, and is the result of numerous metabolic activities operating in a certain balance. These conditions and balance of activities vary with different fungi and even different strains of the same species. The balance of metabolic activities can be altered by manipulation of the culture conditions to favor production of a certain product, often at the expense of the production of other substances and even growth. This has been exploited for the commercial production of certain microbial natural products and has been studied for lipid production by fungi. The rate of lipid synthesis relative to the rate of synthesis of other cellular products determines whether lipids accumulate. The distribution of carbon from glucose into various cellular components can be manipulated by altering the medium composition. For example, the use of nutrient-limited media has proven useful in lipid production studies. For example, fungi grown under nitrogen-limiting conditions accumulate lipid; this occurs because protein synthesis is reduced resulting in more carbon substrate being available for conversion to lipid.

This section is concerned with how culture conditions such as temperature, pH, and oxygen; nutrients such as carbon and nitrogen; and other factors influence growth and lipid production in fungi. The effects of such factors on lipid production should be evaluated relative to optimum conditions for growth.

Temperature. Fungi differ in their responses to different growth temperatures with respect to lipid content (Table 2.8). The lipid content of some yeasts and yeast-like fungi is higher at temperatures below that optimum for growth compared to that at optimum temperatures, such as in *Candida lipolytica* and *Saccharomyces cerevisiae* (Hunter and Rose,

1971). On the other hand, *Rhodotorula gracilis* contains less
than half the lipid when grown at 22 C compared to when grown
at 28 C (Steinberg and Ordal, 1954). Too few species have
been tested for generalizations to be made, but reductions in
temperature below that for optimum growth of mycelial fungi
may result in reduced lipid content.

Fungi are classified into four groups according to
optimum growth temperatures. Mesophilic fungi have optimum
growth temperatures between 25 and 33 C, and psychrophilic
fungi between 10 and 20 C. Thermophilic fungi are defined
on the basis of maximum and minimum i.e., maximum at or above
50 C and minimum at or below 20 C (Cooney and Emerson, 1964).
Thermotolerant fungi can grow maximally near 50 C but mini-
mally below 20 C. There are no consistent differences in
the lipid content of certain mesophilic and thermophilic
Humicola, *Penicillium*, *Sporotrichium*, *Stilbella*, or *Chaetom-
ium* species (Mumma et al., 1970) on the basis of optimum
growth temperatures, but the lipid content of spores and
mycelia of certain mesophilic Mucorales fungi is lower than
that of closely related thermotolerant and thermophilic species (Sumner and Morgan, 1969) (Tables 2.1 and 2.8).

Changes in growth temperature have a more striking and
consistent influence on the degree of lipid unsaturation,
which is usually reflected in the number of double bonds in
the acyl moieties of complex lipids. The degree of lipid
unsaturation is based on the fatty acids obtained by alkaline
hydrolysis of the total lipid extract (total fatty acids),
and is expressed as unsaturation per mole (Δ/mole). The
degree of lipid unsaturation is calculated as follows:
Δ/mole = 1 × (% monoenes)/100 + 2 × (% dienes)/100 + 3 ×
(% trienes)/100.

A relatively high degree of unsaturation has been found
in lipids of higher plants, certain animals (including insects), and certain microorganisms grown at low temperatures.
Since increasing unsaturation alters the physical properties
of lipids, particularly melting point, it has been postulated
that this phenomenon represents an important factor in the
adaptation to cold environments. It follows that the order
of decreasing unsaturation in lipids of fungi having different growth temperature optima should be psychrophile > mesophile > thermophile. It should also follow that if these
organisms are grown at temperatures at either extreme of
their optima, the degree of unsaturation would adjust

TABLE 2.8 The Effect of Growth Temperature on the Degree of Unsaturation in Fungal Lipids

Fungi	Unsaturation Δ/mole[a]	Temp. (°C)	Reference
PHYCOMYCETES			
Mucor mucedo[b]	1.16–1.47	25	Sumner & Morgan, 1969
M. ramannianus[b]	1.33–1.50	25	Sumner & Morgan, 1969
M. racemosus[b]	1.23–1.28	25	Sumner & Morgan, 1969
M. hiemalis[b]	0.96–1.29	25	Sumner & Morgan, 1969
M. miehei	0.95	25	Sumner & Morgan, 1969
	0.76	48	
M. pusillus	1.02	25	Sumner & Morgan, 1969
	0.93	48	
M. pusillus[d]	0.83	45	Mumma et al., 1970
M. globosus[b]	0.96	25	Mumma et al., 1970
M. strictus[c]	1.13–1.26	10	Sumner et al., 1969
	0.98–1.24	20	
M. strictus[c]	1.03–1.18	10	Sumner et al., 1969
	0.94–1.04	20	
M. strictus[c]	0.99–1.19	10	Sumner et al., 1969
	0.89–1.06	20	
M. oblongisporus[c]	0.95–1.12	10	Sumner et al., 1969
	0.84–1.07	25	
M. oblongisporus[c]	1.00–1.29	10	Sumner et al., 1969
	1.17–1.37	25	
Mucor sp. I[d]	0.96–1.04	28	Sumner et al., 1969
	0.91–1.04	48	

Organism			Reference
Mucor sp. II[d]	0.96-1.06	28	Sumner et al., 1969
M. sp. III[d]	0.79-1.00	48	Sumner et al., 1969
(Mucor miehei)			
Rhizopus sp. III[d]	0.93-0.98	28	Sumner et al., 1969
	0.76	48	
Rhizopus sp. I[e]	0.76-0.84	36	Sumner et al., 1969
	0.60-0.86	48	
Rhizopus sp. II[e]	1.02-1.22	28	Sumner et al., 1969
	0.80-0.95	48	
Rhizopus sp.	0.99-1.24	28	Sumner & Morgan, 1969
	0.94-1.29	48	
	1.22	25	
R. arrhizus	0.94	48	Gunasekaran & Weber, 1972
	0.85	15	
	0.93	20	
	0.69	25	
	1.29	30	
Pythium ultimum	0.83	20	Chang & Matson, 1972
	0.82	30	
ASCOMYCETES			
Candida lipolytica[b]	1.00	25	Kates & Baxter, 1962
Candida sp. (#5)[c]	1.38	10	Kates & Baxter, 1962
C. scottii (AL25)	1.72	10	Kates & Baxter, 1962
C. scottii (5AAP2)	1.72	10	Kates & Baxter, 1962
C. lipolytica[b]	0.86	25	Kates & Baxter, 1962
C. lipolytica	1.12	10	Kates & Baxter, 1962
C. utilis[b],[f] NCYC 321	1.40	30	Brown & Rose, 1969
C. utilis[b],[g] NCYC 321	1.27	30	Brown & Rose, 1969
C. utilis[b],[f] NCYC 321	1.51	15	Brown & Rose, 1969

TABLE 2.8   Continued

Fungi	Unsaturation Δ/mole[a]	Temp. (°C)	Reference
C. utilis NCYC 321	1.08	15	Brown & Rose, 1969
Chaetomium thermophile[d]	0.65	45	Mumma et al., 1970
C. globosum[b]	0.96	25	Mumma et al., 1970
Rhodotorula glutinis[b]	0.63	25	Kates & Baxter, 1962
Humicola grisea[d] var. thermoidea	0.97	45	Mumma et al., 1970
H. insolens[d]	1.01	45	Mumma et al., 1970
H. lanuginosa[d]	0.82	45	Mumma et al., 1970
H. grisea[b]	1.54	25	Mumma et al., 1970
H. nigrescens[b]	1.34	25	Mumma et al., 1970
H. brevis[b]	1.17	25	Mumma et al., 1970
M. pulchella[b]	1.27	25	Mumma et al., 1970
Penicillium duponti[d]	0.86	45	Mumma et al., 1970
P. chrysogenum[b]	1.60	25	Mumma et al., 1970
Sporotrichum thermophile	1.00	45	Mumma et al., 1970
S. exile[b]	1.27	25	Mumma et al., 1970
Stilbella thermophila[d]	0.56	45	Mumma et al., 1970
Stilbella sp.[b]	1.47	25	Mumma et al., 1970
Sclerotinia sclerotiorum (14 day old)	2.09	5	Sumner & Colotelo, 1970
S. sclerotiorum (sclerotia)	1.66	20	Colotelo et al., 1971
S. sclerotiorum (sclerotial exudate)	0.44	20	Colotelo et al., 1971

[a] Δ/mole = 1.0 × (% monoene)/100 + 2.0 × (% dienes)/100 + 3.0 × (% trienes)/100; [b] Mesophile; [c] Psychrophile; [d] Thermophile; [e] Thermotolerant; [f] 75 mm HgO_2 tension; [g] 91 mm HgO_2 tension.

accordingly. Increases in the degree of lipid unsaturation
occurs in *Aspergillus niger* and *Rhizopus nigricans* (Pearson
and Raper, 1927), *A. nidulans* (Singh and Walker, 1956), and
Rhodotorula gracilis (Bass and Hospodka, 1952) grown at reduced temperatures, while the lipids of *S. cerevisiae* (Chang
and Matson, 1972) and *C. utilis* (Meyer and Bloch, 1963) become less unsaturated when grown at elevated temperatures.
This has also been demonstrated with thermotolerant *Rhizopus*
species and thermophilic *Mucor* species (Sumner and Morgan,
1969), mesophilic *Candida lipolytica*, and psychrophilic *C.*
scottii (Kates and Baxter, 1962). Similar responses to
changing temperatures have not been shown for other fungi
tested such as *Pythium ultimum* (Bowman and Mumma, 1967),
Cunninghamella blakesleeanus and *Rhizopus arrhizus* (Shaw,
1966). The influence of growth temperature on the degree
of lipid unsaturation in several fungi is given in Table
2.8.

The degree of lipid unsaturation in lipids from mesophilic and thermophilic fungi grown at their optimum growth
temperatures is consistent with the above reasoning. Lipids
of mesophilic fungi tested have a high degree of unsaturation
(0.96 to 1.60) compared to that of the thermophilic fungi
(0.65 to 1.01) (Mumma et al., 1970). The degree of lipid
unsaturation is similar in several mesophilic *Mucor* species
(Sumner and Morgan, 1969) and appears to be typical of most
mesophilic fungi grown at their optimum growth temperatures.
The degree of unsaturation is also similar for certain other
mesophilic and psychrophilic fungi, and for a thermotolerant
Rhizopus species grown at 28 C (Sumner et al., 1969).

Because of the dependence of the desaturase on oxygen,
temperature effects on the degree of lipid unsaturation may
be exerted through its influence on the solubility of oxygen
in the medium (see below for oxygen effects). It has been
pointed out that temperature effects on cellular fatty acid
composition may be due to alterations in the metabolic balance rather than to a specific effect (Brown and Rose, 1969);
or perhaps the temperature effect is primarily on synthesis
of the desaturase (Meyer and Bloch, 1963). It has also been
suggested that the rates of unsaturated fatty acid synthesis
and degradation are both temperature dependent and have different temperature coefficients (Kates and Baxter, 1962). A
reciprocal relationship between $C_{18:1}$ and $C_{18:2}$ during growth
of *C. lipolytica* at 10 and 25 C has been observed (Kates and
Paradis, 1973).

   Carbon Source. Glucose is the most important carbon
substrate for the growth of fungi. Other monosaccharides
such as fructose, mannose, and galactose can support growth
equally well in some species and less in others. Maltose
appears to be the best disaccharide for supporting fungal
growth, but trehalose and sucrose support good growth of
most species.

   Naegeli and Loew (1878) were the first to show that
sugars are converted to lipid by yeast. Smedley-MacLean
(1936) has also shown that reserve carbohydrates of yeast
are converted to lipid in aerated culture. Several comparative studies have been conducted to determine the best carbon
source for fat production by certain fungi. Glucose and
maltose are the best carbon substrates for lipid production
in many fungi (Chesters and Peberdy, 1965), such as *Penicillium lilacinum* (Philip and Walker, 1958) and *Aspergillus
nidulans* (Naquib, 1959). Sucrose and fructose are good carbon sources for lipid production by *P. chrysogenum* (sucrose
> fructose > glucose) (Divakaran and Mod, 1968) and *Rhizopus
arrhizus* (Gunasekaran and Weber, 1972). Medium sucrose content of 22.5% results in maximum lipid content in *P. soppi*
(34.8%) and *Fusarium lini* (28.4%) (Murray et al., 1953), but
17% sucrose is best for lipid production by *P. lilacinum*
(58%) (Osman et al., 1969). Sucrose is also the best carbon
substrate for lipid production by *A. fisheri* (37%) (Gad and
Hassan, 1964), and maximum lipid production by *P. aurentio-
brunneum* occurs at a medium sucrose of 40% (Singh and Singh,
1957). Although lactose has been the carbon source used for
superior lipid-producing strains of *Lipomyces starkeyi* and
L. lipofer (Cullimore and Woodbine, 1961), this disaccharide
and pentose sugars are generally poor substrates for fungal
growth and lipid production. The ability of fungi to convert various carbon substrates to lipid differs with the
species, but the best three for most fungi appear to be
glucose > sucrose and fructose.

   The utilization and conversion of glucose to lipid by
yeasts has been reviewed by Rattray et al. (1975). Two
categories of yeast have been described according to their
metabolic behavior with glucose as the carbon substrate
(DeDeken, 1966). Crabtree-positive yeasts, *Saccharomyces*
species, metabolize glucose primarily via glycolysis, and
the lipid content of these yeasts decreases as the glucose
concentration of the medium is increased from 0.2 to 1.0%
(Johnson et al., 1972). Crabtree-negative yeasts, on the

other hand, generally do not possess enzymes of the glycolytic pathway and oxidize glucose by aerobic metabolism involving mainly the pentose phosphate pathway (Mian et al., 1974).
These yeasts, which include *L. starkeyi*, *Candida lipolytica*,
C. tropicalis, *C. utilis*, *Candida* 107, *Hansenula anomala*,
Rhodotorula glutinis, and *Rhodotorula graminis*, tend to
accumulate lipid with increasing glucose concentrations in
the medium. Triacylglyceride is the principal type of lipid
that accumulates in these yeasts.

   Many fungi (filamentous) have been tested for their ability to convert sugars (mainly glucose) to lipid. Generally,
lipid production increases with increasing glucose content
of the medium. For example, the lipid content of *Aspergillus
fischeri* increases linearily from 10.4 to 36.0% of the fungal
dry weight with increasing medium glucose concentrations from
1 to 70% (Prill et al., 1935). Fungi vary considerably in
the optimum medium glucose concentration for the highest
lipid production. *Penicillium javanicum* has the highest
lipid content (41.5%) with 40% glucose in the medium. This
seems to be the best medium glucose concentration for lipid
production in many fungi (Ward et al., 1935).

   Fungi generally remove between 15 and 18% of the available sugar from the medium by the end of logarithmic growth.
The efficiency of sugar to lipid conversion is known as the
'fat coefficient' or 'lipid yield' and is expressed as grams
lipid produced per 100 g sugar (glucose) depleted from the
medium. Fat coefficients for *P. javanicum* grown in media
containing 20 to 60% glucose ranges between 7 and 8 (Ward
et al., 1935). The fat coefficient for *Rhodotorula gracilis*
grown at pH 6 is 19 prior to nitrogen exhaustion from
the medium, and 14 after nitrogen depletion (Kessell,
1968). A fat coefficient of 22 has been achieved with *Candida* 107 grown under continuous culture conditions (see below). The fat coefficient of a fungus is dependent on culture conditions and medium composition. The reciprocal relationship between lipid production and glucose utilization by
Aspergillus parasiticus is illustrated in Figure 2.3.

   See below for the relation between carbon and nitrogen
contents of the medium and lipid production.

   The potential of an organism as an economic producer of
fat is related to the efficiency of substrate utilization.
In this connection, many fungi have been screened for their

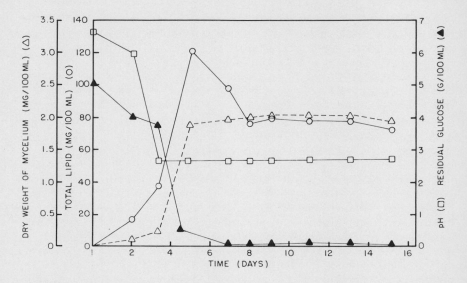

Figure 2.3   Relationship between lipid production and glucose utilization by *Aspergillus parasiticus* grown at pH 6
(Redrawn from Shih and Marth, 1974).

ability to utilize inexpensive industrial waste products,
particularly petroleum products, as carbon sources for both
protein and fat production. *Candida* and *Torulopsis* species
are particularly efficient in utilizing hydrocarbons (C_9 to
C_{18}) as carbon substrates (Thorpe and Ratledge, 1972). The
ability to utilize hydrocarbons for growth is adaptive, but
also involves constitutive enzymes. That is, hydrocarbon-
grown cells have a greater capacity for utilizing hydrocarbons than glucose-grown cells. Higher cellular yield is
obtained on glucose-containing media, and the ability to
utilize hydrocarbons efficiently is dependent on an active
respiratory system and the availability of oxygen. Whether
hydrocarbon-utilizing yeasts produce more or less lipid than
when grown with glucose as the carbon source varies with the
strain and environmental conditions.

    It has also been shown that certain filamentous fungi
degrade alkanes (Pelz and Rehm, 1973; Cerniglia and Perry,
1974; Gerasinova et al., 1975). For example, certain
Mucorales fungi such as *Absidia spinosa*, *Cunninghamella*

echinulata, and *Mortierella isabellina* metabolize C_{13} to C_{17}
n-alkanes in the terminal and subterminal positions (Hoffmann
and Rehm, 1976; 1978).

The mechanisms of alkane oxidation for different organisms is described in Chapter 8.

Inorganic Nutrients. In addition to a carbon source,
a defined medium for the laboratory culture of fungi is composed of mineral salts that generally includes the elements
calcium, magnesium, potassium, phosphorus, a nitrogen source,
and a few minor elements (zinc, iron, and manganese). The
nutritional factors influencing growth and lipid production
have been reviewed by Woodbine (1959). Generally, relatively
little is known about the importance of the amount and composition of various mineral salts for lipid production. Ca,
Na, and Fe do not appear to be required for lipid production
in yeast (Steinberg and Ordal, 1954). Increasing NaCl concentrations from 0 to 10% in the medium results in an almost
linear increase in the lipid content (0.3 to 6.3%) of a *Candida* species (Combs et al., 1968).

Phosphorus is essential for fungal growth, but reports
seem to vary as to whether limiting phosphorus results in
more or less total lipid. Increasing the medium phosphorus
content has little effect on the phospholipid composition of
Saccharomyces pombe (White and Hawthorne, 1970), but significantly alters the fatty acid and polar lipid composition of
Candida utilis (Johnson et al., 1973). *S. cerevisiae* (NCYC
366) grown in continuous culture under phosphorus-limiting
conditions contains only 10% less lipid than cells grown in
a phosphorus-sufficient medium, but about 23% more phospholipid (Ramsey and Douglas, 1979). The cells from the phosphorus-limited medium contain 40% less triacylglycerol, 67%
less sterol ester, 46% less phosphatidylserine, 46% more
phosphatidylcholine, and 38% more phosphatidylethanolamine.
Other reports show that *S. cerevisiae* and *C. utilis* produce
more lipid under phosphorus-limiting conditions, mainly as
triacylglycerol (Johnson et al., 1973).

Candida 107 grown under magnesium-deficient continuous
culture conditions accumulates myristic acid (C_{14}) suggesting
that this element is required for completion of fatty acid
synthesis [palmitic acid (C_{16}) is the main product of the
fatty acid synthetase, see Chapter 3] (Gill et al., 1977).

Other than the carbon source, nitrogen has the most pronounced effect on lipid production by fungi. The best nitrogen source for growth varies depending on the fungal species, but inorganic nitrogen in its most reduced form (NH_4^+) is more efficiently utilized by a greater number of fungi than other forms. Comparison of the effects of several concentrations (0.2 to 10%) of NH_4NO_3 on the growth and lipid production by *A. fischeri* has shown that the highest fat coefficient and mycelial production occur at 0.2%. However, the greatest amount of glucose is removed from the medium at 2% NH_4NO_3.

An organic nitrigen source is best for the growth of some fungal species, but not for lipid production. However, there are some exceptions such as *A. nidulans* which contains more lipid when grown with asparagine as the nitrogen source compared to NO_3^- or NH_4^+ (Naquib and Saddik, 1960). Asparagine is second to $(NH_4)_2HPO_4$ for lipid production by *Mortierella vinacea* (Chesters and Peberdy, 1965). Glycine is a more efficient nitrogen source than $NaNO_3$ for fat production by *P. lilacinum*. Other organic and inorganic nitrogen sources have proven less effective for lipid production.

The most important nutritional parameter for lipid production by fungi is the carbon:nitrogen (C:N) ratio. Within limits, increasing the carbon and nitrogen of the medium, while holding the C:N ratio constant, results in increased growth but not lipid production. High lipid production is associated with low rates of protein synthesis. Low C:N ratios in the medium favor protein synthesis and high ratios favor lipid accumulation in fungi. For example, the best lipid production by *P. lilacinum* with $NaNO_3$ (0.65%) as the nitrogen source is obtained at a C:N ratio of 65:1 (Osman et al., 1969), and the best growth and lipid production by *M. vinacea* grown with $(NH_4)_2HPO_4$ as the nitrogen source is obtained at a C:N ratio of 80:1. When nitrogen is depleted from the medium, carbon substrate continues to be taken up by some species until it is depleted from the medium, and assimulated to lipid. The highest lipid contents of unidentified yeasts from soil have been obtained using nitrogen-deficient media. It has been suggested that the regulation of acetyl-CoA carboxylase, rather than the rate of glucose uptake, may be important in lipid accumulation in *Candida* 107 grown in a medium with a high C:N ratio or nitrogen deficient conditions (Botham and Ratledge, 1978). Figure

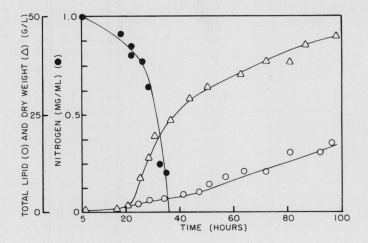

Figure 2.4   Changes in dry weight, total lipids, and
nitrogen content of the medium during the growth
of *Rhodotorula gracilis* (Redrawn from Kessell,
1968).

2.4 illustrates the relation between nitrogen depletion,
growth, and lipid accumulation in *R. gracilis*.

pH. The pH of an unbuffered medium decreases with
fungal growth as shown in Figure 2.3. pH optima for
the growth of most fungi are between 6.0 and 7.0. When am-
monium salts of mineral acids are used as the nitrogen source,
the decrease in pH results mainly from the removal of NH_4^+
ions from the medium by the fungus. Less growth and glucose
utilization by *Aspergillus fischeri* is obtained when grown
in an unbuffered medium compared to a buffered medium (Prill
et al., 1935). Lipid accumulates linearly between pH 3.0
and 8.5 in *Rhodotorula gracilis* at a rate of 2.1 to 3.1 grams
of lipid per 100 g cells per hour (Steinberg and Ordal, 1954).
At constant temperature there are no differences in lipid production by *Pythium vexans* and *P. irregulare* at pH levels of
5.9, 6.8, and 7.5 (Cantrell and Dowler, 1971). There are
other reports that show no correlation between pH and lipid
production within a pH range optimum for growth (Gad and

El-Nockrashy, 1960). For example, maximum growth of *P. lilacinum* can be obtained at pH 4 and maximum fat production at pH values from 4.0 to 6.8. At a constant pH 5.5, a three-fold increase in CO_2 results in 27% more lipid but at pH 6 and the same CO_2 level no increase in lipid occurs (Castelli et al., 1969).

Aeration (Oxygen). The culture method (see next section) determines the degree of aeration for fungal growth. Generally, fungi grow better and take up more carbon substrate from the medium when grown in aerated compared to non-aerated culture; however, in some cases there are no significant differences in the amount of lipid produced (Prill et al., 1935; Starkey, 1946; Kleinzeller, 1944). On the other hand, *Rhodotorula gracilis* grown in aerated culture accumulates more lipid than in non-aerated culture (Enebo et al., 1944).

Oxygen is required for the biosynthesis of unsaturated fatty acids and sterols. Respiratory-deficient yeasts (Crabtree-positive), such as *S. cerevisiae*, contain less lipid when grown under anaerobic compared to aerobic conditions. Growth inhibition of *S. cerevisiae* on unsupplemented medium under anaerobic conditions can be overcome by the addition of unsaturated fatty acids (mainly oleic and linoleic acids) and a sterol.

Growth of *Mucor rouxii* under aerobic conditions results in filamentous growth and under anaerobic conditions cells of this fungus are yeast-like. Anaerobically grown cells contain relatively high proportions of short chain (C_8 to C_{14}) fatty acids and very low degrees of lipid unsaturation (Safe and Duncan, 1974). This is also true for *S. cerevisiae*. Polar lipids comprise over 90% of the total lipid in anaerobically grown *M. rouxii* compared to 60% in aerobically grown cells.

Growth Factors. Many fungi are able to grow and reproduce on a medium containing only a carbon source and mineral salts, while others require such additives as thiamine, biotin, pyridoxine, amino acids, and/or inositol. The relationship between lipid production and vitamin deficiency in fungi is not well established, but generally vitamin deficiencies result in reductions in lipid content. A reduction of lipid content by 40% may be expected in *Hanseniaspora valbyensis* grown in a pyridoxine-deficient medium (Haskell

ans Snell, 1965). Phytosphingosine content is lower in the
deficient cells while lipid phosphorus is unchanged. Pyridoxine levels may also influence the degree of lipid unsaturation. Nicotinic acid deficiency results in an increase
in lipid content while less lipid is found in *S. cerevisiae*
grown in a pantothenic acid-deficient medium (Klein and
Lipmann, 1953). The formation of fatty acids with 18 carbons, particularly $C_{18:1}$, is reduced while there is greater
C_{16} accumulation under these conditions (Suomalainen and
Keranen, 1968). Biotin-deficiency results in decreased
fatty acid synthesis (see Chapter 4).

Since the first report that inositol is required for
normal growth in certain yeasts (Eastcott, 1928), several
studies have been conducted to determine its metabolic role.
S. cerevisiae and *Neurospora crassa* grown in inositol-
deficient media have abnormally high lipid contents (Johnston and Paltauf, 1970; Paltauf and Johnston, 1970; Lewin,
1965; Challinor and Daniels, 1955; Shatkin and Tatum, 1961;
Shafai and Lewin, 1968). Generally, inositol-containing
lipids are low in cells grown under conditions of inositol-
deficiency while other phospholipids are unaltered. Triacylglycerols accumulate in inositol-deficient *S. carlsbergenesis* cells, and can be observed as globules in the cytoplasm. It has been suggested that the increased lipid content of inositol-deficient cells is due to a more active
fatty acid synthetase (Johnson and Paltauf, 1970).

The precise functions of these growth factors, so far
as they are known, are discussed in the appropriate chapters
that follow.

FUNGAL GROWTH AND LIPID PRODUCTION

The overriding influence of genetic factors and the
importance of nutritional parameters in fungal growth and
lipid production has been stressed above. Another important
consideration in the interpretation of physiological and biochemical data involving fungi is the culture method employed.
Most studies with fungi have been conducted using the batch
[surface (stationary) or submerged (shake)] culture technique. In this type of culture, the fungus grows on a
limited amount of medium and progressively modifies the
medium during growth. At any point in time after inoculation, the culture contains mycelia or cells differing in age

and stages of development and are responding to continuously
changing conditions. Measurements of physiological or biochemical activities of fungi in such cultures represents a
composite, or "average," of the population of mycelia or
cells, and cannot be associated with a particular growth or
cellular stage of development at the time of measurement.
Since most fungal research has been carried out using this
type of culture technique, relatively little is known about
the precise details of lipid metabolism and its significance
in relation to the fungal growth cycle (spore germination
through sporulation), or cell cycle in the case of yeasts
and yeast-like organisms. However, surface culture offers
advantages in studying environmental and genetic factors
related to growth, whereas submerged culture has advantages
in nutritional and biochemical studies. Most mycelial fungi
readily sporulate in surface culture, but not in submerged
culture.

Based on studies using surface or submerged culture
techniques, fungal growth is characterized by rapid lipid
production and accumulation proportional to logarithmic
growth (Figure 2.5). This may be preceded by a period of
delayed lipid production not associated with the lag phase
of growth, as with *Rhodotorula gracilis* (Enebo et al., 1946).
Changes in lipid metabolism as the growth rate decreases
due to limiting conditions in the medium depends on several
factors, and appears to vary according to the type of fungus,
or fungal species. In the case of some mycelial fungi grown
as surface cultures, the stationary growth phase is accompanied by rapid lipid utilization which may be associated
with sporulation and may be due to depletion of a medium
component. Lipolytic esterase (lipase) activity is absent
or low in vegetative hyphae but, regardless of the culture
method and mode of induction (sporulation), occurs in mycelial
extracts during conidiation of *Aspergillus niger* (Lloyd et
al., 1972). Esterase activity increases immediately prior
to vesicle and phialide formation, and is present in the
conidiophore tip prior to formation of these structures.
This suggests that lipid serves as a source of carbon and
energy for sporulation in this and perhaps other fungi.
Lipid utilization in *Rhizopus arrhizus* grown in petri plates
for 6 days begins just prior to sporulation at 72 hours after
inoculation (Figure 2.5) (Weete et al., 1973). This pattern
of lipid metabolism has been reported for several other fungi
such as *Phycomyces blakesleeanus* (Bernhard et al., 1958), *R.
arrhizus* (Shaw, 1966), *Aspergillus nidulans* (Singh and

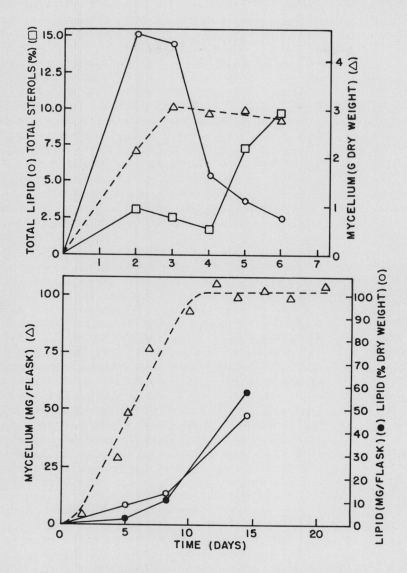

Figure 2.5   (a) Change in the lipid content of *Rhizopus arrhizus* grown in surface culture (Weete et al., 1973). (b) Change in the lipid content of *Pythium ultimum* grown in liquid culture (Bowman and Mumma, 1967).

.lker, 1956), *A. parasiticus* (Shih and Marth, 1974), *Mor-*
.ierella vinacea (Chesters and Peberdy, 1965), and *Penicil-*
lium species (Gylenberg and Raitio, 1952).

When grown in submerged batch culture, some fungi tend
to accumulate rather than utilize lipid during stationary
growth. As illustrated before, this may be due to depletion
of nitrogen from the medium, resulting in the conversion of
remaining sugar to lipid. This may occur until the carbon
source is exhausted from the medium, after which the lipid
content of the mycelium or cells decreases. For example,
the lipid content of the mycelial fungus *Pythium ultimum*
accumulates lipid (up to 48%) up to and after the onset of
stationary growth (Figure 2.5) (Bowman and Mumma, 1967).
The lipid content of *Saccharomyces cerevisiae* ATCC 7755 decreases during the lag phase and first half of logarithmic
growth, and then progressively increases during the second
half of exponential and post-exponential growth (Castelli
et al., 1969). Similar results have been obtained with the
so-called "fat yeasts" *Candida lipolytica* (Nyns et al., 1968)
and *Rhodotorula gracilis* (Figure 2.3) (Enebo et al., 1946).
The lipid content of *C. utilis* reaches maximum during late
stationary growth, and then declines with exhaustion of
glucose from the medium (Dawson and Craig, 1966).

Sporulation in *S. cerevisiae* is accompanied by an increase in lipid which is mainly sterol (81%) initially,
rather than triacylglycerol that usually accompanies vegetative growth (Chassang et al., 1972). Yeast sporulation is
also accompanied initially by increased triacylglycerol and
phospholipid formation, and later only by increased neutral
lipid. Neutral lipid accumulates in the mature ascospores,
but disappears during the transition to vegetative growth
(Henry and Halvorson, 1973; Steele and Miller, 1974).

The water-mold *Achlya* sp. grown in liquid culture requires 27 to 30 hours from spore germination to formation
(Law and Burton, 1976). Lipid is utilized during germination
and initial growth, but the mycelial lipid content increases
at the onset of sporangium formation. This is due to increased activity of the fatty acid synthetase.

The problem of changing medium content with growth has
been solved for some fungi with the development of the continuous culture technique. This technique involves the continuous feeding of medium into a culture maintained at

constant volume. By maintaining culture conditions in this
way, a steady-state growth rate can be achieved for a randomly dividing population of cells. One of the advantages
of this type of culture over batch culture is that a constant
rate of growth and level of product production can be maintained over an extended period of time. Continuous cultures
are particularly valuable in studying the relationship of
certain nutrients or growth factors to growth and metabolism.
Continuous culture usually involves two stages, a cell multi-
plication stage in a complete medium followed by growth at a
constant rate under nutrient-limited conditions.

    Studies on lipid production in fungi grown in continuous
culture are limited and have involved mainly *Candida* and
Saccharomyces species. The most comprehensive of these
studies has been conducted using single- (Gill et al., 1977)
and two-stage (Hall and Ratledge, 1977) continuous cultures
of the oleaginous yeast-like fungus *Candida* 107 grown under
various nutrient-limited (carbon, nitrogen, phosphorus,
nitrogen plus phosphorus, and magnesium) conditions. In the
single-stage continuous culture, a maximum lipid content of
37% and high fat coefficient of 22 grams lipid per gram
glucose utilized could be obtained under nitrogen-limited
conditions, and maintained for several weeks. A high cellular lipid content of 35% could be obtained under nitrogen
plus phosphorus-limited conditions along with the highest
specific rate of lipid formation (0.059 grams lipid per gram
yeast per hour), but low fat coefficient of 5.2. Under car-
bon-limited conditions, an average of about 10% lipid could
be achieved, with increasing fatty acid unsaturation occurring at lower growth rates. Limitations in other nutrients
results in a low degree of unsaturation. A two-stage continuous culture, where the growth phase is separated from
the lipid accumulation phase, offers no advantage over the
single-stage culture. Maximum lipid content of 28% could
be achieved under these conditions with a fat coefficient of
14 (Hall and Ratledge, 1977).

    The continuous cultures discussed above contain cells
of different ages dividing randomly; and no information
about events occurring in relation to the cell cycle can be
ascertained using this culture technique. The problem of
having cells of differing age with respect to cell division
has been solved for yeast-type fungi with the development
synchronously dividing cell cultures. In this type of culture, changes in cellular constituents during the cell cycle

can be readily determined. The total lipid content of *Can-dida utilis* NRRL Y-900 dividing synchronously under glucose
limitation does not change during the first 2 hours of the
3.5 hour cell cycle (Figure 2.6) (Hossack et al., 1979).
However, the total lipid content decreases by 15 to 20% dur-ing the third hour, and then returns to approximately the
original level at the end of the cell cycle. This change
in total lipid is due entirely to corresponding changes in the
phospholipid content, which averages 77% of the total lipid
during the cell cycle. In nitrogen-limited cells, oscilla-tions in the contents of major fatty acids have been observed
(Dawson and Craig, 1966). The triacylglyceride and sterol
(free or esterified) do not change during the cell cycle.
On the other hand, it has been shown that sterols (free and
esterified) and phospholipids accumulate in a continuous and
exponential pattern during the cell cycle of the yeast
Kluyveromyces fragilis (Penman and Duffus, 1976). Increases
in sterol content have been associated with cell division in
Saccharomyces cerevisiae (Cejkova and Jirku, 1978). It is
also well established that sterol esters accumulate with age
in the same yeast dividing non-synchronously (Parks et al.,
1978).

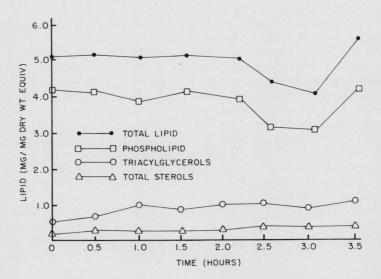

Figure 2.6   Changes in lipid composition during the cell
cycle of *Candida utilis* NRRL Y-900 (Prepared from data of
Hossack et al., 1979).

CHAPTER 3   FATTY ACIDS

INTRODUCTION

Fatty acids are aliphatic monocarboxylic acids, and those considered lipids have 10 or more carbons. Generally, fatty acids occur in nature as a homologous series ranging in chain length from C_{10} to C_{36}. However, fatty acids commonly range from C_{14} to C_{20} in most organisms with the even-numbered homologues being predominant. Fatty acids with 16 and 18 carbon atoms are most common and quantitatively important. Fatty acids may be saturated ($C_nH_{2n+1}COOH$) or unsaturated [mono-($C_nH_{2n-1}COOH$) and poly-($C_nH_{2n-x}COOH$)], they may have odd-numbered carbon chains, or they may be substituted with methyl, oxygen (keto, epoxy), or hydroxyl functions. Cyclic (cyclopropyl- and cyclopentyl-) fatty acids also occur in some organisms. The principal product of the fatty acid synthetase in plants, animals, and microorganisms contains 16 carbons (see Chapter 4). Most other fatty acids are formed by elongation, desaturation, and substitution, or combinations of these modifications. Palmitic acid (C_{16}) is the predominant saturated fatty acid of most organisms, and oleic ($C_{18:1}$), linoleic ($C_{18:2}$) and linolenic ($C_{18:3}$) are the major unsaturated acids. Animals and aquatic organisms produce fatty acids with a higher degree of unsaturation (tetra-, penta-, hexaenes). Alpha (α)-linolenic is the predominant acid of photosynthetic tissues whereas $C_{18:1}$ and $C_{18:2}$ are the major acids of most seed oils. Gamma (γ)-linolenic acid is the $C_{18:3}$ isomer, rather than α-linolenic, produced by animals. Bacteria do not produce polyunsaturated fatty acids; some produce high relative proportions of methyl-branched chain fatty acids and some produce cyclopropyl fatty

acids. Fatty acids are produced by all living organisms and
potentially represent the most abundant class of lipids in
nature. However, fatty acids occur in low amounts in biological systems since they are readily incorporated into
complex lipids. They occur as esters of glycerol (acylglycerols and phosphoacylglycerols), sterols, sugars, sphingolipid bases (N-acyl), and hydroxy fatty acids. Thioesters
(coenzyme A and acyl carrier protein) are the principal
forms of fatty acids involved in lipid metabolism, but do
not accumulate. The hydrocarbon chains from fatty acids
may also be linked to glycerol and phosphoglycerol via ether
and vinyl ether (plasmalogens) linkages. Fatty acids produced by various groups of organisms are summarized in Table
3.1.

     This chapter is concerned with the distribution of fatty
acids in fungi.

NOMENCLATURE AND STRUCTURE

     The systematic naming of fatty acids is based on the
name of the longest hydrocarbon chain containing the carboxyl group (COOH). The final e is dropped from the name
of the corresponding alkane and replaced by oic. For example,
a fatty acid containing 16 carbons is based on the alkane
name hexadecane and is named hexadecanoic acid. The carboxyl
carbon is designated as number 1 and the others increase in
number sequentially toward the terminal carbon. Another
system of designating carbon atoms in a fatty acid molecule
uses the greek letters alpha (α), beta (β), omega (ω), and
ω-1 to indicate the number 2 and 3, terminal, and penultimate
(next to last) carbons, respectively. The presence and number of double bonds in the hydrocarbon chain are indicated
by inserting ene, diene, etc. into the fatty acid name.
Position of the double bond is designated by the number of
the carbon atom nearest the carboxyl carbon involving the
ethylenic bond, i.e. 9,12-octadecadienoic acid. Unsaturated
fatty acids may also be designated by counting from the
terminal end of the molecule, i.e. ω-6,9-octadecadienoic acid.
Polyunsaturated fatty acids are classified according to
position of the double bonds relative to the terminal methyl
(ω-carbon). Most polyenes fall into one of two families,
the ω-6 family or ω-3 family, which are based on γ-linolenic
($\Delta^{6,9,12}$) and α-linolenic ($\Delta^{9,12,15}$) acids, respectively.
These families result from different modes of synthesis (see

Chapter 4). The configuration of the double bond is also designated in the systematic name, i.e. *cis*-9-octadecenoic acid, and the greek letter delta (Δ) sometimes procedes the carbon number when counting from the carboxyl end of the molecule i.e. Δ^9. The position of substituents on the fatty acid molecule are indicated by the carbon number at which they are located, i.e. *cis*-9,10-epoxyoctadecanoic acid. The position of methyl substituents is designated accordingly, but fatty acids with methyl groups at the ω-1 and ω-2 positions are called *iso*- and *anteiso*-branched acids, respectively.

A shorthand system is often used to denote fatty acids. This system employs a letter C accompanied by a subscript figure to denote the number of carbons in the hydrocarbon chain, followed by a colon and a number to indicate the number of double bonds, i.e. $C_{18:2}$, octadecadienoic acid. Position of the double bond is indicated by adding Δ followed by the carbon number of the double bond; i.e. $\Delta^{9,12}$ $C_{18:2}$. Configuration of double bonds may be designated by c or t for *cis* and *trans*, respectively, i.e. $\Delta^{9c,12c}$ $C_{18:2}$. A shorthand system may also be used for unsaturated fatty acids whereby numbering from the terminal carbon is used, i.e. $C_{18:3}$ $\omega 3$, α-linolenic acid. Substituents on the molecule may be denoted accordingly, br C_{18}, iso C_{18}; 2-OH C_{18} or 2h C_{18}.

Trivial names are most often used for the most common fatty acids and are given in Tables 3.2 and 3.3

Fatty acids are often broadly grouped by their quantitative occurrence in nature, major and minor acids. Major fatty acids usually represent >1% of the total fatty acids and minor acids represent <1%. Of course, a fatty acid considered major in one organism may be a minor one in another, or when produced under a different set of conditions. Generally, fatty acids with even-numbered carbon chains from C_{14} to C_{18} are considered major acids, and fatty acids with odd-numbered carbon chains and with more than 18 carbons are considered minor acids. Fatty acids may also be grouped as unusual acids, which include those with structures other than that of the straight chain, saturated or unsaturated monocarboxylic acid. Some of these include acids with non-methylene-interrupted double bond arrangements (see below), and those with substituents.

The carboxyl and terminal methyl carbon atoms are separated by methylene (CH_2) groups which are generally

TABLE 3.1   Fatty Acids Produced by Various Groups of Organisms

Type of Organism	Fatty Acids
Prokaryotes Bacteria	Homologous series commonly ranging from C_{10} to C_{20}; produce no polyunsaturated fatty acids; have vaccenic acid as predominant C_{18} monoene; most species produce high relative amounts of *iso* and *anteiso*-branched fatty acids, mainly *anteiso*-C_{15}; species of certain genera (*Lactobacillus, Escherichia, Vibrio*) produce cyclopropyl fatty acids (C_{17} and C_{19}); certain species produce other unusual fatty acids such as lactobacillic acid, tuberculostearic acid, mycocerosic acids, and sterculic acid.
Blue-Green Algae	Fatty acid distribution intermediate between bacteria and higher algae; homologous series commonly range from C_{10} to C_{18}; some produce no polyunsaturated fatty acids, others produce di- and trienoic C_{18} acids; oleic rather than vaccenic is the principal C_{18} monoene, but the latter fatty acid is produced by some blue-green algae; branched chain fatty acids are not common in blue-green algae; $C_{18}:3$ $\omega 3$ is the predominant C_{18} triene but the $\omega 6$ isomer has been detected.
Eukaryotes Fungi	See text.
Algae	Homologous series commonly ranging from C_{14} to C_{22}; characterized by polyunsaturated C_{18}, C_{20}, and C_{22} fatty acids; polyunsaturated fatty acids are mainly of $\omega 3$ series, but some $\omega 6$ acids present.

Mosses, Ferns — Mosses have fatty acids typical of photosynthetic tissues of higher plants (see below), but also have high proportions of C_{20} polyunsaturated fatty acids; homologous series from C_{12} to C_{24}; ferns have fatty acids typical of photosynthetic tissue but little or no C_{20} polyenes; may have long chain ($>C_{20}$) saturated fatty acids.

Higher Plants — Homologous series commonly from C_{14} to C_{20}; $C_{18}:3$ $\omega3$ only C_{18} triene produced and is major fatty acid of photosynthetic tissue; trans-3-hexadecenoic acid is characteristic of photosynthetic tissue of eukaryotes, and is specifically associated with phosphatidylglycerol; commercially important seed oils have low levels of $C_{18}:3$ but are high in $C_{18}:1$ and $C_{18}:2$; seed oil of some species contain unusual acids (trans fatty acids, substituted acids; acetylenic acids, etc.).

Animals — Diet greatly influences fatty acid composition of animals; homologous series from C_8 to C_{22}, $C_{18}:2\omega9$ and $C_{18}:3\omega6$ produced by animals; C_{20}, C_{24} polyunsaturated fatty acids abundant, particularly in fish.

TABLE 3.2   Some Normal Saturated Fatty Acids

Number of Carbons	Systematic Name	Trivial Name
10	Decanoic	Capric
11	Undecanoic	Undecylic
12	Dodecanoic	Lauric
13	Tridecanoic	-
14	Tetradecanoic	Myristic
15	Pentadecanoic	-
16	Hexadecanoic	Palmitic
17	Heptadecanoic	Margaric
18	Octadecanoic	Stearic
19	Nonadecanoic	-
20	Eicosanoic	Arachidic
22	Docosanoic	Behenic
24	Tetracosanoic	Lignoceric
26	Hexacosanoic	Cerotic
28	Octacosanoic	Montanic
30	Triacontanoic	Melissic
32	Dotriacontanoic	Lacceroic

referred to as being linked together in a straight chain in normal fatty acids. However, the C-C bond angle of $109°$ results in a "zig-zag" shape of the molecule which is in constant thermal motion in living cells (Figure 3.1a). The melting points of saturated fatty acids from 10 to 28 carbons in chain length range from 31.6 to 90.9 C. The addition of double bonds to the molecule decreases the melting point, i.e. oleic acid, 10.5 C; linoleic acid, -5.0 C; α-linolenic acid, -11 C; arachidonic acid, -49.5 C. The most common distribution of ethylenic bonds in polyunsaturated fatty acids is the methylene-interrupted arrangement, or the 1,4-diene non-conjugated arrangement of double bonds (Figure 3.1b). Fatty acids with conjugated double bonds (Figure 3.1c) and the arrangement where double bonds are separated by two or more methylene groups, such as $\Delta^{5,9,12}$-octadecatrienoic acid, are less common in nature.

Double bonds of most naturally occurring fatty acids have the *cis* configuration; however, *trans* fatty acids do occur in some organisms (Figure 3.1d). *Cis* fatty acids are

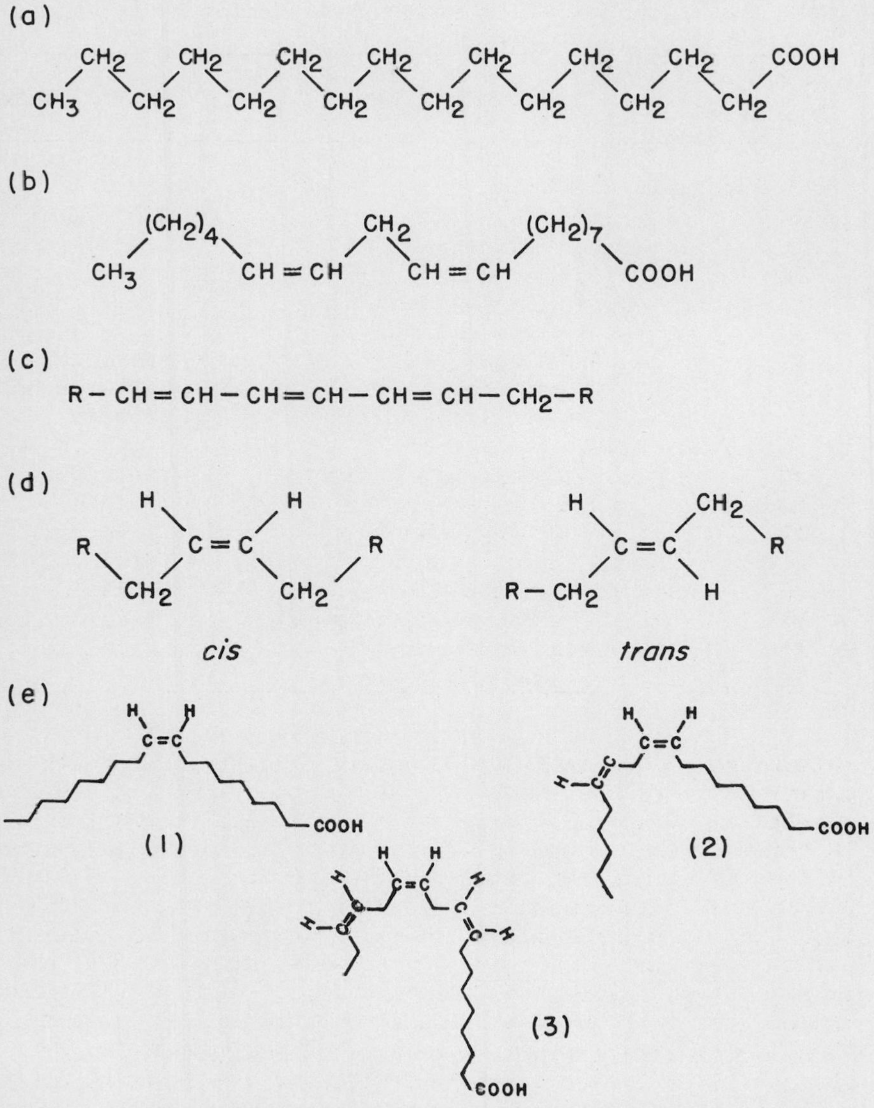

(a)

(b)

(c)

R — CH = CH — CH = CH — CH = CH — CH₂ — R

(d)

cis   *trans*

(e)

(1)   (2)

(3)

Figure 3.1   (a) Structure of palmitic acid illustrating "zig-zag" shape of the molecule resulting from C–C bond angle, (b) Methylene-interrupted arrangement of double bonds in polyunsaturated fatty acids, (c) Conjugated double bond system, (d) *Cis* and *trans* double bond configuration, (e) Structures of oleic (1), linoleic (2), and α-linolenic (3) acids.

TABLE 3.3   Some Mono- and Polyunsaturated Fatty Acids

Number of Carbons	Systematic Name	Trivial Name
Monoenoic Fatty Acids		
14	*cis*-9-tetradecenoic	Myristoleic
16	*cis*-9-hexadecenoic	Palmitoleic
18	*cis*-9-octadecenoic	Oleic
18	*cis*-11-octadecenoic	Vaccenic
20	*cis*-9-eicosenoic	Gladoleic
22	*cis*-13-docosenoic	Erucic
24	*cis*-15-tetracosenoic	Nervonic
Dienoic Fatty Acids		
18	*cis*,*cis*-9,12-octadecadienoic	Linoleic
18	*cis*,*cis*-6,9-octadecadienoic	-
20	*cis*,*cis*-11,14-eicosadienoic	-
Trienoic Fatty Acids		
18	*cis*,*cis*,*cis*-9,12,15-octadeca-trienoic	α-Linolenic
18	*cis*,*cis*,*cis*-6,9,12-octadeca-trienoic	γ-Linolenic
20	*cis*,*cis*,*cis*-8,11,14-eicosatrienoic	-
20	*cis*,*cis*,*cis*-11,14,17-eicosatrienoic	-
Tetra- and Pentaenoic Fatty Acids		
20	*cis*,*cis*,*cis*,*cis*-5,8,11,14-eicosatetraenoic	Arachidonic
20	*cis*,*cis*,*cis*,*cis*,*cis*-5,8,11,14,17-eicosapentaenoic	-

less stable than the corresponding *trans* isomers. A *cis*
double bond introduces a bend in the molecule, and increasing the number of methylene-interrupted double bonds brings
the methyl end of the molecule closer and closer to the
carboxyl end (Figure 3.1e). Increasing unsaturation of
fatty acids is also accompanied by increasing polarity.
Substituents also alter the physical properties of the fatty
acids, i.e. methyl branching lowers the boiling point relative to the corresponding straight chain acid. Some unusual
fatty acids of fungi are given in Table 3.4.

TABLE 3.4   Some Unusual Fatty Acids of Fungi

Systematic Name (Symbol)	Common Name	Source
BRANCHED CHAIN ACIDS		
14-Methyl-pentadecanoic (iso C_{16})	Isopalmitic	*Conidiobolus denaeosporus*
HYDROXY ACIDS		
12-Hydroxy-*cis*-9-octadecenoic (12h $\Delta^{9c}C_{18:1}$)	Ricinoleic	*Claviceps purpurea*
3-Hydroxyhexadecanoic[a] (3h C_{16})	3-Hydroxypalmitic	*Rhodotorula graminis* *R. glutinis*
3-Hydroxyoctadecanoic[a] (3h C_{18})	3-Hydroxystearic	*R. graminis* *R. glutinis*
13-Hydroxydocosanoic[a] (13h C_{22})	-	*Candida bogoriensis*
17-L-Hydroxyocta-decanoic[a] (17h C_{18})	17-Hydroxystearic	*Torulopsis apicola*
17-L-Hydroxyocta-decenoic[a] (17h $\Delta^{9}C_{18:1}$)	17-Hydroxyoleic	*T. apicola*
2-Hydroxyhexacosanoic[b] (2h C_{26}; αh C_{26})	-	*Saccharomyces cerevisiae* mushrooms
9,10-Dihydroxyocta-decanoic (9,10 Dih C_{18})	9,10-Dihydroxy-stearic	*Claviceps sulata*
Erythro-8,9,13-Tri-hydroxydocosanoic[a,c] (8,9,13 Trih C_{22})	-	Yeast NRRL YB-2501, *Rhodotorula* strain 62-506
15,16-Dihydroxyhexa-decanoic[a] (15,16 Dih C_{16})	Ustilic acid A	*Ustilago zeae* PRL-119

TABLE 3.4   Continued

Systematic Name (Symbol)	Common Name	Source
2,15,16-Trihydroxyhexa- deconoic[a] (2,15,16 Trih C_{16})	Ustilic acid B	*U. zeae* PRL-119
KETO ACIDS		
13-Oxo-8,9-dihydroxy- docosanoic[a,c] (13k,8,9 Dih C_{22})		Yeast NRRL YB-2501
EPOXY ACIDS		
9,10-Epoxyoctadecanoic (epoxy C_{18})	Epoxystearic	Rust fungi

[a] Extracelluar; may occur as portion of complex glycolipid, see text.

[b] This and other long chain α-hydroxy fatty acids occur as part sphingolipid molecules, see text.

[c] Occurs as partially or fully acetylated (acetoxy).

FATTY ACIDS IN FUNGI

Since the development of gas chromatography in the early 1950's, the fatty acid composition of numerous fungal species has been reported. Generally, fatty acids produced by fungi are similar to those of other organisms. They consist of a homologous series of saturated and unsaturated aliphatic acids ranging from 10 to 24 carbons in chain length. Without exception, even-numbered carbon chains are predominant. Fatty acids with 16 and 18 carbons are most abundant, with palmitic being the major saturated acid, and oleic and linoleic being the principal unsaturated acids. Generally, fungal fatty acid distributions have not been taxonomically useful; however, there are characteristic features of fatty acids associated with certain groups of fungi. The following

discussion focuses on the unique aspects of fatty acids or
their distributions in members of the major fungal classes:
Phycomycetes, Ascomycetes and Deuteromycetes (Fungi Imperfecti), and Basidiomycetes. Unless indicated otherwise, the
fatty acid distributions discussed below refer to the fatty
acids obtained by hydrolysis of the lipid extract, often
called the total fatty acids.

The fatty acid compositions of fungi have been reviewed
previously by Shaw (1966), Erwin (1973), Weete (1976), and
Wassef (1977).

Phycomycetes. The fatty acid composition of numerous
Phycomycetes has been reported, some of which are given in
Table 3.5. One of the most significant features of fatty
acids from the lower fungi is γ-linolenic acid ($\omega 6$). Higher
fungi (Ascomycetes, Fungi Imperfecti, and Basidiomycetes)
produce only the $\omega 3$ isomer, α-linolenic acid. $\gamma C_{18:3}$ was
first identified as a fungal component in *Phycomyces blakesleeanus* (Bernard and Albrecht, 1948). Zygomycetes and
Mastigomycetes produce $\gamma C_{18:3}$ and some higher homologues of
the $\omega 6$ series such as $\Delta^{8,11,14} C_{20:3}$ and $\Delta^{5,8,11,14} C_{20:4}$,
which are also found in some Mucorales and Entomophthorales
(Haskins et al., 1964; Tyrrell, 1967, 1971). Mucoraceae
and Thamnidaceae fungi produce polyunsaturated fatty acids
with up to 18 carbons and 3 double bonds ($\gamma C_{18:3}$). $\alpha C_{18:3}$
has been reported in a Mucorales fungus, the slightly psychrophilic *Thamnidium elegans*, when grown at reduced temperatures
(Manocha and Campbell, 1978). Another $\omega 3$ polyene, $\Delta^{6,9,12,15}$
$C_{18:4}$, is also produced by this fungus. The Choanephoraceae
and Entomophthoraceae contain fatty acids up to $C_{20:4}$ in chain
length and degree of unsaturation. These fungi also produce
more fatty acids with 20 to 22 carbons than Ascomycetes and
Fungi Imperfecti (Shaw, 1966; Haskins et al., 1964; Tyrrell,
1967; Bowman and Mumma, 1967). The fatty acid composition
of *Smittium culisetae*, a fungus limited to the guts of Diptera
larvae, differs from other Phycomycetes with the absence of
$C_{18:2}$ and $C_{18:3}$, and high relative proportions of C_{16} (34.3%)
and $C_{16:1}$ (38.7%) (Patrick et al., 1973).

It may be phylogenetically significant that some fungi
produce polyunsaturated fatty acids of both the $\omega 3$ and $\omega 6$
series (Brennan et al., 1974). *Blastocladiella emersonii*
contains both α- and $\gamma C_{18:3}$ along with $C_{20:3}$ and $C_{20:4}$
(Sumner, 1970). $\Delta^{5,8,11,14,17} C_{20:5}$ has been detected in
Pythium species and is the principal fatty acid in *Saprolegnia*

TABLE 3.5   Fatty Acids of Phycomycetes (Mycelia)

Fungi	Fatty Acids (%)							
	C_{14}	C_{16}	$C_{16:1}$	C_{18}	$C_{18:1}$	$C_{18:2}$	$C_{18:3}$	$C_{20:4}$
Achlya sp.[x]	9.1	25.5	1.8	7.7	28.9	8.5	-	10.1
Blastocladiella emersonii[a]	0.8	13.3	2.6	2.8	39.1	16.9	10.6	16.4
Basidiobolus haptosporus[b]	5.0	16.6	14.1	4.6	22.5	14.9	22.3	-
Choanophora curcurbitarum[b,c]	2.9	31.2	3.8	7.8	19.9	20.2	10.8	-
Entomophthora muscae[d,e]	5.9	13.8	16.6	2.6	37.9	4.9	1.7	13.5
E. obscura[f]	3.0	5.2	tr	32.2	3.3	tr	tr	4.0
Conidiobolus denaesporus[d,g]	15.2	10.4	5.4	2.5	15.2	1.8	2.1	8.9
C. brefeldianus[h,i]	29.0	12.4	3.2	0.7	18.5	6.1	7.0	-
C. thrombiodes[b,j]	5.3	12.2	14.9	3.0	39.1	5.4	2.3	13.8
Delacroixia coronata[d,k]	26.3	9.3	1.9	3.5	21.9	7.3	5.7	4.8
Mucor strictus[l,m]	5.6	22.4	1.9	9.2	34.8	11.6	14.2	-
M. mucedo[l,n]	1.1	16.8	1.4	11.3	30.5	32.9	6.4	-
Rhizopus arrhizus[o]	1.2	18.4	3.7	11.0	29.4	16.3	0.2	-
Rhizopus nigricans[b]	tr	15.7	3.4	3.9	30.4	19.1	27.5	-
Phytophthora infestans[b,p]	23.5	12.0	3.4	2.1	31.2	4.5	tr	10.2
Phycomyces blakesleeanus[q]	tr	10.9	1.7	20.5	29.8	34.9	2.1	-
Allomyces macrogynus[r,s]	1.0	16.0	4.0	9.0	36.0	6.0	-	-
Rhizidiomyces apophysatus[r,t]	tr	13.0	5.0	4.0	11.0	11.0	4.0	-
Rhizophlyctis rosea[r]	tr	13.0	-	6.0	74.0	3.0	3.0	-
Hypochytrium catenoides[r,u]	2.0	22.0	16.0	-	49.0	4.0	-	-
Pythium debaryanum[b,v]	5.5	13.0	5.6	1.6	13.9	19.7	2.3	12.3
Smittium culisetae[w]	1.1	34.3	38.7	4.3	16.3	-	-	-

Thamnidium elegans[Y]	2.9	14.5	1.0	6.0	31.7	13.4	17.8	-
Saprolegnia parasitica[Z,aa]	6.9	18.3	1.9	2.8	19.6	14.2	2.9	9.6

[a]Sumner, 1970; C_{12}, 0.5%; C_{17}, 0.8%; $\alpha C_{18:3}$, 1.2%; C_{20}, 0.9%; $C_{20:2}$, 0.8%; $C_{20:3}$, 3.2%; [b]White & Powell, 1966; [c]fatty acids of triglyceride fraction; C_{10}, 0.2%; C_{12}, 0.2%; $C_{14:1}$, 0.06%; C_{15}, 0.3%; $C_{16:2}$, 0.6%; C_{17}, 0.2%; C_{20}, 0.3%; $C_{20:1}$, 0.1%; $C_{20:2}$, 0.2%; $C_{20:3}$, 0.06%; C_{22}, 0.2%; [d]Tyrrell, 1967; [e]$C_{14:1}$, trace; $C_{16:2}$, 1.6%; C_{17}, 1.5%; $C_{20:3}$, trace; [f]C_{10}, 16.2%; C_{12}, 8.6%; $C_{14:1}$, 1.0%; $C_{20:1}$, 1.1%; $C_{20:3}$, 1.1%; C_{22}, 24.4%; [g]C_{12}, 1.1%; $C_{20:1}$, 1.0%; $C_{20:2}$, 0.4%; $C_{20:3}$, 0.5%; C_{13}, 1.4%; C_{24}, trace; $C_{24:1}$, 1.6%; [h]Tyrrell, 1971; C_{12}, 6.0%; C_{13}, 0.6%; C_{15}, 2.5%; C_{17}, 0.8%; 35% branched chain fatty acids, iso C_{14}, anteiso C_{15}, iso C_{16}; [j]C_{12}, trace; C_{17}, 1.6%; $C_{20:1}$, 1.2%; $C_{20:3}$, 1.2%; [k]C_{12}, 8.4%; C_{15}, 1.8%; $C_{20:3}$, 4.9%; [l]Sumner et al., 1969; [m]CBS 576-66; psychrophilic fungus, grown at 10 C for 28 days; trace quantities of C_{12}, C_{15}, C_{17}; [n]trace quantities of C_{12}, C_{15}, C_{17}; mesophilic fungus, grown at optimum 25 C for 14 days; [o]Weete et al., 1970; C_{10}, C_{12}, trace; C_{15}, 0.2%; C_{20}, 16.2%; C_{22}, 0.9%; [p]$C_{20:1}$, 3.8%; $C_{20:3}$, 1.6%; C_{22}, 7.7%; [q]Jack, 1966; [r]Rambo & Bean, 1969; [s]C_{15}, 3%; $C_{16:2}$, trace; C_{17}, 6%; C_{20}, 4%; $C_{20:1}$, 1%; [t]C_{15}, 11%; $C_{16:2}$, trace, C_{17}, 7%; $C_{20:3}$, 9%; [u]Unknowns (2), 7%; [v]$C_{16:2}$, 1.5%; $C_{20:1}$, 3.0%; $C_{20:3}$, 2.4%; C_{22}, 19.2%; [w]Patrick et al., 1973; C_8, trace; C_{10}, 1.1%; C_{12}, 1.0%; C_{13}, 0.3%; C_{15}, 0.5%; unknown, 1.1%; [x]Law & Burton, 1976; trace, C_{15}; C_{20}, 2.5%; HO C_{16}, 3.1%; unidentified 2.8%; 27 h growth; [Y]Manocha & Campbell, 1978; 14 day growth; 25 C; $\alpha C_{18:3}$ detected at lower temperatures; $C_{18:4}$, trace; C_{22}, 1.3%; $C_{22:1}$, 2.4%; C_{24}, 8.9%; [z]Gellerman & Schlenk, 1979; [aa]traces of C_{12}; C_{20}; $C_{20:1}$; $C_{20:2}$; odd-numbered; and branched; $C_{20:3}$ $\omega 6$, 2.8%; $C_{20:5}$ $\omega 3$, 20.0%.

parasitica (Gellerman and Schlenk, 1979) along with the $\omega6$ polyenes $\Delta^{8,11,14}$ $C_{20:3}$, $\Delta^{5,8,11,14}$ $C_{20:4}$ and $\Delta^{7,10,13}$ $C_{22:3}$ (Shaw, 1965). Both isomers of $C_{18:3}$ are produced by *Thamnidium elegans* grown at less than 25 C (Manocha and Campbell, 1978).

Several aquatic fungi have distinct fatty acid patterns (Rambo and Bean, 1969). For example, *Rhizidiomyces apothysatus* produces a large number of C_{13} to C_{20} fatty acids with mono-, di- and triunsaturated isomers of almost every chain length. *Rhizophylyctis rosea* contains 74% oleic acid.

Some Phycomycetes produce higher amounts of short chain fatty acids ($<C_{16}$). Some *Conidiobolus* (*Entomophthora*) species can be distinguished from others on the basis of relative amounts of short chain and saturated fatty acids (Tyrrell, 1967).

Branched chain fatty acids are rare in fungi. They were first identified (by GLC retention data) as fungal products in the imperfect fungi *Pithomyces chartarum*, *Cylindrocarpon radicicola*, and *Stemphylium dendriticum* (Hartman et al., 1960; 1962). 12-Methyl-tridecanoic (*iso* C_{14}), 12-methyl-tetradecanoic (*anteiso* C_{15}), and 14-methyl-pentadecanoic (*iso* C_{16}) acids represent up to 50% of the total fatty acids of *Conidiobolus denaeosporus* (Tyrrell, 1968). Branched chain fatty acids are also minor components of *Saprolegnia parasitica* (Gellerman and Schlenk, 1979).

The degree of lipid unsaturation does not appear to differ appreciably according to optimum growth temperature of the fungus, but the relative proportions of unsaturated C_{18} acids differ between thermophilic and other fungi studied (Sumner et al., 1969). Thermophilic Mucorales species contain ca. 50% $C_{18:1}$, ca. 15% $C_{18:2}$ and ca. 4% $C_{18:3}$. Corresponding thermotolerant, mesophilic, and psychrophilic species contain low relative proportions of $C_{18:1}$, 15 to 24% $C_{18:2}$, and 8 to 19% $C_{18:3}$.

The fatty acid composition of mycelia and spores of the same species has been compared in relatively few fungi. The fatty acid composition of sporangiospores from several *Mucor* species and a *Rhizopus* species are similar to the parent mycelia, but the degree of unsaturation is lower in the spore fatty acids (Sumner and Morgan, 1969). This is true for thermotolerant, thermophilic, and mesophilic species studied.

The relative proportions of fatty acids, particularly the C_{18} acids and C_{20}, are different for sporangiospores and mycelium of *R. arrhizus* (Weete et al., 1970). Fatty acids of *Phycomyces blakesleeanus* (Furch et al., 1976) differ from those of mycelium of the same species (Bowman and Mumma, 1967; Jack, 1966) mainly by the low relative amounts of C_{16} (4.3%) and high amounts of $C_{18:3}$ (48.6%) (Table 3.6).

Some other Phycomycetes for which the fatty acid composition has been reported are listed in Table 3.7.

Ascomycetes and Fungi Imperfecti. Ascomycetes are divided into two subclasses: (1) Hemiascomycetes, which include the true yeasts and certain parasitic mycelial forms, and are generally considered the more primative among other Ascomycetes, and (2) Euascomycetes, which include the mycelial forms of the Ascomycetes for which the sexual stage of reproduction has been identified. Fungi Imperfecti are fungi for which no sexual stage has been observed. α-Linolenic acid, rather than the corresponding ω6 isomer, is produced by Ascomycetes and Fungi Imperfecti.

The fatty acid composition of over 55 yeast and yeast-like fungi has been reported (Table 3.8). Major fatty acids range from 14 to 18 carbons in chain length, but low relative proportions of even-numbered fatty acids up to 24 carbons are produced by some species. There are no apparent differences in fatty acid composition that distinguishes ascosporogenous (true yeasts) from asporogenous (yeast-like) yeasts. The most commonly studied genera of the two groups are *Saccharomyces* and *Candida*, respectively, with *S. cerevisiae* and *C. utilis* being the most thoroughly studied species. *Saccharomyces* species are characterized by high relative proportions of $C_{16:1}$ which ranges from 26 to 60% of the total fatty acids. They also have high proportions of $C_{18:1}$ and, with a few exceptions, do not produce polyenoic fatty acids. Generally, $C_{16:1} > C_{18:1} > C_{16}$, and these acids account for about 90% of the total fatty acids. Short chain ($<C_{16}$) saturated fatty acids accummulate in *S. cerevisiae* grown anaerobically. High relative proportions of $C_{16:1}$ are not restricted to *Saccharomyces* species, but also occur in such species as *Hanseniaspora valbyensis*, *Brettanomyces lambricus*, *Debaryomyces nilssonii*, *Kloeckera apiculata*, *Kluyveromyces polysporus*, *Nadsonia fulvescens*, *Torulopsis colliculosa*, *T. dattila*, and *Trignopsis variabilis* (Table 3.8). The relative proportions of fatty acids of *Candida*

TABLE 3.6 Fatty Acids of Phycomycete Spores

Fungi	Fatty Acids (%)						
	C_{14}	C_{16}	$C_{16:1}$	C_{18}	$C_{18:1}$	$C_{18:2}$	$C_{18:3}$
Blastocladiella emersonii[a]	–	28.0	4.1	9.0	32.0	–	12.0
Mucor miehei[b]	5.6	24.4	3.1	11.3	32.6	13.1	9.3
M. pusillus[b]	1.0	25.4	3.0	4.8	42.2	19.0	4.5
M. mucedo[b]	2.5	21.3	3.5	12.6	27.2	21.0	12.0
M. ramannianus[b]	4.1	18.7	3.4	5.5	31.2	14.1	20.9
M. racemosus[b]	4.9	21.8	3.8	9.1	31.4	14.9	14.1
M. hiemalis[b]	2.0	15.0	2.1	17.1	28.0	17.9	18.0
Phycomyces blakesleeanus[c]	0.2	4.3	0.6	0.9	11.2	28.4	48.6
Rhizopus arrhizus[d]	1.0	16.8	1.7	19.9	42.4	7.7	0.1
Rhizopus sp.[b]	6.0	27.3	2.9	12.1	28.8	10.6	12.1
Thamnidium elegans[e]	1.5	9.4	1.6	2.9	33.4	15.7	10.4

[a]Mills & Cantino, 1974; $C_{20:4}$, 7%; zoospores; [b]Sumner & Morgan, 1969; [c]Furch et al., 1976; C_{12}, 0.3%; C_{15}, 0.1%; $C_{16:2}$, 0.1%; C_{17}, 0.1%; $C_{17:1}$, 0.1%; C_{18} polyene, 0.8%; C_{22}, 0.7%; unidentified, 3.6%; [d]Weete et al., 1970; C_{15}, 0.8%; C_{20}, 6.9%; Manocha & Campbell, 1978; $\alpha C_{18:3}$, 1.7%; $C_{18:4}$, 3.9%; $C_{22:1}$, 7.1%; C_{24}, 12.2%.

TABLE 3.7   Some Additional Phycomycete Fungi for Which the
Fatty Acid Composition Has Been Reported

Fungus	Reference
Basidiobolus anarum	White & Powell, 1966
B. meristosporus	White & Powell, 1966
Entomophthora virulenta	Tyrrell, 1967
E. thaxteriana	Tyrrell, 1967
E. conglomerata	Tyrrell, 1967
E. tipula	Tyrrell, 1967
E. ignobilis	Tyrrell, 1967
E. coronata	Tyrrell, 1967
E. apiculata	Tyrrell, 1967
E. megasperma	Tyrrell, 1967
E. exitialis	Tyrrell, 1967
Conidiobolus chamydosporus	Tyrrell, 1971
C. gonimodes	Tyrrell, 1971
C. megalotocus	Tyrrell, 1971
C. polytocus	Tyrrell, 1971
C. humicola	Tyrrell, 1971
C. globuliferus	Tyrrell, 1971
C. lamprauges	Tyrrell, 1971
C. nanodes	Tyrrell, 1971
C. paulus	Tyrrell, 1971
C. undulatus	Tyrrell, 1971
C. heterosporus	Tyrrell, 1971
C. osmoides	Tyrrell, 1971
Mucor strictus	Sumner et al., 1969
M. oblongiosporus	Sumner et al., 1969
M. genevensis	Mantle et al., 1969
M. rammannianus	Sumner et al., 1969
M. racemosus	Sumner et al., 1969
M. hiemalis	Sumner et al., 1969
Mucor sp.	Sumner et al., 1969
Rhizopus sp.	Sumner et al., 1969
Rhizopus arrhizus	Gunasekaran et al., 1972

TABLE 3.8 Fatty Acid Composition of Yeasts and Yeast-Like Fungi

Fungus	Fatty Acids (%)						
	C_{14}	C_{16}	$C_{16:1}$	C_{18}	$C_{18:1}$	$C_{18:2}$	$C_{18:3}$
Brettanomyces lambicus[a,b,c](5*)	2.2	16.3	47.8	–	20.6	11.1	–
Candida sp. 107[d,e]	1.0	23.0	2.0	7.0	34.0	27.0	–
Candida sp. 107[f]	tr	31.0	2.1	7.0	36.0	20.0	1.0
Candida sp. 107[g]	tr	20.0	1.5	7.5	39.8	23.7	tr
Candida sp. #5[h,i]	tr	14.0	18.0	1.0	14.0	40.0	11.0
C. albicans[j]	0.6	11.6	8.0	6.8	35.9	25.0	9.6
C. boidinii[k,l]	0.4	14.7	19.7	2.6	27.4	31.4	–
C. krusei[a,b](3*,5*)	–	10.5	26.7	–	22.1	35.4	5.3
C. krusei[d](3*,5*)	–	13.0	13.0	–	25.0	24.0	23.0
C. krusei[m,n](3*,5*)	tr	15.0	6.2	1.1	48.2	14.9	13.7
C. lipolytica[h,f,q]	tr	19.0	12.0	1.0	45.0	21.0	–
C. mycoderma[m,o](3*)	tr	14.1	17.1	2.3	41.1	18.6	4.7
C. mycoderma[dd](3*)	–	13.0	13.0	–	25.0	24.0	23.0
C. petrophillium[p]	0.7	8.6	9.3	0.6	34.4	44.8	–
C. pulcherrima[m,q](3*)	1.4	21.0	7.4	1.5	41.4	25.6	1.4
C. rugosa[dd]	14.0	13.0	11.0	2.0	33.0	30.0	9.0
C. scottii (AL25)[h,i]	tr	15.0	2.0	3.0	17.0	34.0	28.0
C. scottii[h,i]	tr	12.0	2.0	2.0	16.0	51.0	17.0
C. tropicalis[m,r]	tr	21.8	5.4	9.4	28.6	26.2	4.4
C. tropicalis (IFO 0589)[k,s](3*)	0.4	12.2	5.8	4.3	44.1	22.7	–
C. tropicalis (YO-148)[k,t](3*)	0.3	22.4	11.7	1.3	35.5	21.6	–
C. utilis[u](5*)	–	12.0	6.0	–	14.0	37.3	27.1

C. utilis^V(5*)	-	19.0	6.0	2.5	35.0	24.0	12.0
C. utilis^k,w(5*)	0.2	23.0	3.3	1.7	28.7	31.6	-
C. utilis^a(5*)	-	12.8	10.6	-	20.2	39.1	15.9
C. utilis (7005)^x,Y(5*)	tr	22.1	10.5	3.7	31.5	22.9	6.5
C. utilis^z(5*)	-	13.6	2.6	1.2	14.7	54.1	13.7
Cryptococcus laurentii^m,aa(5*)	tr	17.2	tr	8.5	37.5	36.5	-
C. neoformans^x,q	tr	13.5	tr	1.9	38.1	46.6	-
Debaryomyces hansenii^ee	0.4	23.7	2.5	8.2	50.1	2.5	1.5
D. hansenii^m,bb	tr	18.9	1.0	10.5	43.7	18.7	4.7
D. nilssonii^m,cc	1.3	10.7	33.1	2.5	35.3	17.0	-
D. subglobosus^a,z	10.3	12.2	12.1	-	39.0	22.0	11.8
Hanseniaspora valbyensis^ff	1.4	22.0	60.8	1.5	13.1	-	-
Hansenula anomala(1*,5*)	-	16.0	10.0	-	28.0	24.0	19.0
H. anomala var. anomala^dd(1*,5*)	-	14.0	9.0	1.0	31.0	19.0	25.0
H. anomala var. schneiggii^dd,gg(1*,5*)	-	12.0	3.0	-	28.0	25.0	14.0
H. anomala^m,q(1*,5*)	tr	17.2	1.3	3.8	34.7	35.1	7.5
Kloeckera apiculata (KK3)^x,hh	tr	15.3	67.1	tr	16.9	-	-
Kluyveromyces polysporus (EC12-4)^m,ii(1*)	11.0	12.6	51.3	1.3	21.0	-	-
Lipomyces lipoferJ^j(1*)	-	13.9	2.3	7.3	69.5	6.9	tr
L. starkeyi (IAM-4753)^x,kk	tr	31.3	6.8	1.7	52.4	7.6	-
Nadsonia fulvescens^a(5*)	-	11.7	22.6	-	22.2	41.4	-
Pichia membranaefaciens^dd(1*,5*)	-	15.0	16.0	2.0	25.0	25.0	18.0
P. membranaefaciens^m,ll(1*,5*)	tr	12.3	14.8	tr	40.9	23.6	7.7
P. farinosa^m,mm(1*,5*)	tr	25.5	1.5	1.3	23.9	46.4	-
P. fermentans^a(1*,5*)	tr	9.2	18.8	-	21.6	30.2	15.9
Pullularia pullulans^nn	0.1	30.8	2.6	8.7	41.9	13.1	0.6
Rhodotorula glutinis^a(3*,5*)	1.1	9.8	8.6	-	19.0	43.6	15.7
R. glutinis^m,oo(3*,5*)	tr	12.3	tr	1.9	31.6	53.1	-
R. glutinis^h(3*,5*)	-	9.0	1.0	14.0	58.0	2.0	-

TABLE 3.8 Continued

Fungus	Fatty Acids (%)						
	C_{14}	C_{16}	$C_{16:1}$	C_{18}	$C_{18:1}$	$C_{18:2}$	$C_{18:3}$
R. gracilis[pp](3*,5*)	1.4	22.3	1.4	11.2	50.3	9.3	2.1
R. gracilis[q](3*,5*)	1.1	29.8	1.8	8.8	40.1	11.2	4.8
R. graminis[rr](3*,5*)	3.9	31.9	0.3	3.2	37.2	10.2	4.6
R. rubra AY-2[m,q](3*,5*)	tr	22.2	tr	4.2	61.0	10.2	2.7
Np2-17-4B[m](3*,5*)	2.0	23.8	tr	2.3	56.9	11.5	3.2
Saccharomyces bailii[a](1*,5*)	-	16.5	30.2	-	30.2	23.0	-
S. carlsbergenesis[a,ss](1*,4*)	0.2	6.7	57.5	1.9	33.7	-	-
S. carlsbergenesis[m,tt](1*,4*)	2.8	6.2	50.8	1.2	37.6	-	-
S. cerevisiae[a,ss](1*,4*)	3.2	18.1	52.2	2.3	19.8	-	-
S. cerevisiae (IAM-4274)[k,uu](1*,4*)	0.7	6.2	63.0	6.7	18.3	1.3	-
S. cerevisiae (OC-2)[x,vv](1*,4*)	3.0	6.1	54.9	1.1	34.2	-	-
S. cerevisiae[ww](1*,4*)	0.3	12.7	34.5	4.1	46.1	-	-
S. cerevisiae (NCYC712)[ss,xx](1*,4*)	1.7	15.8	45.3	1.3	16.1	-	-
S. cerevisiae[yy](1*,4*)	27.3	11.3	26.4	6.7	18.8	-	-
S. cerevisiae[dd](1*,4*)	-	8.0	35.6	-	30.0	-	-
S. chevalieri[a](1*,4*)	-	5.7	51.9	-	42.3	-	-
S. delbrueckii[a,ss](1*,4*)	2.6	6.2	60.9	-	20.8	-	-
S. fragilis[zz](1*,5*)	2.5	19.2	11.9	3.4	27.0	25.1	9.6
S. fragilis[q](1*,5*)	0.2	16.7	26.0	-	21.5	31.3	3.1
S. italicus[a](1*,4*)	-	5.8	59.9	-	34.3	-	-
S. lactis[a](1*,5*)	-	6.7	31.0	-	18.1	33.9	10.4
S. oviformis[a,q](1*,4*)	0.9	11.8	63.6	-	23.3	-	-

Species							
S. rosei[a] (1*,5*)	–	10.1	31.7	–	15.5	42.7	–
S. rosei[m,aaa] (1*,5*)	3.7	14.7	47.5	1.0	33.3	–	–
S. rouxii[m,bbb] (1*)	tr	13.6	16.4	3.4	31.6	34.4	–
S. veronae[a,q] (1*,5*)	0.5	9.1	38.1	–	28.2	17.6	6.1
Saccharomycodes ludwigii[m,ccc] (1*)	1.0	13.4	3.0	tr	80.2	–	–
Schizosaccharomyces pombe[a,q] (4*)	1.1	3.5	22.8	–	70.4	–	–
S. pombe (IAM-4332)[x,ggg] (4*)	tr	15.8	1.3	7.7	74.4	–	–
S. pombe[ddd] (4*)	–	8.0	–	–	89.0	–	–
S. pombe var. *liquefaciens*[eee] (4*)	8.0	16.0	5.0	7.0	37.0	2.0	7.9
Schwanniomyces occidentalis[a] (5*)	–	17.3	17.7	–	49.6	15.5	–
S. occidentalis[x,fff] (5*)	tr	15.7	4.0	1.1	52.6	18.5	7.9
Sporobolomyces salmonicolor[m,q] (2*)	tr	24.0	tr	6.1	35.9	36.3	3.4
Torulopsis candida[m,hhh] (3*,5*)	tr	27.9	3.7	9.0	42.5	11.9	2.7
T. colliculosa[m,iii] (3*,5*)	3.4	13.9	46.5	tr	34.5	15.5	10.6
T. dattila[d,q] (3*,5*)	0.3	14.2	22.4	–	23.9	35.8	8.2
T. sphaerica[a] (3*,5*)	0.4	22.0	19.9	–	9.9	36.3	3.4
Trichosporon cutaneum[m,q] (3*)	tr	24.0	tr	6.1	35.9	36.3	–
Trigonopsis variabilis[m,q] (3*)	tr	19.0	34.4	tr	22.9	21.5	–

[a] Johnson & Brown, 1972; [b] crabtree-negative; [c] $C_{14:1}$, <2.2%; [d] Ratledge & Saxton, 1968; [e] 1-3%, C_{20}, C_{22}, C_{24}, traces of C_{10}, C_{11}, C_{12}, C_{13}, C_{15}, C_{17}, and sometimes $C_{20:3}$; [f] Hall & Ratledge, 1977; grown in two-stage continuous culture; average values over period of growth; [g] Gill et al., 1977; average values at end of 21 weeks in single-stage continuous culture; 1.5-2.3%, C_{20}, C_{22}, C_{24}; [h] Kates & Baxter, 1962; [i] <1%, C_{12}, $C_{12:1}$, C_{15}, C_{17}, $C_{17:1}$; [j] Combs et al., 1968; ATCC 10231; C_{10-13}, 0.3%; $C_{14:1}$, trace; C_{15}, 0.4%; C_{17}, trace; [k] Tajima et al., 1976; [l] <2.3%, C_{12}, C_{15}, C_{17}, $C_{17:1}$, C_{20}; strain KM-2; [m] Kaneko et al., 1976; [n] traces of C_{12}, C_{15}, C_{17}, C_{20}, and/or C_{22}; [o] traces of C_{15}, C_{20}, C_{17}, 2.4%; [p] Mizuno et al., 1966; C_{10-13}, 0.3%; $C_{17:1}$, 1.4%; grown on glucose; strain SD-14; [q] traces of C_{15}, C_{17}; $C_{22}+C_{24}$, trace; [r] trace C_{12}, C_{24}; C_{15}, 1.2%; C_{17}, 2.6%; [s] traces, C_{11}, traces of unknowns; C_{12}, C_{15}; C_{17},

TABLE 3.8   Continued

2.9%; $C_{17:1}$, 2.4%; C_{20}, 3.1%; traces C_{11}, C_{12}, C_{15}, C_{17}; $C_{17:1}$, 1.6%; C_{20}, 5.1%; traces of unknowns; [u]McMurrough & Rose, 1971; strain NCYC 321; all acids not reported; [v]Dawson & Craig, 1966; strain Y 900; batch culture, 45 hr; [w]traces of unknowns, C_{11}, C_{12}, C_{13}, C_{15}, C_{17}; C_{20}, 9.1%; [x]Itoh & Kaneko, 1974; [y]traces C_{12}, C_{15}, C_{17}, C_{20}, C_{22}, C_{24}; [z]Johnson et al., 1972; [aa]traces of C_{15}, C_{17}, C_{20}, C_{22}, C_{24}; [bb]traces of C_{15}, C_{22}, C_{24}; C_{17}, 2.0%; C_{20}, 2.8%; [cc]traces C_{10}, C_{12}, C_{15}, C_{17}; [dd]Bracco & Muller, 1969; [ee]Merdinger & Devine, 1965; C_{15}, 0.6%; C_{17}, 4.2%; $C_{18:4}$ or $C_{20:1}$, 3.1%; [ff]Haskell & Snell, 1965; C_{10-13}, 0.5%, $C_{14:1}$, 0.7%; [gg]C_{20}, 1%; [hh]traces, C_{10}, C_{12}; [ii]trace C_{10}, C_{15}; C_{12}, 2.8%; [jj]Jack, 1966; ATCC 10742, $C_{20:2}$, 0.7%; [kk]traces C_{10}, C_{12}, C_{15}, C_{17}, C_{20}; [ll]traces C_{15}, C_{17}, C_{20}; [mm]traces C_{15}, C_{17}, C_{20}, C_{22}, C_{24}; [nn]Merdinger et al., 1968; C_{10-13}, 4.5%; trace, C_{15}; C_{20}, 0.3%; $C_{20:2}$, 0.7%; [oo]traces of C_{12}, C_{15}, C_{17}, C_{20}, C_{24}; [pp]Enebo & Iwanoto, 1966; C_{10-13}, traces; C_{15}, 0.4%; C_{17}, 0.5%; $C_{17:1}$, C_{20}, 0.4%; [qq]Holmberg, 1948; C_{12} & below, 2.4%; [rr]Hartman et al., 1959; C_{10-13}, 0.8%; $C_{14:1}$, 1.1%; C_{20} & above, 0.4%; [ss]$C_{14:1}$, <2.2%; [tt]traces of C_{12}, C_{15}; [uu]traces of unknowns, C_{11}, C_{15}, $C_{15:1}$, C_{17}, C_{20}; C_{12}, 1.8%; [vv]traces C_{10}, C_{12}; [ww]Suomalainen & Keranen, 1963; C_{10-13}, 0.7%; C_{15}, 0.7%; C_{17}, 0.6%; [xx]Brown & Johnson, 1970; [yy]Chang & Matson, 1972; C_{10-13}, 3.8%; trace, $C_{14:1}$; [zz]Noble & Duelschaever, 1973; [aaa]traces of C_{10}, C_{15}; C_{12}, 1.7%; [bbb]traces, C_{12}, C_{17}; [ccc]traces C_{10}, C_{12}, C_{20}; [ddd]White & Hawthorne, 1970; [eee]Baraud et al., 1970; [fff]traces of C_{12}, C_{15}, C_{17}, C_{20}; [ggg]traces of C_{10}, C_{12}, C_{20}, C_{22}, C_{24}; [hhh]trace, C_{15}, C_{24}; C_{17}, 1.8%; [iii]trace, C_{10}; C_{12}, 1.2%; 1*Ascosporogenous yeast; 2*Ballistosporogenous yeast; 3*Asporogenous yeast; 4*Crabtree-positive; 5*Crabtree-negative.

utilis vary with the strain and culture conditions, but C_{16}, $C_{18:1}$ and $C_{18:2}$ are the principal acids in this species. This is true for most asporogenous yeasts (Table 3.8). When yeasts are separated on the basis of being Crabtree-positive or Crabtree-negative (see Chapter 2), there is a conspicuous absence of di- and polyenoic fatty acids in the former group which is composed of true yeasts. However, Crabtree-negative ascosporogenous yeasts, including some *Saccharomyces* species, produce high relative proportions of $C_{18:2}$ and in some cases $C_{18:3}$. Fatty acids with chain lengths greater than 18 carbon atoms are usually not detected in lipid extracts of *Saccharomyces* species; however, they appear to be part of sphingolipid molecules associated with the cell envelope (see below).

Several yeasts and yeast-like fungi produce extracellular hydroxy, acetylated, or long chain fatty acids. For example, 3-D-hydroxypalmitic acid has been identified as a product of *Saccharomyces malanga* (Kurtzman et al., 1974). 8,9,13-Triacetoxydocosanoic and 13-oxo-8,9-diacetoxydocosanoic acids have been isolated as products of a yeast-like fungus closely related to *Torulopsis fujisanesis* (Stodola et al., 1965). Partially acetylated and esterified (with long chain fatty acids) 8,9,13-trihydroxydocosanoic acid has been isolated from a *Rhodotorula* species (Stodola et al., 1967).

Extracellular glycolipids called sophorosides have been identified as products of some yeast-like fungi (see Chapter 5). 17-L-Hydroxyoctadecanoic and 17-L-hydroxyoctadecenoic acids are components of sophorosides from *Torulopsis magnaliae* (*T. apicola*) (Gorin et al., 1961), and 13-hydroxy-docosanoic acid is a component of a sophoroside from *Candida bogoriensis* (Tulloch et al., 1968).

See below for hydroxy fatty acids of the plasmalemma of *S. cerevisiae* and Chapter 7 for hydroxy fatty acids of sphingolipids. See Table 3.4 for unusual fatty acids of fungi.

The fatty acid compositions of mycelial Ascomycetes are similar to those of most other organisms. Palmitic is the principal saturated fatty acid and $C_{18:1}$ and $C_{18:2}$ are the major unsaturated acids. The relative proportion of these acids is variable depending on the species and culture conditions (Table 3.9). Ascomycetous thermophilic fungi

TABLE 3.9   Fatty Acids of Ascomycetes and Fungi Imperfecti (Mycelia)

Fungus	Fatty Acids (%)							
	C_{14}	C_{16}	$C_{16:1}$	C_{18}	$C_{18:1}$	$C_{18:2}$	$C_{18:3}$	C_{20}
Aspergillus daucia[a]	1.8	40.1	4.0	5.9	25.0	17.6	2.0	tr
A. flavus[b,c]	0.4	12.6	-	17.6	37.0	31.9	-	-
A. fumigatus[d]	-	19.0	0.8	3.1	35.2	41.9	-	-
A. nidulans[e]	0.7	20.9	1.2	15.9	40.3	17.0	0.2	3.8
A. niger[f,g]	0.3	15.8	0.7	7.2	21.3	37.8	15.6	-
Beauveria bassiana[h]	-	20.0	-	5.0	33.0	40.0	2.0	-
B. tenella[i,j]	1.2	20.4	-	9.2	25.6	23.7	1.5	1.2
Botrytis cinerea[b,k]	3.4	19.0	1.3	3.8	11.0	16.4	41.7	-
Cephalosporium subverticallatum[b,l]	1.6	21.6	2.7	6.8	28.3	35.0	3.1	-
Ceratocystis fagacearum[m]	6.0	12.0	2.0	13.7	16.0	17.1	3.0	2.4
Chaetomium globosum[b]	0.4	19.2	1.7	8.3	17.0	46.4	7.0	-
C. thermophile[n]	-	57.8	3.1	4.4	8.0	26.8	-	-
Cochiobolus miyabeanus[o]	-	22.0	-	3.0	7.0	59.0	9.0	-
Cylindrocarpon radicola[b,p]	3.1	22.4	1.7	7.0	24.6	25.5	9.2	-
Epicoccum nigrum[q]	-	15.0	3.0	7.0	18.0	34.0	4.0	-
Fusarium aquaeductuum[q]	-	11.0	-	4.0	25.0	58.0	1.0	-
F. moniliforme[b]	2.1	14.4	-	11.0	29.7	41.6	1.2	-
F. oxysporum f. lini[r,s]	tr	18.3	-	12.3	41.0	28.3	tr	-
F. roseum[r]	tr	17.5	-	5.3	29.1	39.7	8.2	-
F. solani[t,u]	0.5	18.5	1.0	13.6	24.0	36.5	5.1	0.9
Glomerella cingulata[v]	1.1	43.7	2.2	5.8	26.4	19.8	1.2	-
Humicola brevis[n]	-	28.8	1.9	3.7	20.4	41.3	4.0	-

Species								
H. langinosa[n]	–	21.4	–	4.5	65.2	8.6	–	–
Israia farinosa[h]	–	17.0	1.0	6.0	41.0	34.0	2.0	–
Malbranchea pulchella[n]	–	11.3	–	11.4	26.6	50.7	–	–
Metarrhizium anisopliae[h]	–	21.0	1.0	9.0	23.0	47.0	–	–
Microsporum gypseum[w]	–	17.2	–	7.7	10.4	64.0	–	–
Nectoria achroleuca[b]	–	19.0	–	14.0	24.0	37.0	5.0	–
Neurospora crassa[x]	5.2	17.8	4.3	3.4	9.8	54.7	9.4	–
Penicillium atrovenetum[y,z]	0.2	14.5	1.2	4.6	30.5	43.0	0.8	1.7
P. chrysogenum[b,aa]	3.1	12.8	–	11.9	18.8	43.1	6.4	–
P. cyaneum[bb]	2.8	18.9	4.8	8.6	14.2	45.8	4.5	–
P. duponti[n]	–	25.2	–	10.8	42.2	21.8	–	–
Pithomyces chartarum[cc]	1.4	29.5	0.7	8.1	15.0	41.3	1.1	–
Pyronema domesticum[b,dd]	0.4	13.6	–	21.2	29.5	35.1	–	–
Sclerotium bataticola[ee]	0.4	18.2	17.7	2.1	51.4	16.6	0.4	0.3
Sepedonium ampullosporum[ff]	–	30.4	–	9.1	16.1	44.3	–	–
Sporedonium epizoum[gg]	0.5	17.8	0.9	3.5	11.5	65.1	–	–
Sporotrichum[n,hh]	1.9	17.0	2.1	8.8	8.3	58.4	–	2.1
Stemphylium dendritium[ii]	0.6	18.6	1.6	3.6	17.2	51.5	6.2	0.1
Stilbella thermophila[n]	2.1	42.5	1.9	13.7	25.4	14.3	–	–
Taphrina deformans[b,ii]	4.8	21.2	6.7	4.6	51.5	7.0	2.1	–

[a]Gunasekaran & Weber, 1972; C_{13}, 1.3%; C_{15}, 1.1%; [b]Shaw, 1965; [c]C_{12}, 0.5%; [d]Kaufman & Viswanathan, 1965; [e]Singh et al., 1955; [f]Sumner, 1970; [g]C_{15}, 0.2%; [h]Tyrrell, 1969; [i]Molitoris, 1963; [j]C_{12}, 0.4%; C_{17}, 1.0%; $C_{20}:1$, 0.02%; C_{22}, 1.3%; C_{24}, 0.8%; [k]C_{12}, 3.4%; [l]C_{12}, 0.9%; [m]Collins & Kalnins, 1968; C_{12}, 8.2%; C_{15}, 3.0%; C_{17}, 2.2%; [n]Mumma et al., 1970; [o]Gribanovski-Sassu & Beljak, 1971; [p]C_{12}, 3.5%; [q]Foppen & Gribanovski-Sassu, 1968; Gribanov-ski-Sassu & Foppen, 1968; [r]Rambo & Bean, 1969; [s]C_{12}, 0.8%; [t]Haskins et al., 1964; [u]C_{15}, 0.2%; [v]Jack, 1966; [w]Wirth et al., 1964; [x]Hardesty & Mitchell, 1963; [y]Van Etten & Gottlieb, 1965; [z]C_{15}, 0.6%; C_{17}, 1.3%; [aa]C_{12}, 3.9%; [bb]Komen et al., 1969; C_{12}, 5.4%; C_{17}, 2.1%; $C_{20}:4$,

TABLE 3.9   Continued

2.1%; [cc]Hartman et al., 1960; C_{12}, 0.5%; C_{15}, 0.2%; C_{17}, trace; [dd]C_{12}, 0.4%; [ee]Gottlieb & Van Etten, 1966; [ff]Ando et al., 1969; [gg]Lopez & Burgos, non-polar soluble lipids; $C_{16:2}$, 0.04%; free fatty acids; [hh]C_{12}, 1.4%; C_{13}, 1.0%; [ii]C_{12}, 2.1%.

generally do not differ in fatty acid composition from mesophilic species. *Iso*-branched isomers of C_{16} and C_{18} have
been detected in lipid extracts of *Penicillium pulvillorum*
(Nakajima and Tanenbaum, 1968), and small quantities of
branched chain acids have been reported for *P. cyaneum*
(Koman et al., 1969).

There are few reports on the fatty acid composition of
ascospores, which is probably due to the difficulty in col-
lecting the sexual spores from most species in sufficient
quantities for analysis. However, the ascospores of *Bys-
sochlamys fulva* contain a high lipid content (23%), 10% of
which is in the spore wall (Herbert et al., 1973). Oleic
acid is the principal unsaturated acid in these spores, but
the major saturated fatty acids have chain lengths with more
than 19 carbons.

The fatty acid composition of conidia from several
species has been reported and is generally similar to that
of the mycelia (Table 3.10). However, the fatty acid composition of spores of two powdery mildew fungi differs considerably from that of other fungi. Both species have relatively high amounts of long chain saturated ($>C_{18}$) fatty
acids which are accompanied by the corresponding monoenes.
Docosanoic acid (C_{22}) comprises 41.7% of the fatty acids in
spore lipids of *Sphaerotheca humili* var. *fuliginea* (Tulloch
and Ledingham, 1960). The surface wax of *S. fuliginea*
spores contains a complex mixture of diol, alkyl, Δ^{2t} alkyl,
and methyl esters along with nonesterified (3 to 11%) fatty
acids (Clark and Watkins, 1978). The principal fatty acid
of wax from this species is also C_{22} (57%) which is accompanied by C_{24} (31%). In addition to the high relative
amounts of long chain saturated fatty acids, the most characteristic feature of fatty acids of *S. fuliginea* is monounsaturated fatty acids with a *trans* double bond at the 2-
position (Δ^{2t}). Forty-four percent of the fatty acids of
Erysiphe graminis spore lipids has not been identified and
apparently is not typical of those found in other fungi
(Tulloch and Ledingham, 1960).

Although there are quantitative differences, the major
fatty acids of conidia and mycelia of *Sporendonema epizoum*
(*Hemispora stellata*) are C_{16}, $C_{18:1}$, and $C_{18:2}$ (Lopez and
Burgos, 1976). The soluble lipids of this fungus contain
high levels (29%) of nonesterified fatty acids. The relative
proportions of fatty acids from *Fusarium roseum* and *F.*

TABLE 3.10 Fatty Acids of Ascomycete and Fungi Imperfecti Spores

Fungus	Fatty Acids (%)							
	C_{14}	C_{16}	$C_{16:1}$	C_{18}	$C_{18:1}$	$C_{18:2}$	$C_{18:3}$	C_{20}
Aspergillus niger[a]	1.7	30.6	0.8	5.3	33.3	27.4	–	tr
Curvularia sp.[b]	0.5	27.9	3.3	9.4	21.7	27.8	1.8	0.7
Erysiphe graminis[c]	0.8	8.8	1.0	2.0	4.3	6.0	5.1	0.8
Fusarium roseum[d,e]	tr	21.6	–	tr	40.9	37.5	tr	–
F. oxysporum[d,e]	tr	13.2	–	44.1	27.2	15.4	tr	–
Pithomyces chartarum[f] (*Sporidesmium bakeri*)	0.8	21.6	1.3	6.4	15.9	53.4	–	–
Sphaerotheca fuliginea[g]	–	–	–	–	–	–	–	4.0
S. humili[h]	1.1	5.3	1.0	2.2	4.5	7.0	4.3	13.0
Sporendonema epizoum[i]	0.6	25.6	0.9	1.8	23.8	45.2	0.5	–
Verticillium albo-atrum[j]	0.6	31.7	0.4	35.1	21.0	6.5	0.4	–

[a]Gunasekaran et al., 1972; C_{12}, 0.8%; $C_{14:1}$, trace; [b]Bharaucha & Gunstone, 1956; $C_{14:1}$, 1.4%; C_{15}, 0.6%; C_{13}, trace; C_{22}, trace; unknowns, 0.6%; [c]Tulloch & Ledingham, 1960; C_{15}, 0.3%; C_{22}, 5.0%; $C_{22:1}$, 7.1%; C_{24}, 7.6%; $C_{24:1}$, 6.5%; unknowns, 44.1%; [d]Rambo & Bean, 1969; [e]traces, C_{14}, $C_{18:3}$; [f]Hartman et al., 1962; C_{11}, 0.6%; C_{15}, trace; [g]Clark & Watkins, 1978; spore surface wax (free) fatty acids; C_{22}, 57%; $\Delta^2 tC_{22}$, 8%; C_{24}, 31%; C_{26}, trace; [h]Tulloch & Ledingham, 1960; C_{12}, 0.2%; C_{22}, 41.7%; $C_{22:1}$, 1.7%; C_{24}, 10.0%; unknowns, 8.0%; [i]Lopez & Burgos; non-polar soluble lipids; C_{12}, 0.2%; $C_{16:2}$, 1.2%; free fatty acids; [j]Walker & Thorneberry, 1971; C_{12}, 2.1%; C_{15}, 0.3%; C_{17}, 0.4%.

oxysporum conidia are different in that C_{18} (44.1%) accumulates in spores of *F. oxysporum* (Rambo and Bean, 1969).
This accumulation does not occur in mycelia of the same
species.

With two notable exceptions, the fatty acid composition
of sclerotia is typical of mycelia and spores of other fungi
(Table 3.11). Sclerotial fatty acids of *Sclerotinia borealis*,
S. sclerotiorum, and *Botrytis tulipae* are similar to the
mycelia of the respective species, and has a higher degree
of unsaturation when grown on their natural hosts compared
to synthetic media (Sumner and Colotelo, 1970; Weete et al.,
1970).

Unlike sclerotia of other fungi, those produced by the
ergot fungus *Claviceps purpurea*, and related species, contain 20 to 30% lipid. Sclerotial lipid of *C. purpurea* contains high relative amounts of ricinoleic acid (D-12-hydroxy-
cis-9-octadecenoic acid). Ergot oil may contain up to 44%
of this hydroxy acid. The hydroxy group of ricinoleic acid,
occurring as an ester with glycerol, may be acylated with
long chain fatty acids to form estolides (see Chapter 1).
An estolide may contain up to six fatty acid residues. Formation of ricinoleic acid occurs only during sclerotium
development and does not occur in sphacelial tissue (asexual
sporulating mycelium) (Mantle et al., 1969). Mycelium of
C. purpurea grown on a synthetic medium rarely produce
sclerotia and have higher relative proportions of $C_{18:2}$ than
sclerotia. The ricinoleic acid content of sclerotial lipid
may vary according to the isolate. Comparison of the ricinoleic acid content of sclerotial lipid from nine *C. purpurea*
isolates showed that two contained 0 and 3.3% of the hydroxy
acid and the other six contained 29 to 36% (Mantle et al.,
1969). Ricinoleic acid does not accumulate to more than 1%
of the total fatty acids in lipid from sclerotia of other
Clavicep species examined (Morris, 1967). However, 9,10-
dihydroxyoctadecanoic acid comprises 63.6% of the fatty acids
in sclerotial lipid of *C. sulcata*. *Cis*-9,10-epoxyoctadecanoic
acid has been tentatively identified in *C. sulcata* and an unidentified *Claviceps* species (Morris, 1967).

Some additional Ascomycetes and Fungi Imperfecti for
which the fatty acid composition has been reported are listed
in Table 3.12.

TABLE 3.11 Fatty Acids of Sclerotia

Fungus	Fatty Acids (%)								
	C_{14}	C_{16}	$C_{16:1}$	C_{18}	$C_{18:1}$	$C_{18:2}$	$C_{18:3}$	C_{20}	$OHC_{18:1}$
Botrytis tulipae[a]									
natural[b]	0.4	13.2	1.1	2.2	15.7	62.8	4.7	tr	–
cultured	2.5	19.6	3.3	6.8	22.2	43.6	1.6	tr	–
Claviceps paspali[c,d]	0.2	19.0	7.4	2.1	56.7	14.5	–	tr	tr
C. gigantea[c,e]	0.2	17.2	2.6	2.7	55.6	19.6	tr	tr	~1
C. purpurea[f]	0.1	19.9	6.5	4.3	22.5	14.3	tr	–	32.3
C. purpurea (S_1)[g,h]	0.7	28.0	3.7	6.4	19.6	17.4	–	–	24.1
C. purpurea (S_2)[g]	0.2	26.0	4.8	4.1	24.1	13.2	–	–	27.5
C. purpurea (S_3)[g]	1.2	25.1	4.3	4.6	20.0	9.8	–	–	34.9
C. purpurea (S_4)[g]	0.4	22.9	3.1	5.4	17.0	15.8	–	–	35.5
Claviceps sp.[c,i]	0.3	34.1	2.4	7.3	38.6	14.4	–	tr	~1
C. sulcata[c,j]	0.1	11.1	0.8	4.9	12.8	5.1	0.1	1.4	tr
Sclerotinia borealis[a]									
natural (SB3)	1.7	15.5	6.0	3.3	18.5	41.7	12.3	tr	–
cultured (SB4)	1.2	20.8	1.8	9.8	15.0	47.1	2.5	tr	–
natural (SB4)	0.3	10.5	1.0	1.1	7.2	56.5	23.3	tr	–
cultured (SB3)	0.4	10.7	0.5	0.9	5.8	80.5	1.5	tr	–
Sclerotium rolfsii[k,l]	3.3	18.6	3.2	5.1	7.3	60.7	–	–	–
S. rolfsii[m]									
16 h photoperiod	3.8	15.8	–	4.0	8.2	60.7	4.0	–	–
dark grown	8.3	16.5	–	4.3	13.3	45.5	3.7	–	–

Sclerotinia sclerotiorum

natural[n,o]	0.6	16.3	1.9	5.9	16.6	28.8	26.7	0.7	—
cultured[n,p]	tr	12.9	0.5	5.0	37.5	37.9	4.7	—	—
natural[a]	0.3	11.2	1.4	1.4	21.2	50.0	15.1	tr	—
cultured[a]	0.3	11.2	0.6	1.0	32.4	50.4	4.3	tr	—

[a]Sumner & Colotelo, 1970; [b]From tulip bulbs; [c]Morris, 1967; [d]di OHC$_{18}$, trace, [e]di OHC$_{18}$, ~1%; [f]Mantle et al., 1969; from *Spartina townsendii*; [g]Morris & Hall, 1966; [h]From rye; [i]From *Pennisetum typhoideum*; epoxy C$_{18}$, trace; di OHC$_{18}$, ~2%; [j]di OHC$_{18}$, 63.6%; [k]Howell & Fergus, 1964; [l]C$_{10-13}$, 2.0%; C$_8$, 0.4%; C$_{10}$, 0.5%; C$_{12}$, 1%; [m]Weete et al., 1970; [n]Weete et al., 1970; [o]From peas; C$_{15}$, 1.4%; C$_{17}$, 1.2%; [p]C$_{15}$, trace; C$_{17}$, 1.3%.

TABLE 3.12   Some Additional Ascomycetes and Fungi Imperfecti
for Which the Fatty Acid Composition Has Been Reported

Fungus	Reference
Aspergillus flavus	Singh, 1957
A. fresenius	Stone & Herming, 1968
A. niger	Salmonowitz & Niewiadomski, 1965
A. versicolor	Shaw, 1966
Cephalosporium acremonium	Huber & Redstone, 1967; Berger et al., 1978
C. diospyri	Ando et al., 1969
Ceratocystis coerulescens	Sprechter & Kubeczka, 1970
Penicilium chrysogenum	Mumma et al., 1970; Bennett & Quackenbush, 1969; Divakaran & Modia, 1968
P. crustosum	Audette et al., 1961; Ando et al., 1969
P. javanicum	Coots, 1962
P. lanosum	Shaw, 1965
P. lilacinum	Singh et al., 1956
P. notatum	Shaw, 1965
P. pulvillorum	Nakajima & Tanenbaum, 1968
P. roqueforti	Kubeczka, 1968
P. sophi	Salmonowitz & Niewiadomski, 1965; Singh et al., 1957
P. sopimulosum	Shimi et al., 1959
P. tardum	Ando et al., 1969
Stilbella sp.	Mumma et al., 1970
Trichophyton rubrum	Kostiw et al., 1966
Botryosphaeria ribis	Shaw, 1965
Ceratocystis coerulescens strain 431	Sprechter & Kubeczka, 1970
Chaetomium globosum	Shaw, 1965
Mycosphaerella musicola	Shaw, 1965
Neurospora crassa	Todd et al., 1957; Shaw, 1965; Brady & Nye, 1970
Botrytis cinerea	Haskins et al., 1964; Henry & Keith, 1971
Claviceps gigantea	Morris, 1967
C. paspali	Morris, 1967
C. purpurea	Morris, 1967; Thiele, 1964; Bharusha & Gunstone, 1957; Morris & Hall, 1966; Mantle et al., 1969

TABLE 3.12   Continued

Fungus	Reference
C. sulcata	Morris, 1967
Fusarium aquaductuum var. *medium*	Foppen & Gribanovski-Sassu, 1968; Gribanovski-Sassu & Foppen, 1968
F. oxysporum	Rambo & Bean, 1969
F. sambucinum	Foppen & Gribanovski-Sassu, 1968; Gribanovski-Sassu & Foppen, 1968
F. solani f. *phaseoli*	Gunasekaran & Weber, 1972
Cephalosporium acremomium corda	Huber & Redstone, 1967
Cylindrocarpon radicicola	Hartman et al., 1962
Humicola grisea	Mumma et al., 1970
H. insoleus	Mumma et al., 1970
H. nigrescens	Mumma et al., 1970
Macrophomina phaseolina	Wassef et al., 1975
Malbranchea pulchella var. *sulfurea*	Mumma et al., 1970
Sclerotium rolfsii	Gunasekaran & Weber, 1972
Sporotrichum thermophile	Mumma et al., 1970
Trichophyton mentagrophytes	Audette et al., 1961
T. rubrum	Wirth & Annand, 1964
Trichoderma viride	Ballance & Crombie, 1961
Cephalosporium falciforme	Sawicki & Pisano, 1977
C. kiliense	Sawicki & Pisano, 1977

    Basidiomycetes. Basidiomycete fungi are divided into
two subclasses: 1) Homobasidiomycetes, which are saprophytic and include mushrooms, puffballs, and related forms;
and 2) Heterobasidiomycetes, which are parasitic and include
the rusts (Uredinales) and smuts (Ustilaginales). The fatty
acid compositions of over 30 homobasidiomycetous fungi collected from their natural habitats have been reported. Generally, there are no characteristics of fatty acid distributions that distinguish homobasidiomycetes from other fungi
(Table 3.13). Chain lengths of fatty acids from these fungi
commonly range from 12 to 20 carbons, with C_{16}, $C_{18:1}$, and
$C_{18:2}$ being the predominant acids.

TABLE 3.13 Fatty Acids of the Mycelia and Fruiting Bodies of Basidiomycete Fungi

Fungus	Fatty Acids (%)								
	C_{12}	C_{14}	C_{16}	$C_{16:1}$	C_{18}	$C_{18:1}$	$C_{18:2}$	$C_{18:3}$	C_{20}
HOMOBASIDIOMYCETES									
Agaricus bisporus[a]	2.6	4.1	20.5	7.6	6.9	15.5	38.0	-	-
A. bisporus[b,c]	-	0.6	16.0	-	4.9	0.9	73.0	-	-
A. campestris[d]	2.5	1.8	11.9	-	5.5	2.6	63.4	-	9.0
A. campestris[e]	2.2	2.8	15.0	4.0	3.0	5.0	65.5	-	2.5
Amanita muscaria[f]	-	0.5	10.1	1.6	8.0	39.6	52.7	-	-
Armillaria mellea[b,g]	-	2.6	44.3	6.0	4.8	29.3	2.0	-	-
Auricularia auriculajadae[h]	-	-	19.0	-	4.0	32.0	42.0	1.0	-
Boletus edulis[i,j]	0.6	0.1	9.6	2.0	0.7	9.2	60.7	1.6	0.4
Calvatia caelata[k,l]	21.8	2.3	9.5	2.0	0.6	3.0	55.8	tr	tr
C. gigantea[k,l]	12.2	2.5	8.7	0.8	0.4	8.5	61.5	tr	0.7
C. gigantium[m]	-	-	-	-	-	-	75.0	-	-
Cantharellus cibarius[i,j,n]	0.4	0.1	8.5	2.0	2.5	15.4	45.8	0.6	0.2
C. cibarius[b,o]	-	-	24.7	-	18.7	-	-	-	1.0
Clathrus cibarius[k,p]	1.6	1.5	10.3	1.2	4.8	22.8	55.2	1.2	-
Clitocybe nebularis[i,j,r']	0.5	0.1	5.8	1.4	3.4	37.3	39.0	1.0	0.1
C. illudens[m]	-	tr	18.6	1.5	1.8	42.8	34.8	-	-
C. tabescens[b]	0.6	0.1	15.9	5.7	1.0	32.3	41.2	1.5	-
Collybia sp.[e]	2.5	3.2	13.1	1.9	1.7	6.3	54.3	17.0	-
C. velutipes[b,q,r]	-	2.2	26.0	0.8	1.8	15.9	48.2	0.8	-
Coprinus comatus[s]	-	0.7	22.5	2.3	9.6	20.4	42.0	2.5	-
Corticium solani[s]	5.5	3.5	13.4	2.2	7.3	22.3	28.4	14.9	-

Cortinarius sp.[k,q]	2.0	0.6	14.0	1.2	0.7	13.5	65.6	0.2	0.2
Daedaleopsis confragosa[h]	–	–	22.0	4.0	7.0	18.0	5.0	–	–
D. tricolor[h]	–	–	21.0	9.0	11.0	16.0	34.0	–	–
Exobasidium vexans[t]	2.4	2.5	16.4	2.6	5.6	23.1	34.2	13.2	–
Fomes annosus[u]	–	–	21.0	–	37.0	10.0	29.0	–	–
F. fomentarius[v]	–	–	35.0	3.0	10.0	17.0	4.0	1.0	–
F. igniarius[w]	–	1.5	12.3	–	12.6	–	–	–	–
Fomes sp.[e]	1.2	2.2	12.4	2.1	3.1	4.5	70.3	4.2	–
Geastrum triplex[k,p]	0.5	1.6	17.7	1.5	3.5	25.4	39.4	9.2	–
Hebeloma sp.[k]	0.4	0.5	10.9	1.7	1.1	31.3	53.0	0.3	0.1
Hydnellum sp.[k]	0.5	0.4	10.2	1.0	3.5	19.5	60.0	0.4	0.5
Hydnum rufescens[b,x]	–	1.2	24.7	7.9	2.0	54.3	6.2	–	–
Hypholoma sublateritium[b,y]	–	1.7	24.9	11.4	3.3	47.9	3.9	0.7	–
Lactarius deliciosus[b,z]	–	–	9.0	–	68.6	–	–	–	–
L. rufus[j,aa]	0.2	0.7	9.1	0.7	5.4	16.5	33.0	0.4	0.2
L. torminosus[i,j,bb]	0.1	0.1	4.7	0.5	27.5	26.6	31.5	0.5	0.1
L. trivialis[i,j,cc]	0.4	0.2	4.5	0.8	3.1	24.4	30.3	0.1	0.2
L. vellereus[b,dd]	–	0.2	7.1	0.3	53.8	21.6	14.3	2.2	–
Lycoperdon sp.[k,gg]									
coat	0.6	1.0	13.7	6.1	0.6	7.3	65.6	2.6	0.9
spores	1.7	1.7	15.7	3.9	2.5	12.6	56.4	2.0	0.4
Marasmius sp.[k]	0.4	1.6	13.3	3.0	5.2	11.4	55.9	0.4	5.9
Polyporus hirsutus[k]									
fruiting body	0.2	2.5	9.5	3.7	0.4	18.7	52.3	12.2	tr
vegetative hyphae	0.9	1.2	19.7	6.6	2.1	21.5	49.9	2.2	tr
P. pubescens[k]									
fruiting body	0.2	1.7	10.4	4.5	0.5	14.7	51.1	16.0	tr
vegetative hyphae	1.2	1.5	20.0	1.7	1.5	11.2	59.7	2.9	tr
P. ramosissimus[b,ee]	–	2.4	19.5	14.6	–	14.4	40.4	–	–
P. sulphureus[h]	–	–	2.0	2.0	3.0	35.0	30.0	–	–

TABLE 3.13　Continued

Fungus	C_{12}	C_{14}	C_{16}	$C_{16:1}$	C_{18}	$C_{18:1}$	$C_{18:2}$	$C_{18:3}$	C_{20}
					Fatty Acids (%)				
Psathyrella candolleana[b,ff]	–	–	12.6	1.6	0.7	4.1	77.6	2.9	–
Rhizoctonia lamellifera[s]	3.6	4.2	21.4	2.6	9.7	12.3	29.5	13.5	–
R. solani[hh]	–	tr	11.7–14.5	tr–3.2	2.4–3.9	15.7–24.9	54.4–66.6	–	–
Stillbum zacallo-canthum[s]	0.6	1.0	21.8	4.3	3.7	21.5	41.4	3.8	0.4
Suillus brevipes[k]	1.2	0.5	10.5	2.3	0.9	29.9	50.7	0.4	0.2
S. granulatus[k]	0.2	0.3	11.3	3.2	0.9	16.2	67.1	0.2	tr
Trametes versicolor[k]	0.5	2.1	12.1	3.9	0.5	9.9	61.0	9.1	1.0
T. cinnabarina[k,ii]	0.2	2.6	12.1	2.8	0.4	11.3	55.5	1.9	tr
T. hispida[k]	0.3	0.8	9.9	3.8	0.3	29.5	59.3	1.0	tr
Tricholoma sp.[k]	1.9	0.3	15.1	2.2	0.8	17.0	62.5	0.3	0.2
T. nudum[r']	–	1.0	29.0	1.6	12.6	33.0	35.2	26.1	–
T. terreum[b,jj]	–	0.8	17.0	1.9	2.7	42.7	27.4	3.9	–
HETEROBASIDIOMYCETES									
Tilletia controversa[kk]	–	1.7	9.2	2.5	0.2	10.6	66.2	–	–
Ustilago scitaminca[s]	1.0	2.0	19.7	3.0	7.3	28.8	32.3	1.7	–
Cronartium fusiforme[ll]	1.9	0.3	18.7	–	4.1	19.0	30.7	13.3	–

[a]Holtz & Schisler, 1971; C_{10}, 4.1%; C_{17}, 2.5%; [b]Prostenik et al., 1978; [c]C_{15}, 2.0%; $C_{15:1}$, 1.8%; C_{17}, 0.7%; [d]Hughes, 1962; C_{17}, 1.1%; [e]Shaw, 1967; [f]Talbot & Vining, 1963; C_{15}, 1.4%,

[g]C_{15}, 3.4%; $C_{16:2}$, 6.9%; C_{17}, 0.7%; [h]yokokawa, 1969; 1970; [i]Aho & Kurkela, 1978; [j]<1.6% C_{10}, $C_{14:1}$, C_{15}, C_{17}, $C_{20:1}$; [k]Sumner, 1973; <1.5% C_{15} and/or C_{17}; fatty acids of pileus; [l]C_8, C_{10}, <2.3%; [m]Bentley et al., 1964; [n]C_{21}, 4%; $C_{21:1}$, 8.4%; C_{24}, 5.3%; [o]$C_{18:x}$, 46.2%; C_{21}, 2.4%; C_{22}, 1.6%; [p]Whole fructification; C_8, C_{10} trace; q<1% C_{15}, C_{17}; [r]$C_{21:1}$, 7.0%; C_{24}, 0.2%; [r]Jack, 1966; [s]Shaw, 1965; [t]Shaw, 1969; [u]Gunasekaran et al., 1970; [v]Vishida & Mitsuhashi, 1970; [w]Epstein et al., 1966; C_{15}, 1.1%; C_{17}, 1.2%; $C_{17:1}$, 1.1%; C_{19}, 1.0%; C_{21}, 1.0%; C_{22}, 24.8%; C_{23}, 1.0%; C_{24}, 26.0%; C_{25}, trace; C_{26}, 15.4%; [x]$C_{16:2}$, 3.7%; [y]C_{15}, 0.5%; $C_{15:1}$, 1.2%; $C_{16:2}$, 3.5%; C_{17}, 1.0%; [z]$C_{18:x}$, 21.5%; [aa]$C_{21:1}$, 11.8%; C_{24}, 1.6%; [bb]$C_{21:1}$, 4.4%; C_{24}, 0.3%; [cc]$C_{21:1}$, 9.7%; C_{24}, 22.5%; [dd]$C_{14:1}$, 0.4%; [ee]C_{15}, 2.8%; $C_{16:2}$, 1.9%; [ff]C_{15}, 0.5%; [gg]C_8, C_{10}, <2.3%; $C_{20:1}$, 0.6%; [hh]Gottlieb & Van Etten, 1966; C_{15}, C_{17}, trace; range of fatty acid levels in rings of culture taken with increasing age;[ii]C_{10}, 11.3%; [jj]$C_{14:1}$, 0.9%; $C_{16:2}$, 2.7%; [kk]Trione & Ching, 1971; C_{13}, 1.0%; C_{15}, 1.7%; [ll]Weete et al., 1979; C_{18} epoxy, 5.9.

Sumner (1973) has compared the fatty acid composition of 19 homobasidiomycetes. They have $C_{18:2}$ as the principal fatty acid (40 to 75%) and C_{16} and $C_{18:1}$ as the other major acids in varying proportions. The fatty acids of the pileus and stipe of these fungi are qualitatively very similar, but the stipe lipids have a higher degree of unsaturation. Linoleic acid is higher in the stipe and, perhaps, has a role in the explosive growth of the developing fruiting bodies of these fungi (Sumner, 1973). The fatty acid composition of cultured vegetative mycelia and fructifications of *Polyporus hirsutus* and *P. pubescens* from natural habitats differ mainly by the fruiting bodies having one-half the relative C_{16} content of vegetative hyphae, and 6 to 8 times more $C_{18:3}$. Generally, homobasidiomycetes have low levels of $C_{18:3}$. *Fomes igriarius* does not contain unsaturated fatty acids, but instead a homologous series of acids from C_{14} to C_{26} with C_{22} (24.8%), C_{24} (26%), and C_{26} (15.4%) representing the major acids (Epstein et al., 1966). Fatty acids of the coat and spores of the puffball *Lycoperdon* sp. are qualitatively and quantitatively similar (Sumner, 1973).

Although the fatty acid compositions of the 19 fungi analyzed by Sumner (1973) are similar, other reports reflect considerable variation among certain homobasidiomycetes collected from their natural habitats (Table 3.13). This is true for the same species and species belonging to the same genera. For example, *Boletus edulis* *Cantharellus cibarius*, *Clitocybe nebularis*, *Lactarius trivialis*, *L. torminosus*, and *L. rufus* have low relative proportions of C_{16} (2.6 to 9.6%), $C_{18:2}$ as the principal fatty acid (31.5 to 60.7%), and high relative proportions of certain long chain ($>C_{18}$) saturated or monounsaturated fatty acids such as $C_{21:1}$ (4.4 to 12.1%) and, in the case of *L. trivialis*, C_{24} (22.5%) (Aho and Kirkela, 1978). The principal fatty acid of *C. cibarius* reported by Prostenik et al. (1978) is $C_{18:x}$ (unknown unsaturation) (46.2%) followed by C_{16} (24.7%). Five *Lactarius* species appear to be distinguished by the fatty acid content. For example, *L. deliciosus* contains only three fatty acids, 68.6% of which is C_{18} and 21.5% $C_{18:x}$ (Prostenik et al., 1978). *L. vellereus* contains 53.8% C_{18}, no $C_{18:x}$, and 14 and 22% $C_{18:1}$ and $C_{18:2}$, respectively. *L. trivialis*, *L. torminosus*, and *L. rufus* contain about the same relative proportions of $C_{18:1}$ and $C_{18:2}$, but *L. trivialis* contains 22.5% C_{24} and *L. torminosus* contains 27.5% C_{18} (Aho and Kurkela, 1978). It is not known if these differences will

hold up for isolates obtained from different locations.
Some other mushrooms appear to have characteristic features
of their fatty acid distributions. For example, *Agaricus
bisporus* and *Psathyrella candolleana* contain 73.1% and 77.6%
$C_{18:2}$, respectively, and *Armillaria mellae* contains 44.3%
C_{16} (Prostenik et al., 1978). Other *Agaricus bisporus* isolates have 29 to 38% $C_{18:2}$ (Holtz and Schisler, 1971).

Hydroxy fatty acids are generally not reported as components of most fungal total fatty acid preparations. However, they are present in sphingolipids of fungi studied
(see next section and Chapter 7). In most cases, hydroxy
fatty acids (as N-acyl esters) are probably relatively minor
components of fungi and go undetected in most fatty acid
preparations. For example, they represent less than 2% of
the fatty acids of yeast cell walls (see below). Saturated
and unsaturated (undetermined unsaturation) hydroxy fatty
acids from 15 to 25 carbons in chain length have been detected in nine mushrooms (Prostenik et al., 1978). Hydroxy
fatty acids in these fungi vary from 0 to 22% of the total
fatty acids. Hydroxy C_{24} (38.2 to 62.1% of the total hydroxy
acids) and, in some cases h C_{22} (30.1%), are the principal
hydroxy fatty acids of the species studied (Table 3.14).

The mycelium of relatively few heterobasidiomycetous
fungi has been analyzed for fatty acid content. The relative
amounts of C_{16}, $C_{18:1}$ and $C_{18:2}$ of *Tilletia controversa* and
Ustilago scitaminca differ considerably (Trione and Ching,
1971; Shaw, 1965) (Table 3.13). The fatty acid composition
of mycelium of the rust fungus *Cronartium fusiforme* is similar to other fungi, except that *cis*-9,10-epoxyoctadecanoic
acid is a minor component (Weete et al., 1979). The epoxy C_{18}
acid is characteristic of rust fungi (see below) and may be
used to confirm rust fungi in culture.

With the exception of rusts, spores of relatively few
Basidiomycetes have been analyzed for fatty acid content.
Like the fruiting bodies (Table 3.13), basidiospores of
several strains of *Agaricus bisporus* contain no $C_{18:3}$, but
have over two times more $C_{18:2}$ (69 to 86%) than fruiting
bodies and low levels of $C_{18:1}$ (2 to 6%) (Holtz and Schisler,
1971).

The spore fatty acids of several smut fungi have been
reported. Lipid extracted from intact teliospores of *Tilletia controversa* and *T. foetida* contain 63.1% and 67.3% C_{16},

TABLE 3.14   Hydroxy Fatty Acids of Fruiting Bodies of Some Basidiomycetes[a]

Fungus	Fatty Acids (%)							
	hC_{16}	hC_{18}	hC_{20}	hC_{21}	hC_{22}	hC_{23}	hC_{24}	$hC_{24:x}$
Agaricus bisporus[b]	32.0	8.0	1.7	1.0	30.1	2.9	22.5	0.8
Armillaria mellea[c]	19.6	2.8	–	1.1	7.6	1.4	40.8	17.8
Cantharellus cibarius	26.9	6.4	–	–	5.4	3.2	55.5	2.6
Clitocybe tabescens[e]	8.1	2.9	–	2.5	16.3	3.0	61.0	5.4
Collibya velutipes[d]	13.7	4.7	–	1.3	9.0	4.2	61.8	1.6
Hypholoma sublateritium[f]	13.3	2.2	–	–	10.5	4.1	62.1	4.2
Lactarius deliciosus[g]	16.2	7.8	4.6	1.1	30.1	7.7	8.9	3.4
L. vellereus	–	23.6	–	–	13.1	3.5	59.8	–
Tricholoma terreum	8.2	3.5	0.9	3.6	37.4	8.2	38.2	–

[a]Prostenik et al., 1978; [b]hC_{15}, 0.3%; hC_{17}, 0.7%; [c]$hC_{16:x}$, 3.0%; $hC_{18:x}$, 0.9%; $hC_{22:x}$, 3.5%; hC_{25}, 1.3%; $hC_{25:x}$, 0.9%; [d]hC_{25}, 3.7%; [e]$hC_{22:x}$, 0.8%; [f]$hC_{16:x}$, 0.6%; $hC_{18:x}$, 1.1%; hC_{22}, 1.9%; [g]$hC_{16:x}$, 1.3%; hC_{17}, 1.4%; $hC_{18:x}$, 1.3%; $hC_{21:x}$, 3.6%.

respectively, whereas the spores of *T. caries* contain 36.9%
C_{16} and 26.6% $C_{16:1}$ (Laseter et al., 1968). This probably
reflects the fatty acid composition of spore wall lipids
since that from fractured spores of *T. controversa* contains
16% C_{16}, 14.4% $C_{18:1}$, and 49.4% $C_{18:2}$ (Trione and Ching,
1971). This is similar to the fatty acid composition of
mycelium of the same species (Table 3.13). The spore fatty
acid composition of several *Ustilago* species is generally
similar to other fungi (Table 3.15). Generally, spores of
smut fungi contain low relative amounts of C_{20} and C_{22}
acids.

The fatty acid compositions of spores of over 20 rust
fungi have been reported (Table 3.16). Lipid from spores
of these fungi is characterized by *cis*-9,10-epoxyoctadecanoic
acid which ranges from 0 to 78% of the total fatty acids.
Lipid from the spores of some rust fungi have high levels
of $C_{18:3}$, which appears to be inversely related to the epoxy
C_{18} content. Rust spores also contain low relative proportions of fatty acids with 20 and 22 carbons. Although there
is variation among species, the fatty acid composition of
rust spores are not sufficiently distinctive to be taxonomically useful (Tulloch and Ledingham, 1960; 1962). In most
cases, the fatty acid composition of the various spore forms
(uredospores, aeciospores, basidiospores, and teliospores)
are similar, even when the spores are produced on different
hosts (Carmack et al., 1976; Weete and Kelley, 1977).

FATTY ACIDS OF CELL WALLS AND MEMBRANES

Cell Wall. Although there are numerous reports on the
chemistry of fungal cell walls, relatively little is known
about their lipid composition. As noted in Chapter 2, the
total lipid content of fungal cell walls ranges from 0.5 to
15% of the dry weight. However, caution should be exercised
in the interpretation of data on cell wall lipid content because of the difficulty in removing all the plasmalemma during cell wall isolation. Approximately 70% of the fatty
acids of *S. cerevisiae* cell wall lipid is comprised of $C_{16:1}$
and $C_{18:1}$ (Table 3.17). About 2% of the yeast cell wall
fatty acids consist of saturated, unsaturated, and 2-hydroxy-
lated homologues ranging from 19 to 26 carbons in length,
which are associated with a glycosphingolipid fraction of
the lipid extract (Nurminen and Susmalainen, 1971). The
principal very long chain fatty acids of the cell walls are
C_{20}, C_{26}, 2-OH C_{26}, and unidentified acids.

TABLE 3.15   Fatty Acids of Spores from Basidiomycete Fungi (Smuts)

Fungus	C$_{14}$	C$_{16}$	C$_{16:1}$	C$_{18}$	C$_{18:1}$	C$_{18:2}$	C$_{18:3}$	C$_{20}$	C$_{22}$
Sphacelotheca reiliana[a]	–	15.9	1.8	8.0	29.0	27.8	2.4	0.9	2.4
Tilletia foetida[b,c]	5.7	67.3	7.6	6.0	7.6	5.8	0.3	–	–
T. foetida[d,e]	0.9	15.0	2.6	1.7	12.3	46.7	1.8	1.1	1.7
T. foeteus[f]	0.5	14.6	3.0	1.0	8.8	63.2	–	0.3	0.4
T. controversa[b,c]	3.4	63.1	9.7	4.2	4.9	11.2	3.6	–	–
T. controversa[d,g]	0.4	11.1	1.9	0.5	15.8	41.2	1.2	0.7	1.7
T. controversa (free)[h]	0.9	16.0	2.1	1.3	14.4	49.4	3.2	–	–
T. caries[b,c]	2.0	36.9	26.6	2.4	4.6	14.3	13.2	–	–
T. caries[d,i]	0.3	13.3	2.5	1.3	12.9	44.9	1.5	0.8	1.6
Urocystis agropyri[j]	–	8.6	–	6.3	18.0	7.7	–	–	–
Ustilago zeae[k,l]	0.4	9.3	5.2	2.6	53.8	17.4	–	2.3	–
U. tritici[k,m]	0.3	26.9	9.7	4.7	30.2	16.4	–	1.0	–
U. nigra[k,n]	1.4	24.9	5.7	5.5	35.0	22.0	1.6	0.5	–
U. levis[k,o]	1.1	19.8	6.8	3.9	42.0	19.0	1.6	–	0.2
U. maydis[p]	1.5	15.0	1.9	6.4	23.1	22.6	3.5	2.8	1.5
U. maydis[q,r]	0.5	18.2	3.2	1.4	39.6	30.7	2.6	3.7	–
U. maydis[d,s]	3.7	28.9	3.9	1.3	12.4	5.7	–	5.8	–
U. bullata[q,t]	3.4	16.9	14.2	4.8	29.7	24.9	0.3	0.9	–

[a]Weete et al., 1969; C$_{17}$, 8.0%; C$_{24}$, 4.6%; [b]Laseter et al., 1968; [c]spore wall fatty acids; [d]Weete, 1970; [e]C$_{17}$, 0.3%; C$_{20:1}$, 1.8%; C$_{22:1}$, 0.8%; C$_{24}$, 2.4%; C$_{24:1}$, 9.0%; [f]Tulloch &

Ledingham, 1960; $C_{16:2}$, 1.2%, $C_{20:1}$, 1.2%, $C_{24:1}$, 6.0%; [g]C_{15}, 0.2%; C_{17}, 0.3%; $C_{20:1}$, 1.5%; $C_{22:1}$, 3.1%; C_{24}, 2.2%; $C_{24:1}$, 8.5%; [h]Trione & Ching, 1971; C_6, 6.3%; C_8, 4.4%; C_{10-13}, 1.1%; [i]C_{15}, 0.3%; C_{17}, 0.4%; $C_{20:1}$, 1.5%; $C_{22:1}$, 1.0%; C_{24}, 2.3%; $C_{24:1}$, 8.7%; [j]Weete et al., 1969; unknowns, 69.4%; [k]Tulloch & Ledingham, 1960; [l]C_{15}, 0.2%; $C_{20:1}$, 7.6%; [m]C_{15}, 2.7%; $C_{20:1}$, 1.9%; unknown, 3.5%; [n]C_{15}, 1.2%; $C_{20:1}$, 1.0%; unknown, 0.8%; [o]C_{15}, 0.9%; $C_{20:1}$, 3.2%; unknown, 1.5%; [p]weete et al., 1969; $C_{14:1}$, 1.2%; C_{15}, 4.1%; C_{17}, 6.7%; $C_{22:1}$, trace; C_{24}, trace; [q]Gunasekaran et al., 1972; [r]C_{12}, 0.2%; $C_{14:1}$, trace; C_{15}, 0.2%; C_{17}, trace; [s]free fatty acids; C_{13}, 1.8%; $C_{14:1}$, 13.8%; C_{15}, 13.9%; $C_{15:1}$, 2.9%; C_{17}, 3.9%; $C_{17:1}$, 1.2%; [t]C_{12}, trace; $C_{14:1}$, 2.4%; C_{15}, 1.8%; C_{17}, 0.5%.

TABLE 3.16 Fatty Acids of Spores of Basidiomycete Fungi (Rusts)

Fungus	Fatty Acids (%)										
	C_{14}	C_{15}	C_{16}	$C_{16:1}$	C_{18}	$C_{18:1}$	$C_{18:2}$	$C_{18:3}$	Epoxy C_{18}	C_{20}	C_{22}
Cronartium commandrae [a,e,f]	1.3	1.5	21.6	2.4	2.8	6.8	6.2	12.1	39.1	0.6	-
C. fusiforme [a,g,h]	1.0	0.5	12.3	0.9	4.3	5.7	6.2	9.2	49.4	0.8	2.2
C. fusiforme [b,i,j]	1.2	0.3	11.7	0.7	3.2	3.5	17.2	16.1	40.7	4.1	-
C. hasknessia [a,e,k]	1.1	0.8	17.7	1.2	4.8	6.9	5.2	12.1	38.0	0.5	1.0
C. ribicola [a,e,l]	0.7	0.9	21.6	1.6	4.0	5.9	8.0	19.1	34.4	1.1	-
Fommea obtusa var. *duchesneae* [a,e]	0.5	0.1	16.3	0.8	73.0	9.6	4.2	9.7	42.9	0.9	1.9
Gymnosporangium juvenescens [m,n]	0.6	0.7	11.6	1.0	9.6	2.9	7.0	54.6	-	2.9	3.3
Hemileia vastatrix [c,e]	-	-	5.0	1.0	3.0	2.0	7.0	1.0	78.0	2.0	1.0
Melampsora lini [c,m,o]	0.6	0.4	11.1	0.3	3.1	2.2	2.0	3.0	74.2	1.1	1.4
M. medusae [c,m,p]	1.3	0.6	32.0	1.3	5.7	7.0	6.7	7.8	30.8	0.7	1.2
Peridermium stalactiforme [a,e,q]	1.0	1.1	19.7	3.2	2.5	8.0	5.0	10.9	39.4	-	-
Phragmidium andersonii [d,e]	0.1	-	6.6	0.7	2.6	20.1	23.0	43.9	1.7	0.3	-
P. speciosum [d,e]	0.4	-	10.4	0.6	3.0	14.7	20.1	42.3	7.5	0.4	0.7
Puccinia aspargi [d,e]	0.6	0.5	18.8	1.9	2.7	18.7	8.6	40.6	4.4	1.0	2.2
P. asteris [d,e]	0.4	0.3	8.5	0.8	0.6	62.3	10.7	12.6	3.4	-	0.4
P. carthami [m]	0.4	0.4	17.5	0.5	3.2	28.4	22.2	22.6	4.8	-	-
P. coronata [a,e]	1.4	0.8	21.4	0.8	5.3	13.1	18.0	2.2	36.5	0.2	0.3
P. coronata [d,e]	0.8	0.6	14.1	0.9	1.4	25.8	22.2	16.3	15.2	1.2	1.5
P. graminis avenae [c,e]	1.9	0.2	35.5	0.5	6.0	11.4	6.1	8.0	29.5	-	-

P. graminis avenae[c,r,s]	2.2	-	31.6	1.0	5.6	1.0	3.6	6.8	8.3	2.8	30.0
P. graminis avenae[a,e]	1.6	0.7	28.2	0.4	6.4	7.8	6.1	8.2	30.1	-	-
P. graminis avenae[d,e]	0.8	0.5	23.1	0.4	3.5	15.0	4.6	8.3	41.0	-	-
P. graminis tritici[c,e,t]	2.8	0.2	42.9	1.4	4.7	5.9	4.5	11.2	26.3	-	-
P. graminis tritici[a,e]	2.1	0.3	33.6	1.8	5.2	16.9	10.0	12.6	17.5	-	-
P. graminis tritici[d,e]	1.4	0.6	20.6	1.4	3.1	17.6	9.6	21.6	23.0	-	-
P. helianthi[d,m]	1.0	0.9	23.3	1.3	4.1	15.0	11.9	23.0	16.4	0.3	-
P. hieracii[c,m]	0.3	0.4	16.3	0.5	5.5	11.7	21.1	33.7	9.5	1.0	-
P. hieracii[d,m]	0.3	0.4	17.1	0.3	4.6	15.1	19.2	34.0	8.5	0.5	-
P. pulsatillae[d,e]	0.2	0.2	11.9	3.7	3.8	6.7	63.7	3.1	0.7	3.1	1.2
P. recondita[a,e]	0.9	0.2	18.8	0.9	3.4	36.2	5.6	6.6	26.8	0.4	0.2
P. sorphi[c,e]	0.7	0.3	25.8	0.9	7.5	18.7	11.3	32.6	-	1.2	1.0
P. sorphi[d,e]	1.7	-	25.6	1.2	9.3	13.2	9.5	37.5	-	1.0	1.0
Ravinedia hobsoni[d,e]	0.8	0.6	23.7	1.6	9.2	46.1	6.4	3.0	-	1.8	1.9
Uromyces											
hedysariobscuri[d,e]	0.5	0.2	15.8	0.5	3.5	7.5	40.9	27.7	-	0.3	2.0
U. phaseoli[c,e,u]	1.1	0.6	25.4	0.7	11.5	32.7	8.3	17.4	-	0.3	2.0
U. psoraleae[a,e,v]	0.9	0.3	19.9	1.4	8.8	40.9	5.2	15.6	1.0	-	1.1

[a] Aeciospore; [b] basidiospore; [c] uredospore; [d] teliospore; [e] Tulloch & Ledingham, 1962; [f] $C_{20:1}$, 3.6%; others, 2.0%; [g] Carmack et al., 1976; [h] C_{12}, 0.2%; $C_{22:1}$, 2.3%; $C_{24:1}$, trace; [i] Weete & Kelley, 1977; [j] C_{17}, 1.3%; [k] $C_{20:1}$, 1.6%; unknowns, 9.1%; [l] $C_{20:1}$, 0.8%; others, 1.9%; [m] Tulloch & Ledingham, 1960; [n] unknowns, 5.1%; [o] $C_{20:1}$, 0.6%; unknowns, 4.2%; [p] C_{17}, 0.4%; unknowns, 4.2%; [q] $C_{20:1}$, 4.6%; unknowns, 4.6%; [r] Laseter & Valle, 1971; [s] C_{17}, 3.1%; $C_{24:1}$, 6.0%; [t] C_{24}, 1.5%; $C_{22:1}$, 6.3%; [u] unknowns, 4.9%; [v] unknowns, 5.9%.

TABLE 3.17   Fatty Acids of Yeast Cell Wall and Membrane
Lipids

Fatty Acid	Cell Wall (A)[1]	Plasma Membrane		Mitochondrial Membranes			
		(A)[2]	(B)[3]	(C)	(D)	(C)[4]	(C)[5]
C_{10}-C_{13}	7	2	0.5	tr	1.6	tr	23.2
C_{14}	6	6	1.4	0.6	3.0	4.5	10.4
$C_{14:1}$	6	2	–	–	–	–	–
C_{16}	6	6	11.4	17.9	19.6	20.5	33.7
$C_{16:1}$	51	24	48.6	43.7	44.6	6.5	12.0
C_{18}	4	1	6.3	3.6	4.1	3.9	13.7
$C_{18:1}$	18	52	31.3	34.2	24.8	61.5	7.0
$C_{18:2}$+$C_{18:3}$	<1	1	–	–	3.2	–	–

(A) Suomalainen, 1969; (B) Longley et al., 1968; (C) Paltauf
and Schatz, 1969; (D) Mudd and Saltzgaber-Müller, 1978.
[1]C_{17}, 1%; [2]C_{17}, 6%; [3]C_{15}, 0.5%; [4]Promitochondria from yeast
cells grown anaerobically in the presence of ergosterol and
Tween 80; [5]Promitochrondria from yeast grown anaerobically
in the absence of added lipids.

    Cell walls of yeast and mycelial forms of *Histoplasma
capsulatum* and *Blastomyces dermatitidis* differ mainly by the
relative proportions of $C_{18:1}$ and $C_{18:2}$. Although the relative proportions of some of these acids are between 75 and
80% of the total fatty acids in both morphological forms,
the yeast forms contain 52 to 66% $C_{18:1}$ and the mycelial
forms contain 30 to 40%.

    Plasmalemma and Mitochondrial Membranes. The composi-
tion of fungal membranes has been described in Chapter 2.
The fatty acid composition of plasmalemma lipids of two
strains of *S. cerevisiae* is similar in that $C_{16:1}$ and $C_{18:1}$
represent 76 to 80% of the total, but the relative proportions of the two acids differ (Table 3.17).

    There are relatively few reports on the fatty acid com-
position of membranes of subcellular particles. The fatty
acid composition of mitochondrial membranes is similar in
two yeast strains grown aerobically, with $C_{16:1}$ and $C_{18:1}$
being the principal acids (Table 3.17) (Paltauf and Schatz,

1969; Mudd and Saltzgaber-Muller, 1978). Promitochondria
from yeast grown anaerobically and in the absence of a lipid
supplement (ergosterol and Tween 80) have a low degree of
unsaturation and high relative amounts of short chain fatty
acids (Table 3.17) (Paltauf and Schatz, 1969).

CHAPTER 4   FATTY ACID METABOLISM

FATTY ACID BIOSYNTHESIS

The realization that carbohydrates can be converted to fat prompted several theories on the formation of fatty acids. For example, Emil Fischer proposed the "hexose condensation theory" where three sugar molecules condense to form a C_{18} fatty acid after the hydroxyl groups have been reduced (Fischer, 1890). Later, it was proposed that acetaldehyde might serve as the initial substrate for fatty acid synthesis (Nencki, 1878). This was supported by the fact that the addition of acetaldehyde to the growth medium stimulates fatty acid production in several fungi (Haehn, 1921; Terroine and Bonnet, 1927; Ottke et al., 1951). Acetate was shown to be the C_2 precursor of fatty acids when radiolabeled acetic acid resulted in greater amounts of labeled fatty acids than carbohydrates (Sonderhoff and Thomas, 1937). Probably the two most important contributions to the elucidation of the fatty acid synthesis pathway was the discovery that acetyl-CoA is the "active" form of the C_2 substrate (Lynen and Reichert, 1951; Lynen et al., 1951), and that malonyl-CoA contributes all the carbons but two to the fatty acid molecule (Wakil, 1958; Wakil and Ganguly, 1959).

The aspects of fatty acid biosynthesis covered in this chapter are *de novo* synthesis, desaturation, elongation, and the formation of some unusual fatty acids. This is preceded by a discussion of the origin of the major fatty acid building block, acetyl-CoA, and the form it takes prior to entering the process of fatty acid synthesis.

Origin of Substrates for Fatty Acid Biosynthesis.
Acetyl-CoA is the principal fatty acid building block in
the sense that it is the fatty acid degradation product. Most acetyl-CoA is provided from hexoses via glycolytic
degradation in the cytosol and subsequent decarboxylation of
pyruvate in the mitochondria (See the next section on fatty
acid oxidation for the mechanisms of "acetate" transport
across mitochondrial membranes). Acetyl-CoA may also be
formed from acetate and coenzyme A in a reaction catalyzed
by acetic thiokinase (Table 4.1a). This enzyme has been
isolated from several organisms, including yeast (Jones et
al., 1953), and partially characterized. Acetyl-CoA is also
a breakdown product of citrate as shown in Table 4.1b.

Although acetyl-CoA is the building block of fatty
acids, all but the first two carbons from this substrate
entering the biosynthetic process do so via malonyl-CoA.
One of the most thoroughly studied enzymes involved in fatty
acid biosynthesis is acetyl-CoA carboxylase which catalyzes
the carboxylation of acetyl-CoA to form malonyl-CoA. The
importance of this bicarbonate-requiring reaction was first
recognized in yeast (Klein, 1957; Lynen, 1959), pigeon liver
(Wakil, 1958; Gibson et al., 1958), and mammalian (Brady,
1958) preparations. The carboxylase is not an integral part
of the multifunctional fatty acid synthetase protein (see
below), and contains biotin. The overall reaction catalyzed
by this enzyme is a two-step process (Table 4.1c). The first
step involves the formation of an N-carboxy-biotinyl intermediate, and the second step is a transcarboxylation reaction whereby acetyl-CoA is converted to malonyl-CoA. Carboxylases from all sources catalyze this two-step reaction.
The mechanism of these reactions has been proposed by Lynen
(1967). This reaction is inhibited by avidin. Biotin-
containing enzymes are divided into two groups: Class I
enzymes utilize bicarbonate as the carboxyl donor and require
ATP. Acetyl-CoA is a Class I biotin enzyme. Class II enzymes
are ATP-independent and require an organic acid carboxyl
donor. This enzyme from S. cerevisiae is composed of four
protomers (Sumper and Riepertinger, 1972), each having a
molecular weight of 189,000. Acetyl-CoA carboxylase from
the yeast-like fungus Candida lipolytica has also been characterized (Mishina et al., 1976a). Like the same enzyme from
S. cerevisiae, acetyl-CoA carboxylase from C. lipolytica
appears to be a multifunctional polypeptide, with a subunit
molecular weight of 230,000. This enzyme from C. lipolytica
is activated by polyethyleneglycol, but unlike S. cerevisiae

TABLE 4.1   Formation of Acetyl- and Malonyl-CoA

$$\text{(1)}$$
a) Acetate + ATP $\rightleftharpoons$ Acetyl-adenylate + PPi

$$\text{(1)}$$
Acetyl-adenylate + CoASH $\rightleftharpoons$ Acetyl-CoA + AMP

$$\text{(2)}$$
b) Citrate + ATP + CoASH $\rightleftharpoons$ Acetyl-CoA + Oxalacetate +
ADP + Pi

$$\text{(3)}$$
c) HCO_3^- + Enzyme + ATP $\rightleftharpoons$ CO_2-ENZ + ADP + Pi

$$\text{(3)}$$
CO_2-ENZ + Acetyl-CoA $\rightleftharpoons$ Malonyl-CoA + Enzyme

---

1. Acetic thiokinase
2. Citrate cleavage enzyme
3. Acetyl-CoA carboxylase

(Rasmussen and Klein, 1967) it is not activated by citrate
(see below for the regulation of acetyl-CoA carboxylase).
Acetyl-CoA carboxylases from various sources have been
reviewed (Wood and Barden, 1977; Bloch and Vance, 1977).

De novo Fatty Acid Synthesis. The enzymes that catalyze the synthesis of fatty acids are referred to collectively as the fatty acid synthetase. The overall process
catalyzed by these enzymes is summarized by the following
expression:

$$Acetyl\text{-}CoA + nMalonyl\text{-}CoA + 2nNADPH + 2nH^+ \rightarrow$$

$$CH_3(CH_2\text{-}CH_2)_n CO\text{-}CoA + nCO_2 + nCoASH + 2nNADP^+ + (n\text{-}1)H_2O$$

Although a homologous series of fatty acids from 12 to 20
carbons in chain length typically occur in most organisms,
the principal product of the fatty acid synthetase has 16
carbons:

$$CH_3CO-CoA + 7HOOCCH_2CO-CoA + 14NADPH + 14H^+ \rightarrow$$
$$CH_3(CH_2-CH_2)_{14}COOH + 7CO_2 + 8CoASH + 14NADP^+ + 6H_2O$$

As indicated by the above expressions, coenzyme A derivatives of acetate and malonate are the initial substrates for fatty acid synthesis. Coenzyme A consists of adenosine -3',5'-diphosphate linked to 4'-phosphopantetheine via the 5'-phosphate diester linkage (Figure 4.1). Coenzyme A derivatives of acetate and malonate are thioesters whereby the carbonyl carbon of the substrate is linked covalently to the sulfhydral sulfur of 4'-phosphopantetheine. Although coenzyme A derivatives serve as initial substrates, the intermediates of fatty acid synthesis are "acyl carrier protein" (ACP) derivatives. ACP is composed of 4'-phosphopantetheine linked to a protein rather than adenosine -3',5'-diphosphate (Figure 4.1). The intermediates are thioesters of ACP. Bacterial ACP has a molecular weight of about 9500 (Wakil et al., 1964) and has been considered a "proteinated" form of coenzyme A (Willecke et al., 1969). In bacteria and eukaryotic photosynthetic plants, the intermediates of fatty acid synthesis occur as free ACP derivatives. This is true even though only one of the enzymes (condensing enzyme, see below) of the fatty acid synthetase shows absolute specificity for the ACP derivative. The thioester derivatives are more reactive and more water-soluble which is particularly important for the more hydrophobic intermediates. However, in animals and fungi, the intermediates are linked to 4'-phosphopantetheine which is a component of a multifunctional protein (fatty acid synthetase) that catalyze the reactions of fatty acid synthesis (see below). The intermediates are not released from the synthetase until an appropriate acyl chain length is reached, signifying completion of fatty acid synthesis process.

There are four types of enzymatically catalyzed reactions involved in fatty acid biosynthesis and appear to be of universal occurrence: transfer, condensation, reduction, and dehydration. As noted above, it is now well-established that the intermediates in fatty acid biosynthesis are the acyl carrier protein (ACP) rather than the coenzyme A derivatives. Thus, the first reactions in fatty acid formation are the transfers of the acetyl and malonyl moieties from CoA to ACP. These reactions are catalyzed by acetyl-CoA:ACP and malonyl-CoA:ACP transacylases, respectively, which constitute part of the fatty acid synthetase.

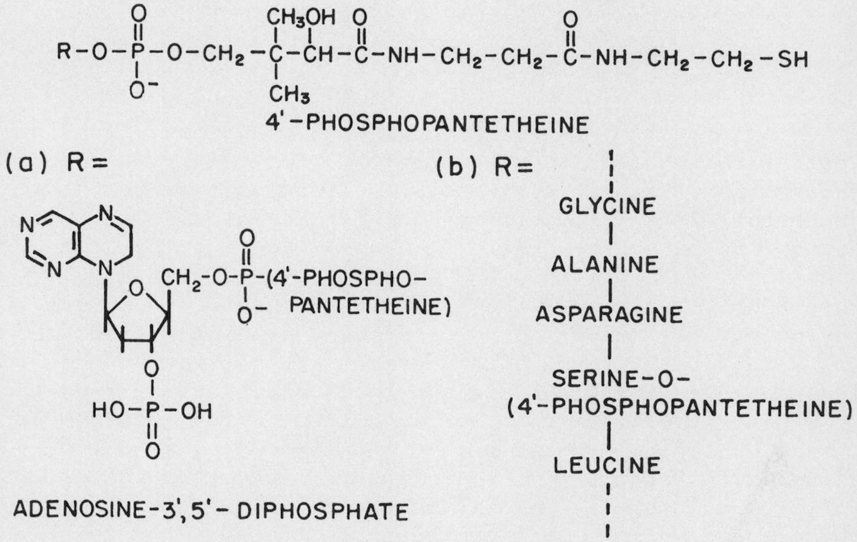

Figure 4.1   Structures of (a) Coenzyme A (CoA) and (b) acyl
carrier protein (ACP).

The active sites of these enzymes from yeast (Schweizer et
al., 1970) and other organisms (Kumar et al., 1972) is the
hydroxyl of serine. The next reaction in fatty acid biosynthesis is the condensation of the acetyl and malonyl moieties
with the loss of CO_2 to form acetoacetyl-ACP. This reaction
is catalyzed by β-ketoacyl-ACP synthetase which is referred
to as the "condensing enzyme." In yeast, the reaction proceeds by first transferring the acetyl moiety from ACP to
the active site of the condensing enzyme which is the SH
group of cysteine. Condensation involves the transfer of
the acetyl moiety from the active site to the C-2 of the
malonyl-ACP coupled with the loss of CO_2, resulting in the
formation of acetoacetyl-S-4'-phosphopantetheine-enzyme (for
fatty acid synthetase complex, see below) or acetoacetyl-ACP.
(It should be recognized that the intermediate will be bound
to the enzyme or a free ACP derivative depending on the
source of the fatty acid synthetase. Only the free derivatives will be noted below.) The importance of malonyl-ACP
rather than acetyl-ACP as the second substrate for condensation is explained on a thermodynamic basis. The loss of CO_2
is believed to shift the equilibrium of the reaction toward

condensation (Lynen, 1961). The condensation reaction is
followed by reduction of acetoacetyl-ACP to D(-)-β-hydroxy-
butyryl-ACP. This reaction is catalyzed by 3-ketoacyl-ACP
reductase and is specific for NADPH and the D(-) stereoisomer
of the substrate. The reduction is followed by the dehydration of the hydroxy intermediate to crotonyl-ACP (trans-2-
butenyl-ACP) catalyzed by D(-)-β-hydroxybutyryl-ACP dehydrase.
This enzyme is highly specific for 3-hydroxy-ACP derivatives,
and several dehydrases with different chain length specificities have been isolated from different sources. The final
chain modification involves the reduction of crotonyl-ACP to
butyryl-ACP in a reaction catalyzed by enoyl-ACP reductase.
This second reduction requires NADPH and FMN (fungi only).
Butyryl-ACP is the result of acetyl- and malonyl-condensa-
tion, and subsequent chain modifications. The above sequence
of reactions are repeated with the butyryl-ACP moiety substi-
tuting for acetyl-ACP until the longer chain lengths (C_{16} and
C_{18}) more prevalent in nature are formed. For example, the
production of an acyl group containing 16 carbons would
require one acetyl and seven malonyl moieties with the
former contributing the two terminal carbons and the latter
contributing the others. The terminal reaction of fatty
acid synthesis appears to vary with the organism and may
include: 1) transfer of the acyl group to free ACP (applies
to fatty acid synthetase complexes, see below), 2) transfer
to CoA, or 3) hydrolysis to the free acid. In yeast, in
vitro products of the fatty acid synthetase are coenzyme A
derivatives of C_{16} and C_{18}. The pathway of fatty acid biosynthesis as described above is outlined in Figure 4.2.

Although the reactions of fatty acid biosynthesis are
of universal occurrence, the type of organization of the
enzymes of the fatty acid synthetase, as indicated above,
differs depending on the organism and in at least one instance
on the growth conditions. In bacteria and plants, the fatty
acid synthetase occurs as seven individual enzymes which
have been comprehensively reviewed (bacteria) (Volpe and
Vagelos, 1973). This has been referred to as Type II fatty
acid synthetase. In animals and fungi, on the other hand,
the catalytic functions of the fatty acid synthetase are not
separable into individual enzymes and have been referred to
as multienzyme complexes. Fatty acid synthetases with this
type of organization are Type I and have been studied extensively in mammalian tissues and avian liver, and in yeast.
The pioneering work on the yeast fatty acid synthetase was
done in Germany by Feodor Lynen and his colleagues. Lynen

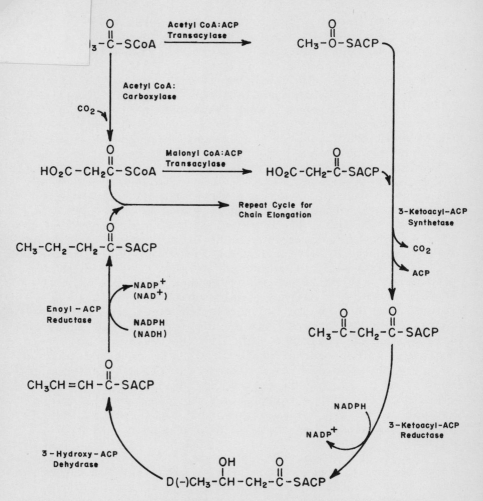

Figure 4.2   Pathway of *de novo* fatty acid biosynthesis.
(Intermediates may be ACP derivatives as shown or acyl-S-
enzyme derivatives if catalyzed by multifunctional proteins.)

visualized the synthetase as being composed of seven distinct
tightly bound enzymes as diagramed in Figure 4.3.   The molecular weight of the yeast fatty acid synthetase is 2.2×10^6
which is similar to that from filamentous fungi studied such
as *Penicillium patulum* (2.2×10^6, Holtermuller et al.,

1970), *Neurospora crassa* (2.3×10^6, Elovson, 1975), and
Pythium debaryanum (4.0×10^6, Law and Burton, 1973).
Fungal fatty acid synthetases are considerably larger than
those from animals which average about 0.5×10^6 in molecular weight. Also, the yeast synthetase contains FMN which
is absent from other synthetases. ACP is also a constituent
of the yeast fatty acid synthetase complex and has a molecular weight of 16,000 which is high compared to that (9,500)
from other systems (Willecke et al., 1969). There may be
two forms of yeast ACP since two protein bands can be separated by electrophoresis. In view of new evidence on the
physical nature of the fatty acid synthetase (see below) it
is unlikely that ACP occurs as a discrete unit in the synthetase. The purified yeast synthetase has outer dimensions of
210×250 angstroms and appears as a trimer (three complexes)
when viewed electron microscopically (Lynen et al., 1964).
The complex is not associated with any particulate component
of the cell (Pirson and Lynen, 1971). Although similar in
many respects, fatty acid synthetases from different sources
vary in amino acid composition (Kumar et al., 1972).

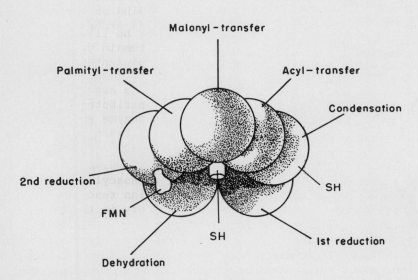

Figure 4.3   Diagrammatic representation of the yeast multienzyme fatty acid synthetase complex (Lynen, 1959).

A new concept of the yeast fatty acid synthetase has
emerged and is based on comprehensive genetic and biochemical
analyses by Schweizer and his colleagues (Schweizer et al.,
1973; Schweizer and Bolling, 1970; Kuhn et al., 1972;
Knobling et al., 1975; Schweizer et al., 1971; Tauro et al.,
1974; Schweizer et al., 1978). The evidence suggests that
the yeast fatty acid synthetase is composed of two multifunctional proteins, designated α and β, with molecular
weights of 185,000 and 180,000 daltons, respectively. The
α and β subunits contain 3 and 5 of the synthetase functions,
respectively, and is believed to exist as six copies of each
subunit, $\alpha_6\beta_6$. This is consistent with the well-known ina-
bility to separate the yeast and animal fatty acid synthetase
complexes into individual monofunctional enzymes, which can
normally be accomplished with multienzyme complexes. Synthesis of the two subunits is coded by two unlinked genes,
fas 2 and fas 1, respectively. The individual catalytic
activities appear to lie at distinct sites and, with the
exception of the acyltransferase activity, they function
independently. The distribution of catalytic functions and
cofactors between the two yeast fatty acid synthetase subunits is given in Table 4.2 as verified by Lynen and his
associates (Wieland et al., 1977). It is suggested that a
multifunctional protein is favored on the grounds of kinetics
and regulation over a complex of the corresponding individual
enzymes (Schweizer et al., 1978).

Lynen (1967) explained the mechanism of fatty acid synthesis with the multienzyme complex in mind on the basis of
a "central" and a "peripheral" sulfhydral group. While the
terms "central" and "peripheral" may not be literally appli-
cable as Lynen intended them, they may remain useful in
explaining the mechanism of fatty acid biosynthesis by the
fatty acid synthetase composed of two multifunctional proteins. The central SH group is contributed by 4'-phospho-
pantetheine (ACP function) now known to be associated with
the α subunit. The peripheral SH is contributed by cysteine
at the active center of the condensing enzyme which is also
located on the α subunit. The "priming" reaction of fatty
acid synthesis is the transfer of the acetyl moiety from the
serine hydroxy group of the acyl transferase (subunit β) to
the peripheral SH group. The first reaction of the chain
elongation process is the transfer of the malonyl moiety from
the serine hydroxyl group of malonyl transacylase (subunit β)
to the central SH group. The condensation reaction involves
transfer of the acyl moiety (acetyl moiety in the case of

TABLE 4.2   Characteristics of Fatty Acid Synthetases from
Various Sources

Characteristic	Source of Fatty Acid Synthetase			
	Bacteria	Plant[a]	Mammals	Fungi
Organization	Individual enzymes	Individual enzymes	Multifunctional protein	Multifunctional protein
Number and molecular weight of subunits	-	-	2 (220,000-270,000)	2 (180,000-200,000
Number of subunit copies	-	-	-[b]	$\alpha_6\beta_6$
Molecular weight	-	-	410,000-547,000	2,300,000
4'-Phosphopantetheine	ACP	ACP	1	~6
FMN	-	-	-	~6
Product	ACP ester	ACP ester	Acid	CoA ester

[a]The fatty acid synthetase of *Euglena gracilis* grown photo-auxotrophically has characteristics similar to those of bacteria and photosynthetic plants, but it has characteristics similar to that from animals when grown heterotrophically.

[b]Not well established whether the two subunits are duplicate copies of same multifunctional protein or different proteins with similar size.

first cycle of fatty acid synthesis) to that of the malonyl residue where the C-1 of the acyl moiety is covalently linked to the C_2 of the malonyl moiety. Subsequent chain modifications (reduction, dehydration, reduction) occur with the

substrate linked to the central SH. The 4'-phosphopante-
theine molecule, which binds acyl chain undergoing modification during the entire process, is visualized as a flexible
arm (20 angstroms in length) that allows optimum juxtaposi-
tion of the substrates and active sites for the reactions
to occur. The saturated acyl moiety may then be transferred
back to the peripheral SH to continue the chain lengthening
process, and to free the central SH for a new malonyl moiety.
Alternatively, when the chain lengthening process is complete
the acyl chain may be transferred to coenzyme A, freeing the
complex for participation in the synthesis of another fatty
acid. See Table 4.3 for a comparison of the fatty acid
synthetases from various organisms.

An important question relating to the mechanism of
fatty acid synthesis is why are fatty acids of 16 and 18
carbons the principal products of the synthetase? Selection
of these chain lengths cannot be explained on the basis of
chain length specificity of the acyl transferase since this

TABLE 4.3   Distribution of Catalytic Functions and Cofactors
between the α and β Subunits of the Yeast Fatty Acid Synthetase

Fas 1 (Subunit β)	Fas 2 (Subunit α)
Acetyl-CoA:ACP transacylase	4'-Phosphopantetheine
Malonyl-CoA (palmityl):ACP	(ACP function)
transacylase[a]	β-Ketoacyl synthetase
D(-)-β-Hydroxybutyryl:ACP	(condensing enzyme)
dehydrase	β-Ketoacyl reductase
Enoyl reductase	
FMN	

[a]In yeast, the malonyl and palmityl transacylase activities
are associated with the same catalytic site (serine hydroxyl)
whereas they are distinct sites in animal fatty acid synthetase complexes. Fatty acid synthetase complexes from various sources have been compared by Vance and Bloch (1977).

enzyme shows no preference for C_2 to C_{20} acids (Schweizer et al., 1970). This phenomenon may be related to the mechanisms of transfer and condensing reactions, the relative rates of these reactions, and the relative affinities of the transfer-binding sites for the intermediates and products of the synthetase. Before chain lengthening can continue, the saturated acyl group must be transferred from the central SH group to the peripheral SH group so that the incoming malonyl moiety can link to the central SH. The terminal reaction requires that the acyl group be transferred to the active site of the transacylase [malonyl (palmityl) transacylase] and then to coenzyme A. Transfer of the substrate to the peripheral SH rather than the terminal transacylase prior to reaching chain lengths of 16 or 18 carbons may be explained on the basis of either relative spatial separation between the substrates and competing active sites, or greater affinity of peripheral SH for the shorter chain lengths. In either case, the preference for the C_{16} and C_{18} chain lengths may be due to an interaction between the growing chain, which is becoming increasingly lipophilic, resulting in conformational changes in the quarternary protein structure. The two competing functions reside on different subunits of the synthetase complex. The conformational changes may bring the transacylase closer to the substrate making it more available, or decrease the affinity of the peripheral SH for the substrate. This is supported by evidence that the growing acyl chain interacts with the enzyme only after a chain length of 13 carbons is reached, and this interaction changes the relative velocities of the transferring and condensing activities in favor of product formation by an energy increment of -0.9 kcal per methylene (Sumper et al., 1969).

Regulation of *de novo* Fatty Acid Synthesis. The regulation of fatty acid synthesis should be considered on the basis of short-term control relating to substrate and cofactor concentrations or activation-deactivation of preformed enzymes, and long-term control involving changes in enzyme content. These control mechanisms in various systems have been reviewed by Vance and Block (1977).

Much attention has been focused on acetyl-CoA carboxylase as an important factor in the regulation of fatty acid synthesis. The enzyme is subject to both short-term and long term control. Acetate carboxylation is generally considered the rate-limiting step of fatty acid biosynthesis. Yeast carboxylase activity is stimulated by several metabolites,

particularly citrate (Rasmussen and Klein, 1967); this is
also true for *Candida* 107 (Gill and Ratledge, 1973). Citrate
acts as an allosteric effector by changing the conformation
of the protein in the region of the biotin prosthetic group,
allowing association of inactive protomers into functional
active units. Short-term regulation of acetyl-CoA carboxylase also occurs through feedback inhibition. The carboxylase is more sensitive to palmityl-CoA than the fatty acid
synthetase (Lust and Lynen, 1968; White and Klein, 1966;
Volpe and Vagelos, 1973). Palmityl-CoA not only antagonizes
the positive effects of citrate but also influences the production of citrate in the mitochondria and the transport of
citrate into the cytoplasm. A dual role for acetyl-CoA
carboxylase relating to the control of fatty acid synthesis
has been recently described by Vance and Bloch (1977). In
addition to providing malonyl-CoA for fatty acid synthetase,
the carboxylase may serve as a temporary acceptor for acyl-
CoA and deliver the acyl-CoA to the membrane site of phospholipid synthesis. Thus, cytoplasmic acyl-CoA levels are
kept low. If the acyl-CoA level exceeds the capacity to
synthesize phospholipids then the acyl-CoA remains bound to
the carboxylase resulting in inhibition of malonyl-CoA, and
hence fatty acid synthesis. In animals, there is some evidence that the control of acetyl-CoA carboxylase activity
may be due to chemical modification of the protein, i.e.
phosphorylation-dephosphorylation (Carlson and Kim, 1974).

The cellular content of acetyl-CoA carboxylase of *Candida lipolytica* varies according to the carbon substrate
available for growth. Cells grown on n-alkanes or fatty
acids exhibit lower levels of the carboxylase than cells
grown on glucose. This reduced carboxylase activity is due
to diminished synthesis of the enzyme rather than increased
degradation (Mishina et al., 1976). Growth of *Candida* 107
on n-alkanes completely represses formation of acetyl-CoA
carboxylase (Gill and Ratledge, 1973).

Although difficult to measure, the rate-limiting step
of the fatty acid synthetase *per se* appears to be that catalyzed by the condensing enzyme (Lynen, 1969; Schweizer and
Bolling, 1970). Under the conditions employed, the rate of
condensation was 10 times less than the next slowest reaction
which was dehydration (Table 4.4). The condensation reaction
also appears to be rate-limiting in animal fatty acid synthetases, which are multifunctional proteins, but not in the
fatty acid synthetase of bacteria which is composed of

TABLE 4.4   Rates of Individual Reactions Catalyzed by Fatty
Acid Synthetases from a Wild-Type Strain and FAS-14 Mutant
Yeast (Schweizer and Bolling, 1970)

Reaction	Specific Activity[a]	
	Wild-Type (×2180, mat 1-α)	FAS-15
Malonyl transfer	11,000	10,500
Acetyl transfer	32	33
Condensation	0.36	--[b]
First reduction	5,000	4,800
Dehydration	3.3	3.9
Second reduction	33,000	32,500
Palmityl transfer	370	370
Fatty acid synthetase	2,050	--[b]

[a]Activities are expressed as units/mg; a unit is defined as
the turnover of 1 μmole of substrate per minute.

[b]No detectable activity.

individual enzymes. In this later case, the acetyl and
terminal transacylase reactions appear rate-limiting (Kumar
et al., 1972).

Palmityl-CoA inhibits the fatty acid synthetase and
also causes dissociation of the synthetase (Willecke and
Lynen, 1969).

In yeast, synthesis of the synthetase subunits appears
to be coordinated in that the formation of one stops in the
absence of the other, preventing overproduction of either
subunit (Schweizer et al., 1978). In S. cerevisiae, which
cannot utilize exogenous fatty acids, the synthetase is
formed constitutively and is not repressed by fatty acids in
the growth medium (Meyer and Schweizer, 1976). The reverse
is true for Candida lipolytica which can use exogenous fatty
acids as a carbon source and Candida 107 where the synthetase
is partially repressed by growth on n-alkanes (Gill and Ratledge, 1973). In rat liver, the fatty acid synthetase is

assembled in three stages (Vance and Bloch, 1977). First
the multifunctional polypeptide chains are synthesized from
amino acids. The next step involves association of the subunits lacking the 4-phosphopantetheine prosthetic group.
The third step involves the enzymatically catalyzed addition
of the prosthetic group to the protein to form a functional
fatty acid synthetase. It is reasonable to expect a similar
sequence of steps in the formation of the fungal fatty acid
synthetase and perhaps one or more of the steps are involved
in the adaptive regulation of fatty acid synthesis.

The principal characteristics of fatty acid biosynthesis and β-oxidation are compared in Table 4.5.

Fatty Acid Elongation. As described above, *de novo*
fatty acid biosynthesis results in the formation of fatty
acids with chain lengths up to 16 and perhaps 18 carbons.
However, as shown in Chapter 3, fatty acids with greater
than 16 carbons are abundant in nature, including some fungi.
Therefore, the production of fatty acids with more than 16
carbons must occur independent of the fatty acid synthetase.
This is illustrated by a mutant of *S. cerevisiae* deficient
in fatty acid synthetase that can elongate dietary fatty
acids (C_{13} to C_{17}) by one or more C_2 units depending on the
fatty acid provided (Orme et al., 1972).

Fatty acid elongation has been demonstrated in plants
(Martin and Stumpf, 1959) and animals (Fulco and Mead, 1961).
There appears to be two types of fatty acid elongating systems. One involves the condensation of acetyl-CoA with a
preformed acyl-CoA, and the other requires malonyl-CoA as
the elongating unit.

Outer membranes of mammalian mitochondria contain
enzymes that catalyze the elongation of medium and long
chain fatty acids (Mooney and Barron, 1970; Colli et al.,
1969; Whereat et al., 1969). Acetyl-CoA is the C_2 donor
for this elongation system, and NADPH and NADH are required
for maximum activity. 3-Hydroxy and Δ^2 fatty acids appear
to be intermediates in the elongating process (Barron and
Mooney, 1970). There is no evidence of a keto intermediate,
and the unfavorable energetics for the addition of C_2 units
via acetyl-CoA are believed to be overcome by rapid reduction
to the 3-hydroxy intermediate.

The second type of fatty acid elongation is associated
with the microsomal cell fraction. Malonyl-CoA is the C_2
donor and NADPH is required for the elongation process.
Saturated and unsaturated fatty acids are substrates for
this system. 3-Keto, 3-hydroxy, and Δ^2-unsaturated intermediates have been detected.

Long chain fatty acids of 20 to 26 carbons are not
detected in *Candida utilis* or *S. cerevisiae* grown on lipid-
free media, but they are detected when C_{22} is added to the
medium (Fulco, 1967). The chain elongation process in *C.
utilis* is specific for C_{20} to C_{24} acids. When these acids
are added to the medium, C_{26} is the principal product of the
elongation reactions. When C_{21} or C_{23} acids are added, C_{25}
and C_{27} acids are the products of elongation.

Biosynthesis of Unsaturated Fatty Acids. The majority
of the fatty acids that accumulate in biological systems
possess one or more carbon-carbon double bonds. See Chapter
3 for details of the structure, nomenclature, and distribution of unsaturated fatty acids. This section includes a
description of the pathways of mono- and polyunsaturated
fatty acid biosynthesis, the desaturases, the mechanism of
desaturation, and the formation of some unusual fatty acids.

There are two pathways of monoenoic fatty acid formation, one is anaerobic and the other is aerobic. The anaerobic pathway appears to be restricted to bacteria and is a
nonoxidative pathway rather than oxygen-requiring as described below. In this pathway, the formation of saturated
fatty acids proceed as described above until the 10 carbon
level is reached. At this point, β-hydroxydecanoyl thioester
dehydrase catalyzes the dehydration and double bond isomerization leading to the formation of *cis*-3-decenoyl-ACP(β,γ)
and *trans*-2-decenoyl-ACP(α,β) (Figure 4.4). The *trans* isomer
is the normal intermediate in saturated fatty acid synthesis
and undergoes reduction and chain elongation to the preferred
saturated product. The *cis* isomer cannot be reduced by
enoyl-ACP reductase but instead continues in the elongation
process as an unsaturated intermediate. The intermediates
are ACP derivatives. The principal product of anaerobic
monounsaturated fatty acid synthesis is vaccenic acid,
$C_{18}\Delta^{11c}$. The anaerobic pathway has been reviewed by Bloch
(1969).

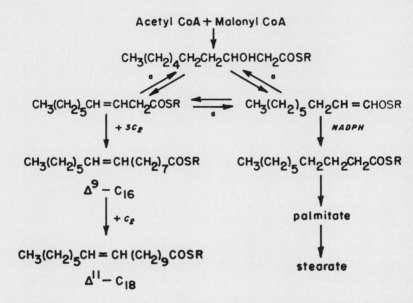

Figure 4.4   Anaerobic pathway of monounsaturated fatty acid
   biosynthesis in *E. coli* (Bloch, 1969).   Reactions catalyzed by β-hydroxydecanoyl-thioester dehydrase are indicated by the letter a.

The aerobic synthesis of monounsaturated fatty acids
occurs in all eukaryotic organisms and involves the removal
of two hydrogens from a preformed acyl chain.   This reaction requires molecular oxygen and NADPH or NADH.   However,
the substrate for fatty acid desaturases differs depending
on the source of the enzyme.   Desaturases from animals,
fungi, and bacteria are specific for the coenzyme A derivative of the fatty acid and plant desaturases are specific
for the ACP derivative.   Desaturases requiring the coenzyme
A derivatives are associated with the microsomal cell fraction and those in plants are soluble and present in the
chloroplast.

Desaturases from animals (liver) have been studied
extensively and have been characterized as multicomponent
systems containing NADH-cytochrome b_5 reductase, cytochrome
b_5, and a cyanide sensitive factor which is a non-heme protein (Holloway and Wakil, 1970; Shimakata et al., 1972;
Holloway, 1971; Strittmatter et al., 1974).   Phospholipid

is an essential component for the NADH-dependent electron
transport system (Jones et al., 1969; Holloway, 1971). The
sequence of electron transfer has been confirmed by recon-
stitution of the stearyl-CoA desaturase from the resolved
components. The corresponding desaturase from *S. cerevisiae*
(Bloomfield and Bloch, 1960) contains essentially the same
components as the hepatic system (Yoshida and Kumaoka, 1969;
Yoshida et al., 1974a; Yoshida et al., 1974b; Tamura et al.,
1976. Present evidence suggests that the yeast microsomal
desaturase catalyzes the formation of monoemoic fatty acids
by the reaction sequence shown in Figure 4.5. Desaturase
systems similar to yeast have been reported for the filamentous fungi *Neurospora crassa* (Baker and Lynen, 1971) and
Penicillium chrysogenium (Bennett and Quackenbush, 1969).

Desaturase systems from plants also contain electron
transport components, one of which appears to be ferredoxin,
and are specific for NADPH and C_{18}-ACP (Nagai and Bloch,
1965; 1966; 1968; Jaworski and Stumpf, 1974).

Although the desaturase systems have characteristics
similar to the mixed function oxygenase systems that catalyze the hydroxylation of hydrocarbons, fatty acids, steroids,
etc. (see Chapter 8), no hydroxy intermediate has been detected in fatty acid desaturation (Nagai and Bloch, 1968).
Also, attempts to convert ACP or coenzyme A derivatives of
9- or 10-hydroxystearate to oleate have been unsuccessful
(Light et al., 1962).

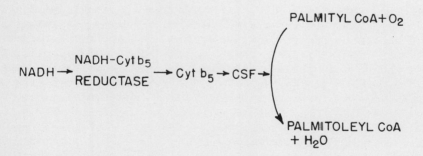

Figure 4.5   NADH-dependent fatty acid desaturase system of
  S. cerevisiae microsomes (Tamura et al., 1976). Cyto b_5,
  cytochrome b_5; CSF, cyanide sensitive factor.

Regardless of the source, desaturases are highly specific for removing hydrogens from the 9 and 10 carbons resulting in the formation of the Δ^9 double bond. This is true regardless of chain length. This probably arises from the enzyme binding the substrate at the carboxyl end, allowing the 9 and 10 methylene groups to be properly positioned at the active center (Howling et al., 1972; Brett et al., 1971). Although $C_{16:1}\Delta^9$ and $C_{18:1}\Delta^9$ are the principal monoenes of yeast, the desaturase from this system converts C_{10} to $C_{10:1}\Delta^9$ (Schultz and Lynen, 1971). It has been suggested that the appearance of certain monoenes in a species is governed by the acyl-thioester substrates provided by the fatty acid synthetase rather than by the chain length specificity of the desaturase. Other than binding the substrate, the requirement for thioesters cannot be readily explained. The thioester does little to "activate" the 9 and 10 positions of the acyl group. It has been proposed that the hydrocarbon chain may not be fully extended in the enzyme-substrate complex but may assume a pseudoannular conformation allowing the reacting carbon atoms to approach the thioester function (Richards and Hendrickson, 1964). Furthermore, the oxygen may attack the sulfur and not a carbon atom, resulting in a perthioester that would provide an "active" oxygen for withdrawal of hydrogens from the 9 and 10 methylene groups.

The stereochemistry of hydrogen removal appears to be the same in plants, animals, and bacteria. Using stereospecifically labelled stearate, it has been shown that the D-9- and D-10-hydrogens (tritium) are lost during desaturation of oleate and the L-9- and L-10-atoms are retained (Schroepfer and Bloch, 1965; Morris et al., 1967; Morris, 1970). Although the stereochemistry of hydrogen removal is established, the mechanism remains unclear. In *Corynebacterium diphtheriae*, there appears to be a stepwise removal beginning with the D-9 hydrogen (Schroepfer and Bloch, 1965). In algae (*Chlorella*) and mammalian systems, on the other hand, oleate formation appears to involve a simultaneous concerted hydrogen removal from stearate (Morris, 1970). This eliminates the requirement of an oxygenated intermediate in fatty acid desaturation.

Unsaturated fatty acids with more than one double bond are classified according to the number of carbons from the terminal methyl to the nearest double bond, i.e. ω^9, ω^6, ω^3. The biosynthetic relationships between these families of

polyunsaturated fatty acids are shown in Figure 4.6. Higher
polyunsaturated fatty acids are produced by desaturation,
elongation, desaturation. Plants and animals are both capable of producing polyenes, but differ in the way they accom-
plish the desaturation. With the thioester of $C_{18:1}\Delta^{9c}$ serving
as the substrate, plants form the corresponding di- and
trienes via a terminal methyl-directed desaturation, result-
ing in the formation of the ω^6 acid linoleate ($C_{18:2}\Delta^{9,12}$)
and ω^3 acid α-linolenate ($C_{18:3}\Delta^{9,12,15}$). Animals, on the
other hand, carry out the desaturation in the carboxyl-
direction resulting in the formation of the ω^9 acid $C_{18:2}\Delta^{6,9}$.
Animals produce the ω^6 acid $C_{18:3}\Delta^{6,9,12}$ (γ-linolenate) from
dietary $C_{18:2}\Delta^{9,12}$. This also occurs in the phycomycetous
fungi. In animals, dietary linoleate is a precursor of the
hormones prostaglandins via arachidonic acid ($C_{20:4}\Delta^{5,8,11,14}$).
Since the corresponding diene $C_{18:2}\Delta^{6,9}$ produced by animals
will not substitute in this pathway, linoleate is considered
an essential fatty acid and is required in the diet of animals.

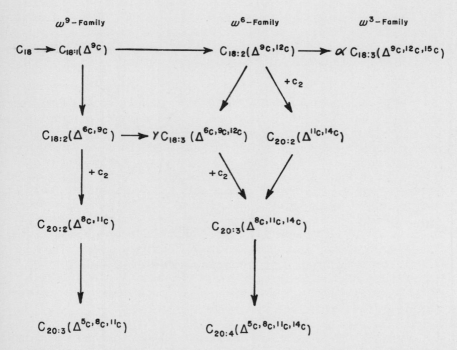

Figure 4.6   Pathways of polyunsaturated fatty acid biosynthesis in plants and animals.

The formation of α-linolenic acid in plants and fungi
is believed to occur via sequential desaturation:

oleate → linoleate → α-linolenate

Typical data illustrating this sequence of reactions is
shown in Figure 4.7. In *Penicillium chrysogenium* exposed
to radioactive acetate during growth, radioactivity appears
rapidly in oleate, and this is followed by a decrease in
radioactivity of oleate with a corresponding increase in
linoleate. Radioactivity subsequently appears in linolenate
(Bennett and Quackenbush, 1969).

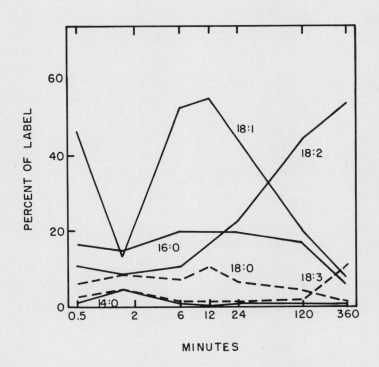

Figure 4.7   Incorporation of radiolabeled substrate into
   oleate, linoleate, and linolenate by *Penicillium chrysogenium* (Redrawn from Bennett and Quackenbush, 1969).

As described above, the conversion of stearate to
oleate involves the coenzyme A derivatives in animals, fungi,
and bacteria, and the ACP derivatives in plants. This has
been shown in *S. cerevisiae* (Yuan and Bloch, 1961) and *C.
lipolytica* (Kates and Paradis, 1975). There is considerable
evidence that complex lipids containing oleate are the substrates for the formation of linoleate rather than thioesters. In fungi, the oleyl moiety of $C_{18:1}$CoA is rapidly
incorporated into phospholipids, particularly phosphatidylcholine. This has been demonstrated with microsomal fractions from the following fungi: *Candida lipolytica* (Pugh
and Kates, 1973), *Torulopsis utilis* (Talamo et al., 1973),
and *Neurospora crassa* (Baker and Lynen, 1971). Phosphatidylcholine has been shown to be involved in linoleate synthesis
in vivo in *Aspergillus ochraceus* (Chavant et al., 1978) and
Mucor mucedo (Chavant et al., 1979), but only oleyl-CoA
could be converted to linoleate by microsomes from these
fungi. In *C. lipolytica*, the desaturation of oleate to
linoleate occurs in all the phospholipids but is more pronounced in phosphatidylcholine. Desaturation occurs at both
the 1- and 2-positions of phosphatidylcholine but only the
2-position in phosphatidylethanolamine (Kates and Paradis,
1973). Microsomal desaturases that catalyze the conversion
of stearyl-CoA to oleyl-CoA have been given considerable
attention whereas relatively little is known about the
desaturase(s) that catalyze the formation of di- and trienes.
However, one such study has partially characterized the
desaturase of *C. lipolytica* that catalyzes the oleate to
linoleate conversion (Pugh and Kates, 1975). The desaturase
from this fungus catalyzes the desaturation of both 1-acyl-
2-[^{14}C-]oleyl-*sn*-glycero-3-phosphorylcholine and 1,2-di-
[^{14}C]oleyl-*sn*-glycero-3-phosphorylcholine. The desaturation
by the microsomal preparation requires molecular oxygen,
either NADPH or NADH, and is inhibited by cyanide suggesting
the involvement of cytochrome b_5 as in the stearyl-CoA desaturase (see above). Using microsomes from cells grown at
different temperatures (10 C and 25 C) evidence for two
desaturases has been obtained, one specific for oleyl-CoA
and the other for the phospholipid. The oleyl-CoA desaturase
appears more active at lower temperatures and may be responsible for the increased degree of unsaturation in cold grown
cells that has been observed in *C. lipolytica* and many other
organisms (see Chapter 2).

Little is known about the enzymes and process of
polyunsaturated fatty acid biosynthesis. With possible

exceptions, $C_{18:3}$ is formed from $C_{18:2}\Delta^{9,12}$ in a methyl-directed desaturation for the $\Delta^{9,12,15}$ isomer and carboxyl-directed desaturation for the $\Delta^{6,9,12}$ isomer. As noted above, methyl-directed desaturation is characteristic of photosynthetic plants and higher fungi and carboxyl-directed desaturation is characteristic of animals and lower fungi. However, it appears that some lower organisms are capable of desaturation in both directions on the fatty acid molecule (Brennen et al., 1975). For example, the lower fungus *Saprolegnia parasitica* contains arachidonic acid ($C_{20:4}\Delta^{5,8,11,14}$) which is produced by a carboxyl-directed desaturation $C_{20:3}\Delta^{8,11,14}$ formed by elongation of γ-linolenate (Gellerman and Schlenk, 1979). This fungus also contains eicosapentaenoic acid ($C_{20:5}\Delta^{5,8,11,14,17}$) an ω^3 acid which is produced by the methyl-directed desaturation of arachidonic acid.

An alternate pathway of α-linolenic acid synthesis has been suggested for chloroplasts, which involves the desaturation of short chain acids (i.e. $C_{12:3}$) followed by elongation to the $C_{18:3}$ level (Kannangara and Stumpf, 1972; Kannangara et al., 1973; Jacobson et al., 1973). The existence of this pathway has been recently questioned. *Penicillium chrysogenium* converts linoleic acid to linolenic, but an alternate route of $\alpha C_{18:3}$ formation suggestive of the desaturation-elongation pathway has been proposed (Richards and Quackenbush, 1974). This is based on the fact that this fungus can convert hexadecatrienoic acid to linolenic acid. Rather than an alternate pathway, perhaps this result reflects only the activity and low specificity of the enzymes for elongation of an unnatural substrate provided exogenously, particularly since no shorter chain trienes could be detected in this fungus.

Little is known about the specific regulation of the desaturases. As noted above, it is well-documented that an adaptive response to cold temperatures by poikilothermic organisms is an increased degree of lipid unsaturation. There is evidence obtained with *Tetrahymena* that fatty acid desaturase activity is regulated by the degree of membrane fluidity rather than cell temperature *per se*, e.g. membrane fluidity is self-regulating and the activity of the desaturases are modulated by the physical state of the membrane (Martin et al., 1976).

Biosynthesis of Unusual Fatty Acids. As defined in
Chapter 3, unusual fatty acids have some structural feature in addition to the usual straight-chain, saturated
or unsaturated (methylene-interrupted ethylenic bonds)
monocarboxylic acid. They may be in high or low amounts
but restricted to certain taxonomic groups, or they may be
widely distributed but present in low relative proportions.
There are numerous unusual fatty acids in nature, but only
the acetylenic acids and substituted acids having hydroxyl,
epoxy, and methyl groups will be discussed here. In most
cases, their biosynthesis has not been widely studied in
fungi.

Acetylenic fatty acids are secondary products, and
their distribution was not considered in Chapter 3. Over
400 naturally occurring acetylenic compounds have been
identified as products of various plants and fungi (Bu'lock,
1966). About one-fifth of these are fungal products, mostly
from Basidiomycetes. Mono-, di-, and triacetylenes from 6
to 18 carbons in chain length occur in fungi, with C_9 and
C_{10} acetylenes being most common. Most acetylenes do not
occur as simple acids, but they may contain various oxygen-
containing groups, i.e. epoxy, hydroxy, aldehyde or a second
carboxyl group. The biosynthesis of acetylenes is not well
established but has been discussed by Turner (1971) and
reviewed by Bu'lock (1966). A common monoacetylenic acid
is crepenynic acid ($C_{18:2}\Delta^{9c,12a}$). Acetylenic bonds are
apparently formed by successive dehydrogenations beginning
at a saturated portion of the precursor molecule. For example, [10-^{14}C] oleic acid is converted to [10-^{14}C] crepenynic
acid via linoleic acid by *Tricholoma grammopodium* (Bu'lock
and Smith, 1967). Also, a possible precursor to polyacety-
lenes, dehydrocrepenynic acid ($C_{18:3}\Delta^{9c,12a,14c}$), has been
isolated from polyacetylene-producing funti. Acetylenes
with relatively short chains may be formed by β-oxidation
of those with longer chains (see below).

There are several hydroxy fatty acids that are fungal
products, some of which include 17-hydroxy-hexadecanoate
17-hydroxy-octadecanoate, 2-hydroxy-hexadecanoate, 2-hydroxy-
hexacosanoate, 9,10-dihydroxy-octadecanoate, and others.
The most extensively studied of the fungal hydroxy acids is
D-12-hydroxy-octadec-9-enoate (ricinoleic acid). This fatty
acid is a major component of the oil from immature sclerotia
or mycelia of the ergot fungus *Claviceps purpurea* and the
castor bean (*Ricinus communis*). The formation ricinoleic

acid appears to occur by different mechanisms in the different organisms (Morris, 1970). In castor bean, ricinoleic acid is formed from oleic acid apparently by a mixed function oxygenase (see Chapter 8) in a reaction requiring molecular oxygen and NADPH (Yamada and Stumpf, 1964; Morris et al., 1966). In the ergot fungus, ricinoleic acid is formed by the hydration of linoleic acid in a reaction that does not require molecular oxygen.

Extracellular glycolipids are produced by several yeast-like fungi (see Chapter 5). Glycolipids from *Torulopsis* species contain L-17-hydroxy-C_{18} acids linked glycosidically to the disaccharide sophorose. The hydroxy acids are formed first and then linked to the sugar. The hydroxyl oxygen is retained in the sophoroside. Hydroxylation is catalyzed by a mixed function oxygenase (see Chapter 8) in a reaction requiring molecular oxygen and NADPH, and the position (terminal or penultimate carbon) of the hydroxy group depends on the chain length and degree of unsaturation of the fatty acid substrate. Hydroxylation activity is present in a 48,000 g particulate cell fraction of this fungus (Heinz et al., 1969; Tulloch et al., 1962; Heinz et al., 1970).

Long chain 2-hydroxy fatty acids have been detected in *Candida utilis* and *Saccharomyces cerevisiae* and are most commonly the acyl component of membrane sphingolipids. Their formation has not been investigated in fungi but the formation of 2-hydroxy-hexacosanoic acid in *C. utilis* may be associated with the α-oxidation pathway (see below) (Fulco, 1967).

An oxygenated fatty acid that appears to be restricted to the rust fungi is L-*cis*-9,10-epoxy-octadecanoic acid. The C_{18} epoxy acid can be produced by *Puccinia graminis* from acetate, stearate, and oleate in a reaction requiring molecular oxygen but not light (Knoche, 1968; 1971). It appears that $C_{18:1}$ is produced from C_{18} as described above and the epoxide bond is formed across the 9 and 10 carbon atoms of oleate. The actual substrate for epoxide formation is not established. Oleic acid is rapidly incorporated into glycerolipids, thus the oleate-containing lipid may be the substrate for epoxidation rather than the thioester or free acid as in the formation of linoleic acid (see above).

When *P. graminis* and *Melampsora lini* uredospores are incubated in water, the C_{18} epoxy acid is enzymatically hydrated to *threo*-9,10-dihydroxyoctadecanoic acid (Tulloch, 1963; Jackson and Frear, 1967; Hartmann and Frear, 1963). The L-9, D-10 configuration of the dihydroxy acid suggests that the 10-hydroxyl comes from water and the oxygen of the 9-hydroxyl comes from the epoxide function. Hydration of the dihydroxy acid occurs very rapidly upon the initiation of germination, and appears to be the first step in the degradation of the C_{18} epoxy acid (Morris, 1970).

Methyl-substituted fatty acids are widely distributed in nature, but appear to be rare in fungi. Both monomethyl and multiple methyl-branched fatty acids have been detected in various organisms. Branched fatty acids may have a single methyl group in the *iso* or *anteiso* positions, or an internal methyl branch. Methyl groups of multiple branched acids may have isoprenoid (see Chapter 9) or non-isoprenoid spacing. The most common methyl-branched fatty acids have the substituent in the *iso* (ω-1) or *anteiso* (ω-2) positions. They are formed by the fatty acid synthetase when a branched precursor with the appropriate structure substitutes for acetate as the "primer" molecule. For example, 2-methyl and 3-methyl-butyrate formed from the amino acids leucine and isoleucine, respectively, give rise to *iso* and *anteiso* fatty acids, respectively, upon elongation. Internally methyl-branched fatty acids may be formed if 2-methyl-malonate substitutes for malonate during fatty acid synthesis. Place-ment of the methyl-branch depends on the chain length of the intermediate fatty acid that condenses with 2-methyl-malonate and the degree of subsequent elongation.

Methyl branches may also be introduced into a preformed fatty acid through alkylation at a point of unsaturation. The mechanism of alkylation (methylation) has been discussed in detail in Chapter 10.

FATTY ACID DEGRADATION

Triacylglycerides represent an important energy reserve in fungi and may be the primary source of carbon and energy for reproduction when the medium sugar content is exhausted. The degradation of triacylglycerides is catalyzed by lipases, and results in the liberation of fatty acids and glycerol upon complete hydrolysis (see Chapter 5). Fatty acids are

then subject to degradation by one of several pathways that
include α-oxidation, β-oxidation, and ω-oxidation.

The degradation of fatty acids by α-oxidation involves
an oxidative decarboxylation with the liberation of CO_2 and
an acid with one less carbon atom than the substrate. The
2- or α-carbon of the original substrate becomes the carboxyl
carbon of the newly formed fatty acid. This pathway is
widely distributed in nature, having been demonstrated in
plants (cotyledons and leaf tissue), mammalian tissue, and
possibly in fungi. α-Oxidation has been reviewed by Stumpf
(1969) and discussed by Hitchcock and Nichols (1971). The
exact mechanisms of α-oxidation are not well-established.
Although they have several properties in common, α-oxidation
seems to differ in the various organisms studied. For example, the α-oxidation process in cotyledons of germinating
peanut seed involves a peroxidase which is involved in the
decarboxylation reaction where an aldehyde is produced, and
an aldehyde dehydrogenase which requires NAD^+ and catalyzes
formation of the acid (Martin and Stumpf, 1959). Hydrogen
peroxide is required and may be generated through the action
of glycolic acid oxidase. The enzymes that catalyze the
α-oxidation of fatty acids in peanut cotelydons are found
in the mitochondrial, microsomal, and soluble cell fractions.

α-Oxidation in leaf tissue, on the other hand, requires
molecular oxygen rather than hydrogen peroxide and a stable
α-hydroxy intermediate is involved (Hitchcock and James,
1966). NAD^+ is also required. Apparently both the D- and
L-hydroxy isomers are formed during the oxidation process,
but only the L-isomer is converted to the corresponding
acid. The D-isomer tends to accumulate and appears to be
incorporated into sphingolipids.

α-Oxidation in mammalian systems differs from the two
plant pathways. In rat brain, enzymes solubilized from
microsomes require ATP, NAD^+, molecular oxygen, and a ferrous
ion (Lippel and Mead, 1968). In this system, an α-hydroxy
acid is formed as before but it is oxidized to the corresponding α-keto prior to decarboxylation directly to the
acid.

α-Oxidation has not been studied extensively in fungi,
but long chain fatty acids can be decarboxylated by *Candida
utilis* (Fulco, 1967). As in peanut cotyledons, α-hydroxy
fatty acids and aldehydes appear to be intermediates of the

α-oxidation process. Maximum decarboxylation activity is with α-OH C_{18}.

The quantitatively most important and widely distributed pathway of fatty acid degradation is β-oxidation. This process occurs in the mitochondria but also occurs in cer tain microbodies and the cytosol of some organisms under appropriate conditions (see below). This process involves the successive removal of C_2 units from an acyl-CoA sub-strate. Unlike in α-oxidation, the substrates for β-oxida-tion are coenzyme A derivatives rather than the acid; there-fore the acids released from acyl lipids (triacylglycerides, phospholipids, etc.) through the action of lipases, require activation. In animals, fatty acids are activated in the cytosol and transported into the mitochondria via the carni-tine derivative. Fatty acid activation is catalyzed by a fatty acid thiokinase on the mitochondrial membrane, and then the acyl group is transferred to carnitine in a reaction catalyzed by carnitine acyl transferase in the intermembrane space. Several thiokinases with different chain length specificities have been detected. At the inner membrane, a carnitine acyl transferase catalyzes the reverse reaction regenerating acyl-CoA in the mitochondrial matrix. The pre-cise process of acyl moiety transport across the mitochon-drial membrane may vary for different animal tissues and perhaps in other organisms. There is evidence for the par-ticipation of carnitine as a "carrier" of acyl groups across mitochondrial membranes in plants and fungi (see below).

The process of β-oxidation is catalyzed by four soluble enzymes located in the mitochondrial matrix. The first reac-tion of β-oxidation is catalyzed by acyl-CoA dehydrogenase which catalyzes the α,β-dehydrogenation of a saturated acyl chain (N) to an enoyl-CoA (trans-α,β-acyl-CoA). The reaction is specific for FAD. Dehydrogenases with different chain length specificities have been detected and each has 2 moles FAD per mole of enzyme protein. Enoyl-CoA is then hydrated to L(+)-β-hydroxyacyl-CoA in a reaction catalyzed by enoyl-CoA hydratase. This enzyme has a broad chain length speci-ficity (C_8 to C_{18}). The third reaction is catalyzed by L-3-hydroxyacyl-CoA dehydrogenase and gives rise to the corresponding 3-keto thioester (3-ketoacyl-CoA). This enzyme is specific for the L-isomer and NAD^+. The fourth reaction in the β-oxidation reaction sequence is the thiolytic cleav-age (between carbons 2 and 3) of 3-ketoacyl-CoA to acetyl-CoA and an acyl-CoA molecule with two carbons less than the

original substrate. This reaction is catalyzed by acetyl-
CoA:acetyl transferase (thiolase) and requires coenzyme A.
This enzyme also occurs in multiple forms with different
chain length specificities. A thiolase having a molecular
weight of 170,000, and consisting of 4 subunits of 42,000
molecular weight each has been isolated from yeast. Repetition of these reactions with N-2, N-4, etc. as substrates
results in N/2 acetyl-CoA. The following expression sum-
marizes the reactions of β-oxidation with palmitic acid as
the initial substrate:

$$\text{Palmitic acid} + \text{ATP} + 8 \text{ CoASH} + 7 \text{ FAD} + 7 \text{ NAD}^+ \rightarrow$$

$$8 \text{ Acetyl-CoA} + 7 \text{ FADH}_2 + 7 \text{ NADH} + 7 \text{ H}^+ + \text{AMP} + \text{PP}_i$$

The process of acyl transport across mitochondrial membranes
and the reactions of β-oxidation are outlined in Figure 4.8.
The principal characteristics of fatty acid biosynthesis and
β-oxidation are compared in Table 4.5.

Re-oxidation of $FADH_2$ with oxygen via the electron
transport system is coupled to the phosphorylation of 2 ADP
molecules and similar reoxidation of NADH results in the
formation of 3 ATP molecules. Thus, each repetition of the
β-oxidation cycle results in the formation of 5 ATP molecules; therefore, the degradation of palmitic acid to 8
acetyl-CoA molecules would result in the formation 35-1
(one ATP used in the initial activation) = 34 ATP molecules.
Thus, with 12 ATP molecules being generated with complete
oxidation of one acetyl-CoA molecule via the TCA cycle and
electron transport system, 130 ATP molecules are generated
through the complete oxidation of palmitic acid to CO_2 and
H_2O $[34 + (12 \times 8) = 130]$. This results in the generation of
988 kcal free energy/mole fatty acid oxidized (7.6 kcal/
mole ATP×130); and if there are 2330.5 kcal/mole released
on the complete combustion of palmitic acid, the efficiency
of palmitic acid oxidation is about 42% (988÷2330.5×100 =
42%) which is similar to that of glucose.

The oxidation of unsaturated fatty acids via β-oxidation
proceeds as with saturated acids until the acyl chain is
shortened to the double bond. Since enoyl-CoA hydratase
is specific for the Δ^{2t} rather than the Δ^{2c} isomer and double
bonds of most unsaturated fatty acids have the *cis*-configura-
tion, an isomerase catalyzes the $\Delta^{2c} \rightarrow \Delta^{2t}$ reaction. Isomeration is followed by hydration and the remaining reactions
proceed as usual (Stumpf, 1969).

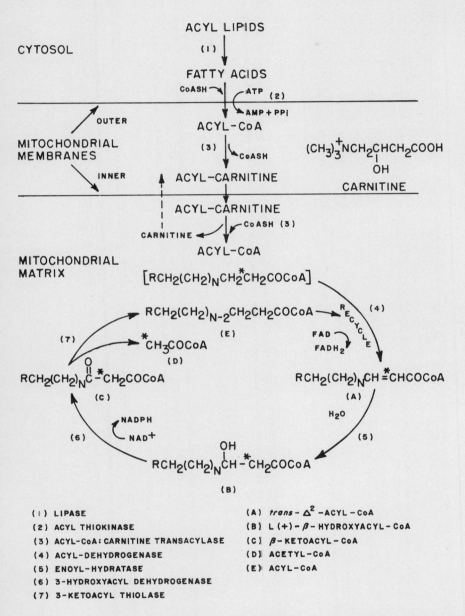

(1) LIPASE
(2) ACYL THIOKINASE
(3) ACYL-CoA: CARNITINE TRANSACYLASE
(4) ACYL-DEHYDROGENASE
(5) ENOYL-HYDRATASE
(6) 3-HYDROXYACYL DEHYDROGENASE
(7) 3-KETOACYL THIOLASE

(A) *trans*-Δ^2-ACYL-CoA
(B) L(+)-β-HYDROXYACYL-CoA
(C) β-KETOACYL-CoA
(D) ACETYL-CoA
(E) ACYL-CoA

Figure 4.8   Transport of acyl moieties across mitochondrial membranes and the β-oxidation of fatty acids (CoA derivatives).

TABLE 4.5   Comparison of Fatty Acid Biosynthesis and β-Oxidation

Characteristic	Biosynthesis	β-Oxidation
Thioester	ACP	CoA
Principal C_2 Unit	Malonate	Acetate
Coenzymes	NADPH (NADP$^+$) (plus FMN in yeast)	NAD$^+$ (NADH)
Stereochemistry of 3-Hydroxy Intermediate	L(+)	D(−)
Enzymes	Soluble (plants, bacteria) Multifunctional protein (animals, fungi)	Soluble
Cellular Location	Cytosol (non-photosynthetic organisms and tissues) Chloroplasts (photosynthetic tissues)	Mitochondria Glyoxysome (plants, fungi) Peroxisome (fungi)

Acetyl-CoA produced by the β-oxidation of fatty acids
or through the decarboxylation of pyruvate is available to
the TCA cycle in the mitochondria, or the acetate moiety may
be transported into the cytosol. Acetate may be transferred
to carnitine and the resulting ester transported into the
cytosol. The acetate moiety may then be transferred back to
coenzyme A. A second alternative is that the acetate moiety
may inter the TCA cycle. TCA intermediates such as citrate
or glutamate may leave the mitochondria and acetate regener-
ated in the cytosol from these substances (Table 4.1). A
third alternative is that acetyl-CoA may be hydrolyzed and
acetate diffuse from the mitochondria. In the cytosol,
acetate may be reactivated in a reaction catalyzed by acetic
acid thiokinase (acetyl-CoA synthetase).

Under certain conditions, plants and microbes utilize
fat for the production of sugars. This involves the participation mitochondrial and soluble enzymes as well as two
enzymes located in microbodies called glyoxysomes. These
enzymes are isocitrate lyase and malate synthetase which
catalyze reactions of the glyoxylate bypass (pathway). The
enzymes of β-oxidation are also associated with glyoxysomes.
The process of fat conversion generally involves the degradation of fatty acids to acetyl-CoA in the glyoxysomes. The
first reaction of the glyoxylate pathway is catalyzed by
isocitrate lyase and involves the breakdown of isocitrate,
from the mitochondria, to succinate and glyoxylate. The
second reaction is catalyzed by malate synthetase, and involves the formation of malate from glyoxylate and acetyl-
CoA resulting from the breakdown of fatty acids. Malate,
and succinate, return to the mitochondria where they are
converted to oxalacetate. Oxalacetate leaves the mitochondria and is then decarboxylated to phosphoenolpyruvate (PEP)
in a reaction catalyzed by PEP carboxykinase. PEP is then
incorporated into sugars by reverse glycolysis in the cytosol.

Although all the details of the glyoxylate pathway and
gluconeogenesis is not well-established in fungi, the sub-
jects have been reviewed by Casselton (1976). The glyoxylate
pathway is active in fungi grown on either acetate or alkanes
as the sole carbon source, and perhaps during the transition
from vegetative to reproductive growth. The key enzymes of
the glyoxylate pathway have been detected in microbodies
from *Candida utilis* grown on n-alkanes believed to be
peroxisomes based on the presence of catalase which is
absent from glyoxysomes. Enzymes of these organelles also

catalyze the breakdown of fatty acids via β-oxidation (Kawamoto et al., 1978). β-Oxidation could not be detected in mitochondria of the alkane grown *Candida* and there is evidence for the transport of acetyl-CoA from peroxisomes to mitochondria of these organisms via an "acetyl-CoA shuttle" (Kawamoto et al., 1978). The interrelation between mitochondria, glyoxysomes (peroxisomes) and enzymes of the cytosol in β-oxidation, the glyoxylate pathway, and gluconeogenesis is outlined in Figure 4.9.

A third pathway of fatty acid degradation is called ω-oxidation. This process of fatty acid oxidation involves mixed function oxygenase enzymes that are discussed in Chapter 8.

Unsaturated fatty acids are also subject to oxidation in a reaction catalyzed by lipoxygenase (lipoxidase). Fatty acids containing two or more *cis* methylene-interrupted double bonds are converted to the corresponding hydroperoxy (-OOH) derivative. For example, $C_{20}\Delta^{8,11,14}$ is converted to 15-hydroperoxy $C_{20}\Delta^{8,11}$. After lipoxygenase action, autocatalytic oxidations result in a number of oxygenated fatty acids. Lipoxygenase is found in plant tissues, particularly seeds.

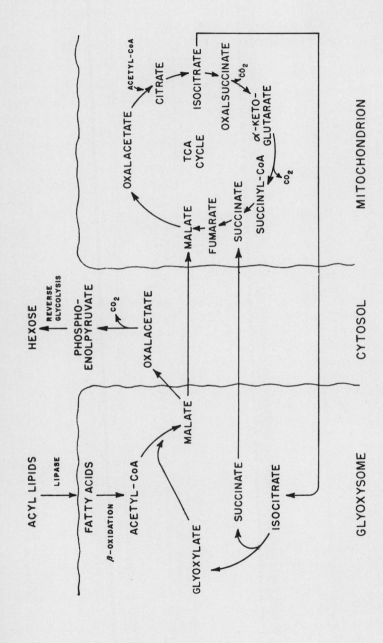

Figure 4.9 Conversion of fats to sugars via lipase action, β-oxidation, the glyoxylate pathway, and reverse glycolysis.

CHAPTER 5   ACYLGLYCEROLS AND RELATED LIPIDS

INTRODUCTION TO ACYLGLYCEROLS

Glycerol serves as the alcohol backbone for a wide
variety of lipids. Fatty acids generally do not occur in
nature as free acids, but primarily as esters with glycerol
called acylglycerols (acylglycerides). Glycerol may be fully
acylated (triacylglycerols or triglycerides) or only partially acylated (mono- and diacylglycerols or mono- and
diglycerides). In some cases a mono- or diglyceride may be
substituted at the third carbon with phosphate or a phosphorylated substance (phosphorylcholine, phosphorylserine,
etc.) via a phosphodiester bond (phosphoglyceride or phospholipid), or with a sugar (glycosylglycerol). Phospholipids are the subject of the next chapter (Chapter 6). Some
glycerides may contain hydroxylated acyl groups that may in
turn be esterified to long chain fatty acids to form estolides (tetra-, penta-, and hexaglycerides). Aliphatic chains
may also be linked to a diglyceride via an ether (alkyl) or
vinyl ether (alkenyl) linkage. The latter type of lipid is
called a neutral plasmalogen. Structures of these lipids are
given in Figure 5.1.

Triglycerides serve mainly as a storage material; fatty
acids are a major source of energy in living systems, yielding over two times more calories per gram upon oxidation
than either carbohydrates of proteins. They are the major
components of natural fats (solid) and oils (liquid). For
example, triglycerides represent up to 95% of vegetable oils
from soybean, peanut, and other high oil content seed. In
fungi, these lipids are the major constituents of oil

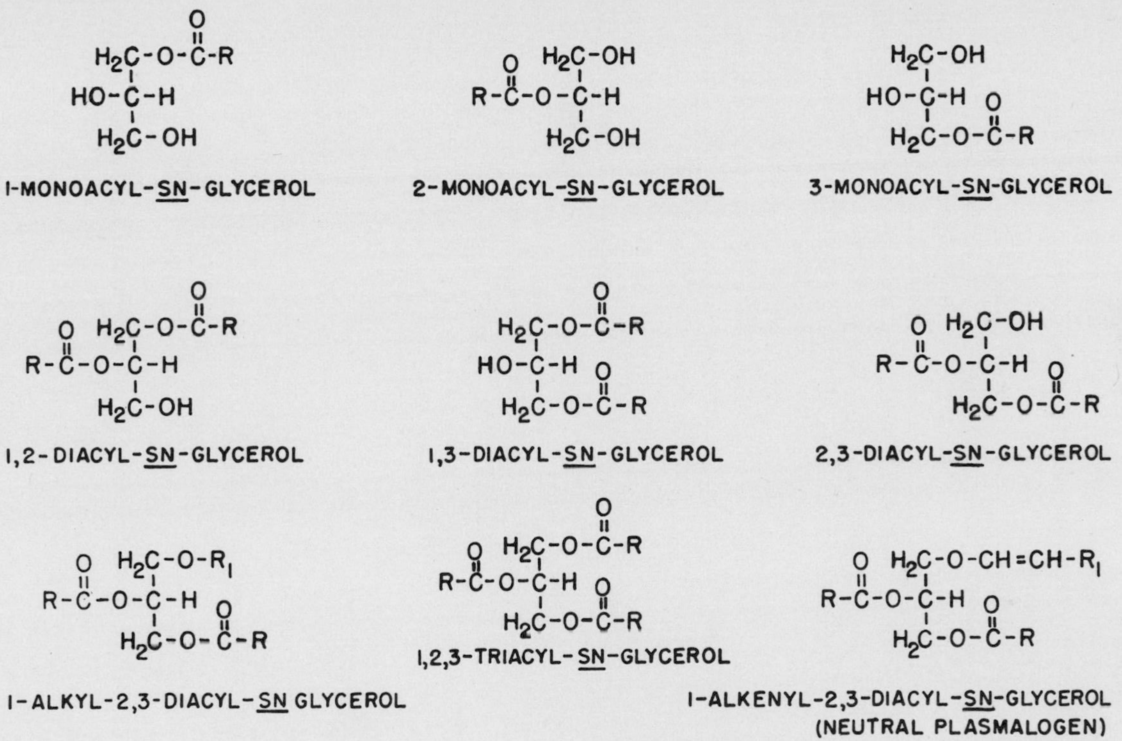

H₂C-O-C-R (structures)

1-MONOACYL-SN-GLYCEROL

2-MONOACYL-SN-GLYCEROL

3-MONOACYL-SN-GLYCEROL

1,2-DIACYL-SN-GLYCEROL

1,3-DIACYL-SN-GLYCEROL

2,3-DIACYL-SN-GLYCEROL

1,2,3-TRIACYL-SN-GLYCEROL

1-ALKYL-2,3-DIACYL-SN GLYCEROL

1-ALKYL-2,3-DIACYL-SN-GLYCEROL

1-ALKENYL-2,3-DIACYL-SN-GLYCEROL
(NEUTRAL PLASMALOGEN)

Figure 5.1 Structures of mono-, di-, and triacylglycerides, alkyl ethers, and alkenyl ethers.

droplets suspended in the mycelial and spore cytoplasm.
These lipids may also be minor components of membranes and
cell walls of fungi.

STRUCTURE AND NOMENCLATURE

Several systems of nomenclature have been applied to the
naming of acylglycerols, including the L-α, D/L, and R/S systems. Each system has its advantages and disadvantages, but
none of them have been universally accepted. Upon formation
of an ester bond, or phosphorylation, at one of the primary
hydroxyl groups of glycerol, the C-2 *meso* carbon becomes
asymmetric and the molecule is optically active. Asymmetry
and optical activity is maintained with substitution at the
second primary hydroxyl group provided the substituent is not
identical with that at the first primary hydroxyl. The main
problem in naming substituted glycerols lies in the fact that
the two primary hydroxyl groups are not identical in their
reactions with enzymes. A system of nomenclature capable of
distinguishing between stereoisomers, or between the C-1 and
C-3 carbinols is required. Such a system called "stereospecific numbering" was developed and appears to be widely accepted (Hirshmann, 1960). The fixed confirmation of L-glyceraldehyde (S-glyceraldehyde) is applied to glycerol (Figure
5.2). The hydroxyl group at C-2 is drawn to the left of a
Fischer projection formula and the carbon above the central
carbon is assigned C-1 and the one below C-3 (Figure 5.2).

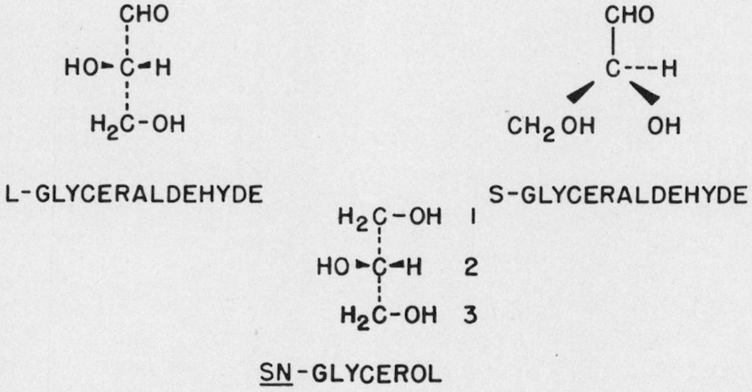

$$CHO$$
$$HO-C-H$$
$$H_2C-OH$$

L-GLYCERALDEHYDE

$$CHO$$
$$C---H$$
$$CH_2OH \quad OH$$

S-GLYCERALDEHYDE

$$H_2C-OH \quad 1$$
$$HO-C-H \quad 2$$
$$H_2C-OH \quad 3$$

SN-GLYCEROL

Figure 5.2   Stereochemistry of glyceraldehyde and glycerol.

The use of "stereospecific numbering" is indicated by "*sn*"
before the root name of the compound. Accordingly, a triglyceride would be named 1,2,3-triacyl-*sn*-glycerol. The same
system of nomenclature can be applied to phospholipids (see
Chapter 6). See the Nomenclature of Lipids (1968) and Burton
and Guerra (1974) for a full description of the "stereospecific numbering" system of nomenclature.

     Triglycerides yield three fatty acids upon hydrolysis.
The number of molecular species of triglycerides depends on
the number of different fatty acids available for esterification to glycerol. If *n* represents the number of different
fatty acids, the number of possible triglyceride species
would be $(n^3 + n^2)/2$. Generally, fungi produce 17 different
medium to long chain fatty acids and theoretically 2,546 individual triglycerides. This number is doubled when the
optical enantiomers are taken into account. Stereospecific
analysis (see below) of triglycerides shows that the fatty
acids are not randomly distributed among the three possible
positions of glycerol, which would be expected because the
C-1 and C-3 carbons of glycerol are not equivalent. The C-2
position generally contains a higher proportion of saturated
fatty acids.

     The stereospecific analysis of most triglycerides can be
accomplished by the use of thin-layer (silver-ion argenta-
tion) and gas-liquid chromatography combined with the action
of certain enzymes with known reaction specificities. One
approach employs the combined action of lipase that attacks
triglycerides at the C-1 and C-3 positions, and phospholipase
A (Chapter 6) that specifically catalyzes hydrolysis at the
C-2 position (Figure 5.3b) (Brockerhoff, 1965; 1967). Partial hydrolysis of a triglyceride with the lipase yields a
mixture of three acylglycerol products (2-monoacylglyceride,
1,2-diacylglyceride, and 2,3-diacylglyceride) and fatty acids
originally at positions C-1 and C-3. This step is followed
by analysis of the fatty acids by GLC and separation of the
glycerides by TLC. The acyl group at C-2 is determined by
GLC after hydrolysis of the monoglyceride. The diglycerides
are converted to their phosphorylphenol (PPh) derivatives
which are subjected to attack by a C-2 specific phospholipase
A. Identification of the fatty acid at the C-1 position of
the resulting lysophosphatide is determined by GLC as before.
The fatty acid in the C-3 position is then calculated by difference from the known fatty acid composition of the triglyceride determined by GLC after acid hydrolysis.

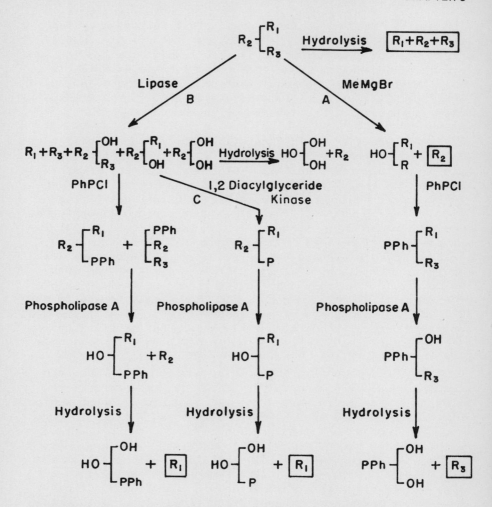

Figure 5.3   Stereospecific analysis of triacylglycerides
using the combined action of pancreatic lipase and phospholipase A.

A similar but more precise method of stereospecific
analysis of triglycerides involves deacylation at C-2 using
methyl magnesium bromide and conversion of the resulting
diglyceride to the 2-PPh derivative, which is then subjected
to attack by a C-1 specific phospholipase A. Analysis of the
fatty acid released during each of the first two steps and
after acid hydrolysis of the lysophosphatide produced in the

third step reveals the identity of the acyl group at each
carbon of the glycerol in the triglyceride (Figure 5.3a).

A third approach used in the stereospecific analysis of
triglycerides employs diacylglyceride kinase which catalyzes
the phosphorylation of 1,2-diacyl-sn-glycerols (Lands et al.,
1966). The 1,2-diacyl-sn-glycerols produced by the action of
lipase on a triglyceride, as before, is converted to its
phosphate ester by the enzyme. The resulting 1,2-diacyl-sn-
glycerol-3-phosphate is hydrolyzed by a phospholipase A specific for the C-2 position. Identities of acyl groups at the
carbon atom of the triglyceride are determined by fatty acid
analysis of the products of enzyme action and after hydrolysis of the 1-monoacyl-sn-glycerol-3-phosphate (Figure 5.3c).

There are certain mixtures of triglycerides in which
all of the enantiomers present cannot be determined by the
above methods. An example of such a triglyceride mixture
is one containing palmitic, stearic, and oleic acids.
Methods for the stereospecific analysis of such mixtures have
been proposed and tested. They employ some of the reactions
discussed above and a lipase from the fungus *Geotrichum
candidum* that is specific for Δ^9 unsaturated fatty acids
(Jensen et al., 1966; Lands et al., 1966; Sampugna and
Jensen, 1968).

OCCURRENCE IN FUNGI

Triglycerides may comprise over 90% of the lipid in
fungi, varying considerably according to the species, stage
of development, and growth conditions. They are the most
abundant type of lipid in many fungi, including mycelia of
Phycomyces blakesleeanus, *Lipomyces lipoferus*, *Glomerella
cingulata*, and *Coprinus comatus* (Jack, 1965). Triglycerides
of *G. cingulata* can be separated into five groups by TLC
according to the degree of unsaturation. Four of these
groups include one containing saturated and monoenoic fatty
acids, another containing saturated, mono-, and dienoic acids,
and two groups with varying proportions of these acids plus a
triene. Triglycerides are also the predominant class of
lipid in *G. cingulata* conidia (Jack, 1965; Bianchi and Turian,
1967), but the conidia of *Neurospora crassa* contain less than
0.6% triglyceride (Bianchi and Turian, 1967).

Neutral lipid of *Blastocladiella emersonii* zoospores represent 29 to 46% of the total lipid, depending on whether the organism is cultured on liquid or solid media. Thirty percent of the neutral lipid are triglycerides, 9% diglycerides, and 12% monoglycerides (Mills and Cantino, 1974a; 1974b).

Most analyses of fungal triglycerides consider fatty acid composition rather than the specific triglyceride structure (see Table 5.1a). Since triglycerides are often the most abundant lipid, the fatty acids of the total lipid may reflect the general fatty acid composition of these acyl lipids. Fatty acids with 16 and 18 carbons are the most abundant in these lipids and are present in a ratio of about 1:3. Palmitic (5-51%), oleic (2-77%) and linoleic (28-50%) acids are usually predominant. The fatty acid composition of triglycerides from several fungi is given in Table 5.1.

Oil of *Claviceps purpurea* may contain up to 44% ricinoleic acid (12-hydroxy-*cis*-9-octadecenoic acid) (Morris and Hall, 1966). The carboxyl group of this acid is esterified with glycerol and the hydroxyl is esterified with a long chain non-hydroxy fatty acid. Depending on the number of hydroxy fatty acids present in the molecule, tetra-, penta-, or hexaglycerides may occur and are called estolides. Non-hydroxy fatty acids of estolides are generally more unsaturated than those of triglycerides. Non-hydroxy acids at the C-2 position of glycerol in estolides are also generally more unsaturated than those at C-1 and C-3. As in typical triglycerides, oleic and linoleic acids are most often found in the C-2 position of estolides and saturated acids are usually at the C-1 and C-3 positions.

It appears that alkyl and alkenyl ethers (plasmalogens) are not common constituents of yeast (Letters and Snell, 1963; Hunter and Rose, 1971) or mycelial fungi, but the latter fungi have not been sufficiently examined for these lipids.

Mono- and diglycerides are generally present as minor constituents of fungal lipid extracts. Monoglycerides of *Rhizopus arrhigus* contain about 96% saturated fatty acids, 88% of which are C_{14} (38.6%) and C_{16} (49.0%) (Gunasekasan et al., 1972). Fatty acids of di- and triglycerides are similar, except that linoleic and linolenic acids are present in relative proportions of 13.2% and 8.1% in the triester, respectively, and of the two acids only linoleic (1.5%) is

present in the diester. Monoglycerides of *Candida* sp. are
probably 1-acyl-*sn*-glycerols and diglycerides mainly mixtures
of 1,2-diacyl-*sn*-glycerols with small amounts of 1,3-diacyl-
sn-glycerols (Kates and Baxter, 1962).

Some unusual neutral lipids from *Lipomyces starkeyi* are
analogues of acylglycerides and neutral plasmalogens (Bergel'son et al., 1966). They are diesters and 1-alkenyl
ethers of dihydric alcohols such as ethylene glycol, pro-
pane-1,2-diol, propane-1,3-diol, butane-1,3-diol, and butane-
1,4-diol. A C_5 diol is the major dihydric alcohol of this
yeast, in a ratio of 1:5 to glycerol (Vaver et al., 1967).
This type of lipid appears to be rare, but has been identified in phospholipid fraction of mammalian tissues (Vaver
et al., 1967).

BIOSYNTHESIS OF TRIACYLGLYCERIDES

A key interemediate in the biosynthesis of triglycerides
and phospholipids (see Chapter 6) is phosphatidic acid (1,2-
diacyl-*sn*-glycero-3-phosphate). Since phosphatidic acid is
a phospholipid, triglyceride biosynthesis may be viewed as a
branch off the phospholipid biosynthetic pathway. Glycerol
can serve as a carbon source for certain yeasts. It is converted to the initial acyl acceptor in phosphatidic acid
snythesis, *sn*-glycero-3-phosphate, in a reaction catalyzed
by glycerol kinase which appears to be an inducible enzyme
(Wieland and Suyter, 1957; Gaucedo et al., 1968). Under most
circumstances, glycerol probably is not an important intermediate in phosphatidic acid formation. The first two reactions in the pathway of phosphatidic acid synthesis involve
the acylation of *sn*-glycero-3-phosphate, resulting in the
formation of phosphatidic acid (Figure 5.4). The first reaction is catalyzed by *sn*-glycero-3-phosphate acyltransferase
which is associated with the endoplasmic reticulum (Schlossman
and Bell, 1977). Acyl-CoA is the acyl donor (Kuhn and Lynen,
1965). This pathway appears to occur universally in eukaryotic organisms. The acylation of *sn*-glycero-3-phosphate by
cell-free extracts of yeast requires ATP and CoA, is stimulated by Mg^{+2}, Mn^{+2}, albumin, and cysteine, and has an optimum pH of 7.4 (Steiner and Lester, 1972; White and Hawthorne,
1970).

A second pathway of phosphatidic acid synthesis has
been described for a particulate cell fraction of rat liver.

TABLE 5.1   Fatty Acid Composition of Fungal Acylglycerides

Fungus	Fatty Acid (%)								Reference
	C_{14}	C_{16}	$C_{16:1}$	C_{18}	$C_{18:1}$	$C_{18:2}$	$C_{18:3}$	C_{20}	
Phycomyces blakesleeanus[a]	tr	28.8	tr	15.7	29.2	15.2	tr	10.8	Jack, 1965
Lipomyces lipoferus[a]	tr	11.9	2.8	5.6	76.7	2.8	tr	tr	Jack, 1965
Glomerella cingulata	tr	40.8	1.0	4.2	30.1	20.7	3.2	--	Jack, 1965
Coprinus comatus[a]	4.2	28.1	1.8	17.3	23.4	25.7	tr	tr	Jack, 1965
Fusarium culmorum[b]	0.3	24.0	0.5	11.0	31.0	33.2	--	--	Marchant and White, 1967
Claviceps purpurea[c]	0.7	28.0	3.7	6.4	19.6	17.4	--	--	Morris and Hall, 1966
C. purpurea[d]	0.2	19.5	6.6	3.3	38.0	32.4	--	--	Morris and Hall, 1966
Tricholoma nudum[e]	--	24.3	--	12.5	34.2	26.5	--	--	Leegwater et al., 1962
Rhizopus arrhizus[f]	0.3	20.5	0.8	24.5	43.1	5.5	--	3.9	Weete et al., 1970
R. arrhizus[g]	8.2	22.9	tr	31.3	16.4	13.2	8.1	--	Gunasekaran et al., 1972
R. arrhizus[h]	38.6	49.0	8.8	1.0	1.8	0.9	--	--	Gunasekaran et al., 1972
R. arrhizus[i]	17.4	22.0	1.0	34.8	23.2	1.5	--	--	Gunasekaran et al., 1972
Choanephora curcubitarum[j]	2.9	31.2	3.8	7.8	19.9	20.2	10.8	0.3	White and Powell, 1966

Species									Reference
Alternaria dauci[k,l]	1.2	62.0	tr	2.8	9.0	23.9	--	tr	Gunasekaran and Weber, 1972
Fusarium solani[k,m]	1.6	45.2	4.1	0.8	15.0	30.5	2.9	--	Gunasekaran and Weber, 1972
Sclerotium rolfsii[k,n]	1.3	51.5	tr	4.5	6.6	31.5	tr	tr	Gunasekaran and Weber, 1972
Candida petrophillum[o]	0.8	5.0	8.2	1.6	41.9	39.6	--	--	Mizuno et al., 1966
Ceratocystis coerulescens[p]	--	8.0	3.0	18.0	9.6	14.4	11.0	--	Sprecher and Kubeczka, 1970
C. coerulescens[q]	--	10.0	1.1	8.1	24.7	35.2	11.6	--	Sprecher and Kubeczka, 1970
C. coerulescens[r]	--	5.1	0.7	14.0	5.3	13.7	7.5	--	Sprecher and Kubeczka, 1970
Pithomyces chartarum[s]	2.6	33.9	0.4	8.6	17.6	35.6	0.9	--	Hartman et al., 1962
Stemphylium dentriticum[s,t]	0.7	21.9	2.0	2.9	21.7	47.7	2.7	--	Hartman et al., 1962
Cylindrocarpon radicicola[s]	0.3	23.5	0.6	8.2	28.6	27.9	10.9	--	Hartman et al., 1962
Sclerotinia sclerotiorum[u]	2.1	22.7	1.8	2.6	20.9	39.3	10.1	--	Sumner and Colotelo, 1970
S. sclerotiorum[v]	0.3	16.1	0.8	1.9	20.9	45.0	15.6	--	Sumner and Colotelo, 1970
S. sclerotiorum[w]	0.2	13.0	0.5	4.0	28.0	47.3	6.2	--	Sumner and Colotelo, 1970

TABLE 5.1 Continued

Fungus	Fatty Acid (%)								Reference
	C_{14}	C_{16}	$C_{16:1}$	C_{18}	$C_{18:1}$	$C_{18:2}$	$C_{18:3}$	C_{20}	
S. sclerotiorum[x]	0.4	7.4	1.0	1.1	31.1	49.6	9.3	--	Sumner and Colotelo, 1970
Neurospora crassa	1.1	28.4	4.9	4.4	9.1	46.5	1.7	--	Kushwaha et al., 1976

[a]Either $C_{20:0}$ and Y$C_{18:3}$ or both. Stereospecific analysis of triacylglycerols of *P. blakesleeanus* mycelium and sporangiophores: *sn*-1 positions have >65% C_{16} and $C_{18:1}$; *sn*-2 positions have 85-90% $C_{18:1}$, $C_{18:2}$ and $C_{18:3}$; and *sn*-3 positions have 40% C_{16}, 30% $C_{18:2}$ and 21% $C_{18:3}$ (Debell and Jack, 1975).
[b]Conidia.
[c]Sclerotia, OH--$C_{18:1}$ (24.1-35.5%); varies with isolate.
[d]Mycelia, OH--$C_{18:1}$ (0-41.8%); varies with isolate.
[e]4 Day old culture.
[f]C_{15} (tr.); C_{22} (1.0%).
[g]96 Hr culture.
[h]Monoglycerides, 72-hr culture (monoglyceride fraction below detectable limits at 96 hr).
[i]Diglycerides, 96-hr culture.
[j]C_{10} (0.2%), C_{12} (0.2%), $C_{14:1}$ (0.06%), C_{15} (0.3%), $C_{16:2}$ (0.6%), $C_{16:3}$ (0.2%), C_{17} (0.2%), $C_{20:1}$ (0.1%), $C_{20:2}$ (0.2%), $C_{20:3}$ (0.06%), C_{22} (0.2%), $C_{22:1}$ (0.09%), $C_{22:2}$ (0.03%), C_{24} (0.3%), $C_{24:1}$ (0.1%), $C_{22:2}$ (0.06%).
[k]Fatty acid constituents of neutral lipids.

^{l}C_{12} (1.0%), $C_{14:1}$ (tr).

^{m}C_{12} (tr), $C_{14:1}$ (tr).

^{n}C_{12} (0.8%), C_{15} (0.8%).

^{o}Grown on glucose substrate, C_{15} (0.8%), $C_{17:1}$ (2.1%).

^{p}Submerged culture, <C_{16} (including C_{17}) (14.2%), >C_{18} (20.0%).

^{q}Surface culture, <C_{16} (including C_{17}) (2.8%), >C_{18} (5.6%).

^{r}Diglycerides, surface culture, <C_{16} (including C_{17}) (7.4%), >C_{18} (46.1%).

^{s}Neutral "glycerides" C_{15} (0.1-0.4%).

^{t}C_{12} (0.3%).

^{u}Neutral lipids (21% of total lipid), mycelia from laboratory culture, C_{17} (0.2%).

^{v}Neutral lipids (87% of total lipid), natural sclerotia, C_{17} (0.2%).

^{w}Neutral lipids (70% of total lipid), cultured sclerotia, 7 days old, C_{17} (0.4%).

^{x}Neutral lipids (27% of total lipid), cultured sclerotia, 84 days old, C_{17} (0.2%).

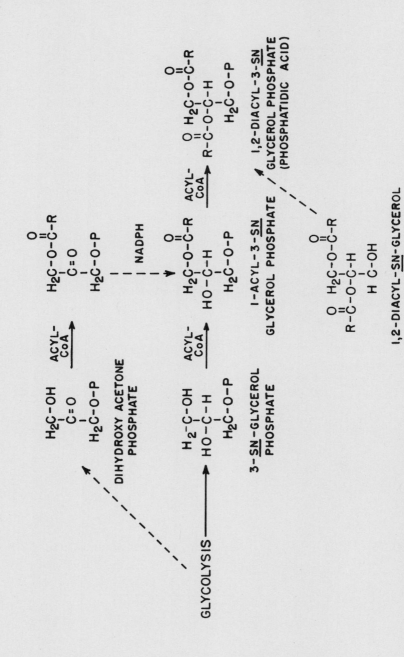

Figure 5.4   Biosynthesis of phosphatidic acid.

This pathway involves the acylation and reduction, followed
by another acylation, of dihydroxy acetone phosphate (Hajra
and Agranoff, 1968a; 1968b) (Figure 5.4). The reaction requires NADPH. The relative contribution of the two pathways
to phosphatidic acid synthesis in the hepatic system is uncertain. Dihydroxy acetone phosphate acyltransferase activity has been detected in two *Saccharomyces* species (Johnson
and Paltauf, 1970; Schlossman and Bell, 1978). However there
is considerable evidence in both mammalian (Schlossman and
Bell, 1977) and yeast (Schlossman and Bell, 1978) systems
that activities of the two acyltransferases constitute dual
catalytic functions of a single enzyme. The second pathway
may not operate in *S. cerevisiae*, since acyl dihydroxy acetone phosphate oxidoreductase activity could not be detected.
However, based on calculations of cellular concentrations of
sn-glycero-3-phosphate and dihydroxy acetone phosphate, the
ratio of acylation of these substrates would be 12:1 and the
first pathway of phosphatidic acid synthesis would be expected to predominate in yeast (Schlossman and Bell, 1978).

In plants, phosphatidic acid formation may also occur
through the direct phosphorylation of a 1,2-diacylglyceride
(Bradbeer and Stumpf, 1960) (Figure 5.4).

Final reactions in triglyceride biosynthesis are also
associated with the endoplasmic reticulum and involves dephosphorylation of phosphatidic acid and acylation of the
resulting diglyceride (Figure 5.5). The enzymes catalyzing

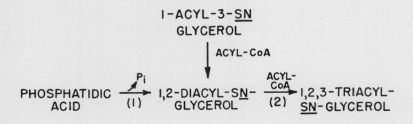

(1) PHOSPHATIDATE PHOSPHATASE

(2) DIGLYCERIDE ACYLTRANSFERASE

Figure 5.5   Biosynthesis of triacylglycerides.

these reactions have been detected in several organisms, but not in fungi (Kuhn and Lynen, 1965; Goldman and Vagelos, 1961). The regulation of triglyceride biosynthesis is influenced by factors that control glycolysis, gluconeogenesis, and lipogenesis (Gurr and James, 1975).

Another pathway of triglyceride biosynthesis has been detected in animals, the monoglyceride pathway (Figure 5.5). This pathway involves the stepwise acylation of a monoacyl- and a diacylglyceride by an endoplasmic reticulum-bound triglyceride snythetase complex (Hubscher, 1970). The preferred substrate in the first acylation reaction is a 1-monoacyl-*sn*-glycerol containing short chain saturated or long chain unsaturated fatty acids. The preferred substrate for the second transferase is a 1,2-diacyl-*sn*-glycerol with a high degree of unsaturation.

See Chapter 6 for the biosynthesis of alkyl and alkenyl ethers.

LIPASES

Lipase is a common name for glycerol ester hydrolases that catalyze the deacylation of acylglycerides and phospholipids. Lipases are distinguished from esterases, as defined by the International Union of Biochemistry (1961), by the physical state of the substrate. Lipases are specific types of esterases that act on glycerides at the interface of emulsions suspended in an aqueous medium.

Lipases are widely distributed in nature, and there are numerous reports on the properties of these enzymes from different sources. Pancreatic lipase has been the most extensively studied. This enzyme has a high reactivity to ester bonds at the C-1 and C-3 positions of triglycerides. After removal of the first acyl group, further deacylation proceeds at a slower rate (Figure 5.6). Lipase action begins with an attack on the carbonyl carbon by a nucleophilic group of the enzyme. The higher rate of lipase action at the C-1 of a triglyceride may be due to an activation effect by the neighboring electrophilic groups. A free hydroxyl group at C-2 may interfere with the formation of the lipase-substrate complex. Short chain fatty acids are more readily cleaved by pancreatic lipase than medium and long chain acids, and polyenoic acids are even less readily cleaved. Bile salts

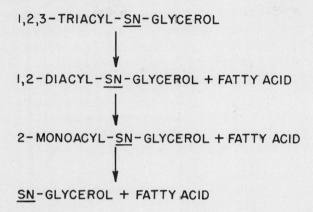

Figure 5.6   Lipase catalyzed degradation of a triacylglyceride.

stimulate lipolysis by pancreatic lipase by solubilizing the
partial glycerides and fatty acids.

2-Monoacyl-*sn*-glycerols produced by the action of pancreatic lipase may serve as the initial substrate for the
monoglyceride pathway of triglyceride biosynthesis (Figure
5.5). However, under certain conditions triglycerides may
be completely degraded. This is due to isomerization of the
2-acyl to the 1 or 3-acyl monoglycerides in the presence of
the enzyme (Borgstrom and Ory, 1970).

It would be expected that all fungi produce lipases.
Nine of numerous fungal species[*] screened were considered
active lipase producers by Alford et al. (1964). Most lipases
from fungi preferentially cleave fatty acids from the C-1
position of trigylcerides. However, a lipase from *Aspergillus flavus* shows no positional specificity. A lipase from
Geotrichum candidum, on the other hand, shows absolute specificity for Δ^9-unsaturation regardless of its position on the
triglyceride molecule. This enzyme shows the greatest specificity of lipases from any source. Lipase activity has also

---

[*]
*Candida lipolytica, Phycomyces nitens, Mucor sufu, Pencil-
lium roquefortii, Rhizopus oligosporus, Chaetostylum fresnei,
Thamnidium elegans, Aspergillus flavus*, and *Geotrichum candidum.*

been detected in *Mucor pusillus* (Somkuti and Babel, 1968;
Somkuti et al., 1969), *Torulopsis ernobii* (Yoshida et al.,
1968), *Candida paralipolytica* (Ota et al., 1970; Ota and
Yamada, 1966a; 1966b; 1967), *C. lipolytica* (Lloyd et al.,
1971; Peters and Nelson, 1948a; 1948b), *C. cylindricaceae*
(Yamada and Machida, 1962), *C. humicola* (Bours and Mossel,
1969), *Torulopsis* sp. (Motai et al., 1966), *Aspergillus niger*
(Lloyd et al., 1971), and *Puccinia graminis tritici* (Knoche
and Horner, 1970). Several fungi produce extracellular
lipases; in yeasts extracellular lipases appear to be restricted to anascosporogenous species (Hunter and Rose,
1971). Lipase activity in *S. cerevisiae* is associated with
the plasma membrane isolated by enzymatic digestion of the
cell wall (Nurminen and Suomalainen, 1970). Generally,
lipases of fungal origin are relatively stable compared to
pancreatic lipase and some are inducible. Most fungal
lipases have a pH optimum of about 8.

     The positional specificity of lipase from *Mucor javani-*
cus is very similar to that of pancreatic lipase (Ogiso and
Sugiura, 1971). Lipase from this source preferentially catalyzes cleavage of ester bonds of the C-1 and C-3 positions
of a triglyceride substrate, and the rate of hydrolysis at
C-1 is higher than at the same position of 1,3-diacylglyc-
erides. The slower rate of hydrolysis of the diglyceride
can probably be explained as before; the diglyceride is less
electrophilic than the triglyceride due to the absence of an
acyl group at the C-2 position. The rate of *Mucor* lipase
catalyzed lipolysis decreases as deacylation occurs: triglyceride > 2,3-diglyceride = monoglyceride. The *Mucor*
lipase more readily catalyzes the hydrolysis of 2-monoglyc-
erides than pancreatic lipase. Bile salts stimulate *Mucor*
lipase activity in the order triglyceride > 2,3-diglyceride
> monoglyceride. The stimulation is due to increased solu-
bility of the substrate rather than enzyme activation.
Stimulation decreases in the later stages of the reaction
due to micelle formation between the salt, monoglyceride,
and fatty acids. Micelles interfere with the approach of
the enzyme (Ogiso and Sugiura, 1971). The rate of hydrolysis
catalyzed by the *Mucor* lipase is also influenced by the degree of unsaturation in the substrate. Triglycerides containing oleic and linoleic acids are more readily hydrolyzed
than those with high levels of linolenic acid. This may be
due to the change in configuration of the fatty acid chain
due to increasing number of methylene-interrupted double
bonds. For example, the chain of linolenic bends so that

the terminal methyl group is near the carbonyl function of
the ester (Figure 3.1, Chapter 3), making development of the
enzyme-substrate complex difficult.

An extracellular lipase has also been isolated from *M.
pusillus* (Somkuti and Babel, 1968) and partially characterized (Somkuti et al., 1969). Production of the enzyme
reaches maximum after 6 days incubation of the fungus in
submerged culture. Using methyl ester substrates, maximal
activity of the enzyme is with a C_{12} substrate at pH 5.5.
Results of inhibitor studies suggest that serine is involved
in substrate binding at the enzyme active site.

The lipase of *Puccinia graminis tritici* uredospores is
believed to be involved in germination. It has temperature
and pH optima of 15 C and 6.7, respectively, and is inhibited
by EDTA, p-chloromercuribenzoic acid, and Hg^{+2}. With triolein
as substrate for the lipase, 1,3- and 1,2-diglycerides are
the products (Knoche and Horner, 1970).

The most extensively studied fungal lipase is that of
Rhizopus arrhizus (Laboureur and Labrousse, 1964; 1966;
1968). The *Rhizopus* lipase has properties similar to that
of pancreatic lipase, but it is extracellular and more stable.
Purified *Rhizopus* lipase (Lipase I) is slowly converted to
another form (Lipase II) in the cold. Lipase II is almost
as active as Lipase I, is more cationic, and even more stable
(Semeriva et al., 1967). Lipase I is a glycoprotein with a
molecular weight of 43,000 and contains 13 to 14 molecules
of mannose, two molecules of hexosamine, and a single N-
terminal aspartic acid (or asparagine) residue. It consists
of two apparently non-covalently linked subunits, a glycoprotein of 8,500 molecular weight, and the enzyme. The glycopeptide does not play a role in the catalytic function of
the enzyme, since, upon storage, it separates resulting in
the conversion of Lipase I to Lipase II. The role of the
glycopeptide is not known, but it may facilitate passage of
the enzyme through the plasma membrane (Semeriva et al.,
1969). A lipase from *C. paralipolytica* appears to be similar
to the *Rhizopus* lipase since a 260 nm absorbing substance can
be separated from the enzyme (Ota and Yamada, 1967). The
Rhizopus lipase also shows preferential activity at the C-1
and C-3 positions of triglycerides (Semeriva et al., 1967).
Migration of the C-2 acyl group to the C-1 or C-3 position
occurs under conditions of optimum activity for the *Rhizopus*
lipase, suggesting that the specificity for the C-1 and C-3

positions is practically absolute. The *Rhizopus* lipase can
also catalyze the hydrolysis of chylomicrons (Infante et al.,
1967; 1968).

GLYCOSYLGLYCERIDES AND OTHER GLYCOLIPIDS

The most abundant glycosylglycerides in nature are the
galactolipids mono- (MGDG) and digalactosyl diglycerides
(DGDG) of photosynthetic plants and bacteria. These two
glycolipids represent about 70% of chloroplast lipid, and
are minor components of brain tissue. The sugar moiety is
linked to the 3-position of *sn*-glycerol. The linkage between galactose and the diacylglycerol is $\beta 1' \rightarrow 3$, and that
between the two galactose units of DGDG is $\alpha 1' \rightarrow 6'$ (Carter
et al., 1961) (Figure 5.7). In plants, acyl groups of
these lipids tend to be more unsaturated than most glycerolipids, with high levels of α-linolenic acid.

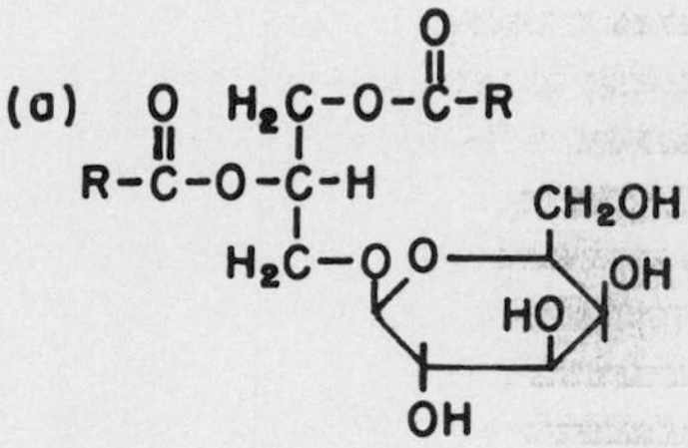

(a) 1,2-DIACYL-β-D-GALACTOPYRANOSYL-
(1'→3)-SN-GLYCEROL

(MONOGALACTOSYLDIGLYCERIDE)

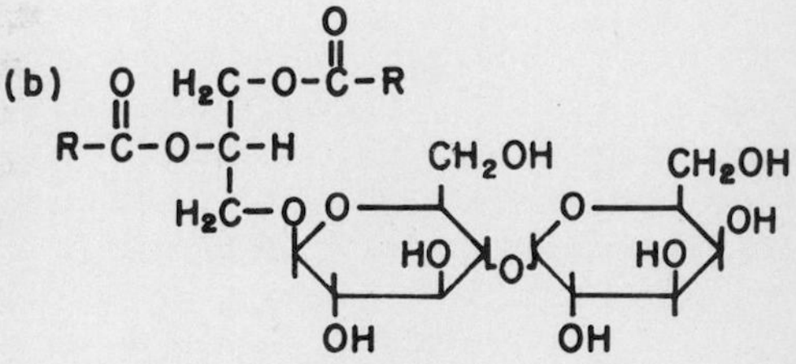

(b) 1,2-DIACYL-β-D-GALACTOPYRANOSYL-
(1'→6')-α-D-GALACTOPYRANOSYL-
(1'→3)-SN-GLYCEROL

(DIGALACTOSYLDIGLYCERIDE)

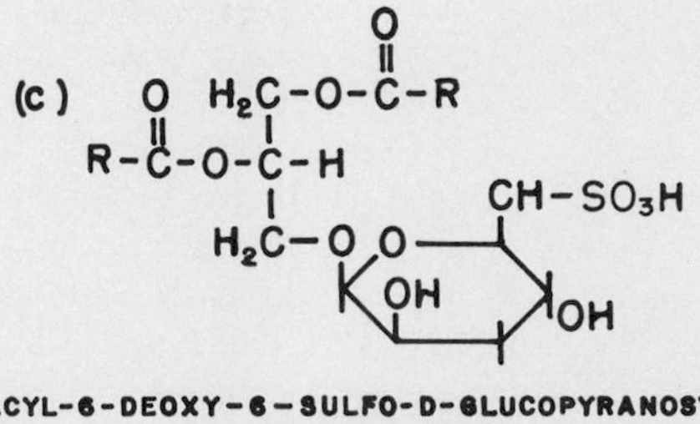

(c) 1,2-DIACYL-6-DEOXY-6-SULFO-D-GLUCOPYRANOSYL-(1'→3)-
SN-GLYCEROL

(SULFOQUINOVOSYLDIACYLDIGLYCERIDE)

Figure 5.7   Plant glycolipids.

Generally, glycolipids have not been widely studied in fungi, and relatively little is known about their occurrence in these organisms. Glycosphingolipids are discussed in Chapter 7. MGDG, sterol glycosides, and sulfatides have been detected in *S. cerevisiae* (Baraud et al., 1970). The most common sulfolipid in plant tissue is sulphoquinovosyl-diacylglyceride found exclusively in the chloroplast (Figure 5.7). Sterol glycosides are sterols linked glycosidically at the C-3 (sterol) position with sugars such as glucose, galactose, or mannose. These lipids have been detected in several fungi but their structures, distribution, and formation have not been fully explored.

Unlike most fungi, *Blastocladiella emersonii* contains a relatively high glycolipid content, 11% to 16%, depending on the stage of development (Mills and Cantino, 1974). The glycolipid fraction from zoospores contains some molybdate and ninhydrin positive substances, but the principal component is DGDG accompanied by MGDG. DGDG is associated with a subcellular organelle of this fungus called a gamma particle which contains chitin synthetase (Mills and Cantino, 1978). DGDG appears to be the carrier intermediate in chitin synthesis rather than a phosphorylated polyisoprenoid as in the formation of some other carbohydrates (see Chapter 10) (Mills and Cantino, 1978).

The biosynthesis of MGDG and DGDG has been studied in higher plants (Harwood, 1977). The formation of MGDG involves the transfer of galactose from its uridine diphosphate (UDP-gal) derivative to a 1,2-diacylglyceride. DGDG is formed by galactosylation of MGDG in a similar manner. There is evidence that the two galactosylation reactions are catalyzed by different enzymes.

Fungi, particularly yeasts and yeast-like organisms, produce a variety of unusual lipids, mostly extracellular, that include free hydroxy fatty acids (see Chapter 3), acetylated sphingolipids (see Chapter 7), polyol fatty acid esters, sophorosides of hydroxy fatty acids, and acetylated hydroxy fatty acids. In some cases, these extracellular lipids may reach a concentration of 1 to 2 g per liter of the growth medium. The occurrence of these lipids in fungi has been reviewed by Stodola et al. (1967), Brennan et al. (1974), and Weete (1976).

Fungal extracellular glycolipids (other than sphingolipids) usually contain hydroxy fatty acids linked glycosidically, or via an ester bond, to a carbohydrate moiety.
Polyol esters of *Rhodotorula graminis* and *R. glutinis* contain D-mannitol, D-arabinol, and xylitol linked to 3-D-
hydroxypalmitic (85%) and 3-D-hydroxystearic (15%) acids
(Tulloch and Spencer, 1964). The two acids are present in
molar ratios of 5-7:1, respectively, and the sugar alcohols
in 1:1 molar ratios with the hydroxy acids. Hydroxyl groups
not linked to acyl groups may be acetylated. Acylglucose
and other acylated polyols have been isolated from *S. cerevisiae* (Brennan et al., 1970) and *Agaricus bisporus* (cited
in Brennan et al., 1974); acylated trehalose has been isolated from *Pullularia pullulans* (Merdinger et al., 1968) and
Claviceps purpurea (Cook and Mitchell, 1970). The basidiomycete *Ustilago maydis* produces 4-0-(2,3,4,6-tetra-0-acyl-
β-0-mannopyranosyl)-D-erythritol which is esterified to the
usual fungal fatty acids (C_{12} to C_{18}) or acetic acid
(Fluharty and O'Brien, 1969) (Figure 5.8a).

A yeast form of the corn smut fungus *Ustilago zeae* (and
U. nuda) produces another group of esterified polyols called
ustilagic acids which are fatty acid esters of β-cellobiose
(Lemieux, 1951; Lemieux and Charanduck, 1951; Lemieux et al.,
1951) (Figure 5.8b). These glycolipids show antibiotic activity. Typical fatty acids and the hydroxylated acids
15-D-16-dihydroxyhexdecanoic acid (ustilic acid A) and
2,15,16-trihydroxyhexadecanoic acid (ustilic acid B) are
linked to cellobiose. The hydroxy acid may be acetylated,
or acylated with L-β-hydroxyhexanoic, L-β-hydroxyoctanoic,
or small relative amounts of hexanoic acid.

A monoglucosyloxyoctadecenoic acid is produced by the
mycelial fungus *Aspergillus niger* and represents 6% of the
total lipid (Laine et al., 1972) (Figure 5.8c).

Another group of fungal extracellular glycolipids are
the sophorosides, which are hydroxy fatty acids linked glycosidically to the disaccharide sophorose (2-0-β-D-gluco-
pyranosyl-D-glucopyranose). Sophorosides of *Torulopsis
magnoliae* (*T. apicola*) contain mainly 17-L-hydroxyocta-
decanoic and 17-L-hydroxyoctadecenoic acids (Gorin et al.,
1961). The fatty acid composition of the sophorosides is
a function of supplement fatty acids added to the growth
medium containing glucose as the primary carbon source.
The structure of a sophoroside from the yeast-like fungus

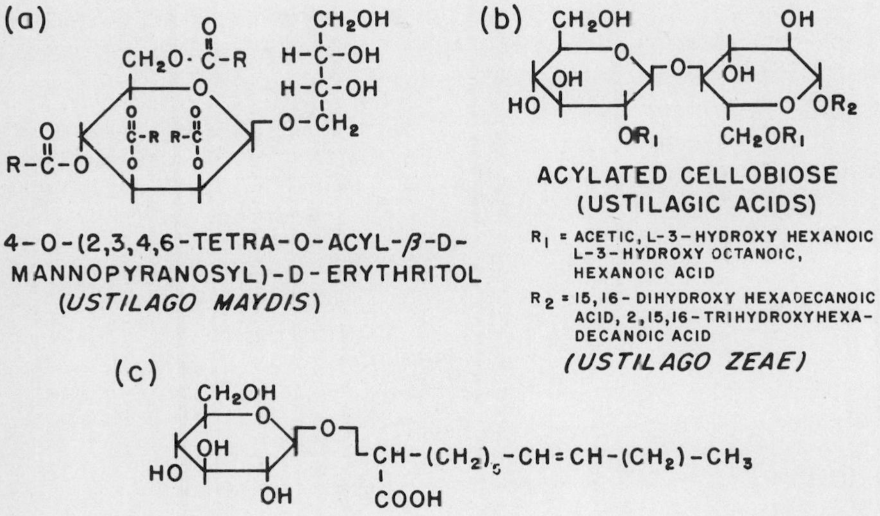

(a)

4-O-(2,3,4,6-TETRA-O-ACYL-β-D-
MANNOPYRANOSYL)-D-ERYTHRITOL
(*USTILAGO MAYDIS*)

(b)

ACYLATED CELLOBIOSE
(USTILAGIC ACIDS)

R₁ = ACETIC, L-3-HYDROXY HEXANOIC
L-3-HYDROXY OCTANOIC,
HEXANOIC ACID

R₂ = 15,16-DIHYDROXY HEXADECANOIC
ACID, 2,15,16-TRIHYDROXYHEXA-
DECANOIC ACID

(*USTILAGO ZEAE*)

(c)

Figure 5.8   Fungal acylated polyols.

grown on a medium containing octadecane is 17-L-[(2'-O-β-D-
glucopyranosyl-β-D-glucopyranosyl)oxy]-octadecanoic acid
6',6"-diacetate (Figure 5.9a). This glycolipid may be accompanied by the two lactones 17-L-[(2'-O-β-D-glucopyranosyl-
β-D-glucopyranosyl)oxy]-octadecanoic acid 1,4"-lactone 6',6"-
diacetate and the corresponding 6"-monoacetate (Figure 5.9b).
Another sophoroside, 13-[(2'-O-β-D-glucopyranosyl-β-D-gluco-
pyranosyl)oxy] docosanoic acid 6',6"-diacetate, and probably
the corresponding monoacetate, are products of *Candida
bogoriensis* (Tulloch et al., 1968) (Figure 5.9c). This
lipid represents about 50% of the extracellular glycolipid
produced by this yeast-like organism. The non-acetylated
form of this glycolipid has also been identified (Esders and
Light, 1972a).

The biosynthesis of fungal extracellular glycolipids
has not been thoroughly investigated. However, two gluco-
syltransferases have been detected in crude extracts of *C.
bogoriensis*, one (most active) that transfers [14C] glucose

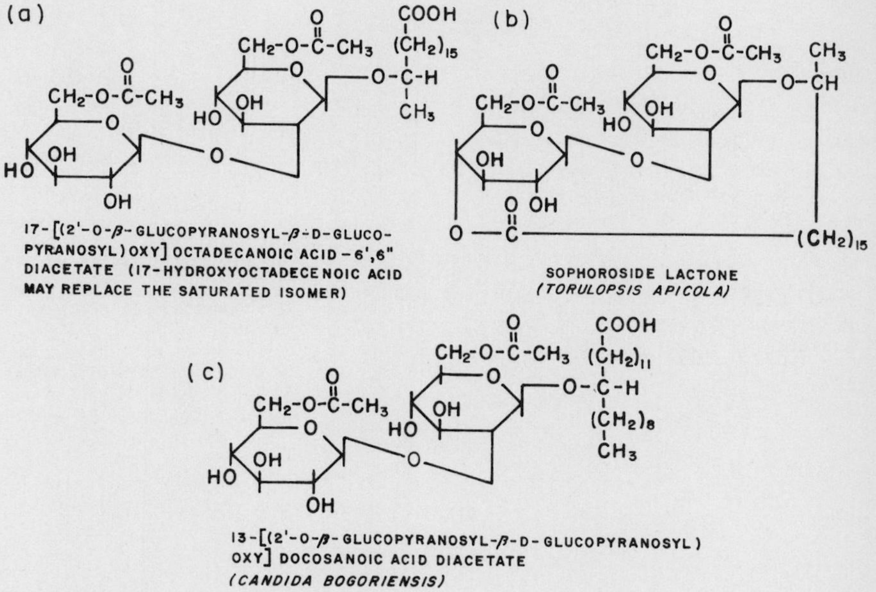

(a)

(b)

17-[(2'-O-β-GLUCOPYRANOSYL-β-D-GLUCO-
PYRANOSYL)OXY]OCTADECANOIC ACID-6',6"
DIACETATE (17-HYDROXYOCTADECENOIC ACID
MAY REPLACE THE SATURATED ISOMER)

SOPHOROSIDE LACTONE
(TORULOPSIS APICOLA)

(c)

13-[(2'-O-β-GLUCOPYRANOSYL-β-D-GLUCOPYRANOSYL)
OXY] DOCOSANOIC ACID DIACETATE
(CANDIDA BOGORIENSIS)

Figure 5.9   Fungal sophorosides.

from UDP-glucose to 13-hydroxydocosanoic acid and one that
transfers glucose to methyl-13-glycopyranosyl-oxydocosanoate
(Esders and Light, 1972b; 1972c). Extracts of this yeast
also contain an acetyltransferase(s) which catalyzes the
acetylation of 13-[(2'-0-β-glucopyranosyl-β-glucopyranosyl)
oxy]-docosanoic acid.

The pathway of sophoroside biosynthesis in *C. bogoriensis* begins with the glucosylation of 13-L-hydroxydocosanoic
acid followed by a second glucosylation and two acetylation
reactions. Sophoroside formation occurs in actively growing
cultures; degradation occurs during the stationary growth
phase beginning with deacetylation (Esders and Light, 1972a).
Complete disappearance of the glycolipid occurs in older
aerated cultures (Ruinen and Deinema, 1964).

Another group of glycolipids that has not been fully
explored in fungi are sterol glycosides. They have been
detected in yeast (Tyorinoja et al., 1974), and there is

evidence for sterol glycosylation in these organisms (Parks et al., 1978).

Sterols may also be non-covalently associated with poly-saccharides, rendering them water-soluble. The "water-soluble" forms of sterols have been detected in *Euglena gracilis* Z. (Brandt et al., 1969) and some higher plants (Anding et al., 1972; Bryce, 1971). The sterol-binding sub-stance of *S. cerevisiae* has been isolated (Adams and Parks, 1967; 1968) and identified as a cell wall mannan (Thompson et al., 1973). Similar results were obtained with *S. carls-bergensis*, but a cell wall glucan(s) is the principal poly-saccharide from *Rhizopus arrhizus* and *Penicillium roque-fortii* that binds sterols (Pillai and Weete, 1975). Very little sterol seems to be associated with the polysaccharides from these fungi *in situ*.

METHYL, ETHYL, AND STEROL ESTERS

As noted earlier in this chapter, the most abundant forms of fatty acids are as acylglycerides. One of the major non-glyceride forms of fatty acids are wax esters, which are esters of long chain fatty acids and fatty alco-hols ranging in chain length from C_{28} to C_{36}. Wax esters may comprise a major portion of leaf wax, fish oils, insect cuticular wax, and lipids of certain algae such as *Euglena gracilis* (Kolattukudy, 1976). Smaller esters, such as methyl, ethyl, and butyl esters of long chain fatty acids, have been reported for a variety of organisms, but they do not repre-sent major components of the total lipid. Fatty acid methyl esters have been reported for mammalian tissue including that of man (Saladin and Napier, 1967; Skorepa et al., 1968), corn pollen (Fathipour et al., 1967), bees (Calam, 1969), beetles (Aitken and Sanford, 1969), and the protozoan *Tetra-hymena pyriformis* (Chu et al., 1972). Methyl and ethyl esters of the major long chain fatty acids from male and female beetles (*Trogoderma granarium*) serve as attractants, but as repellants for red flour beetle (Ikan et al., 1969). Ethyl laurate and ethyl myristoleate (Calam, 1969), along with isoprenoid alcohols and esters, may act as tracking scents of male bumblebees (Kullenberg et al., 1970). Butyl esters of penta- and hexadecanoic acids are produced by the actinomycete, *Streptomyces* sp. (Weete et al., 1979).

Methyl esters of long chain fatty acids are components
of *Ustilago maydis* chlamydospores (Laseter et al., 1968) and
Rhizopus arrhizus mycelium (Weete et al., 1970) (Table 5.2).
Ethyl palmitoleate, ethyl oleate, and ethyl linoleate are
also products of *R. arrhizus* (Laseter and Weete, 1971) (Figure 5.10). The role of these esters in fungi is not known;
however, when placed in the growth medium ethyl esters of
long chain fatty acids increase mycelial growth of *Agaricus
campestris* (Wardle and Schisler, 1969). They also promote
auxin- and gibberellin-induced elongation of etiolated pea
stem sections (Stowe, 1958; 1960).

Enzymatic methylation of a variety of natural products
is well-known and includes pectins, polysaccharides, nucleic
acids, sterols, and proteins. S-Adenosylmethionine (SAM) is
the methyl donor in these reactions. The biosynthesis of

TABLE 5.2   Naturally Occurring Free Fatty Acids and Methyl
and Ethyl Esters of Fatty Acids from *Ustilago maydis* and
Rhizopus arrhizus

Carbon Chain Number	*U. maydis* (%)		*R. arrhizus* (%)		*R. arrhizus* (%)	
	Free	Methyl	Free	Methyl	Methyl	Ethyl
C_{13}	1.8	0.7	--	--	--	--
C_{14}	3.7	1.9	trace	3.0	0.3	--
$C_{14:1}$	13.8	3.3	--	--	--	--
C_{15}	13.9	6.3	--	0.2	0.3	--
$C_{15:1}$	2.9	1.2	--	--	--	--
C_{16}	28.9	20.7	22.8	20.1	11.6	2.4
$C_{16:1}$	3.9	7.9	2.6	4.4	0.5	--
C_{17}	3.9	3.3	--	--	1.6	--
$C_{17:1}$	1.2	1.0	--	--	--	--
C_{18}	1.3	6.2	21.6	6.3	9.0	1.5
$C_{18:1}$	12.4	32.7	41.5	34.4	30.3	21.2
$C_{18:2}$	5.7	9.4	6.8	21.2	3.9	--
$C_{18:3}$	--	--	--	--	--	--
C_{19}	--	--	--	--	--	--
C_{20}	5.8	4.9	3.0	--	2.7	--
C_{22}	--	--	--	--	4.0	--
C_{24}	--	--	--	3.8	--	--
Unidentified	--	--	--	--	11.3	--

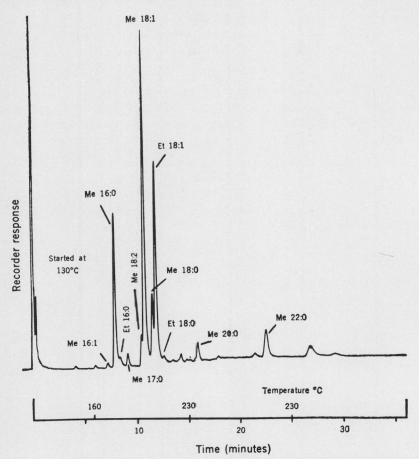

Figure 5.10   GLC separation of methyl and ethyl esters of
   fatty acids of *Rhizopus arrhizus* mycelium (Laseter and
   Weete, 1971).

fatty acid methyl esters has been studied in the bacterium
Mycobacterium phlei (Akamatsu and Law, 1970). A soluble
enzyme from this bacterium catalyzes the transfer of a methyl
group from SAM to the preferred acceptor, oleic acid.

Sterol esters have been detected in several fungi, including yeast (Maguigan and Walker, 1940). Acyl groups include the usual yeast fatty acids, but principally oleic and
palmitoleic acids (Madyastha and Parks, 1960). The sterol
ester content of yeast increases rapidly as the culture goes

into the stationary growth phase (Parks et al., 1978), which corresponds to an increase in sterol esterase activity (Parks and Stromberg, 1978). Sterol esters may represent 80% of the sterols present and represent a pool that can be readily converted to free sterols (Parks et al., 1978). The enzyme that catalyzes the esterification of sterols is associated with the particulate portions of the cell and has a pH optimum of 7 (Parks and Stromberg, 1978).

A purified pancreatic sterol ester hydrolyase preparation has both esterification and hydrolytic activities. The reaction does not require ATP or the CoA derivative of the fatty acid. A second sterol esterifying enzyme, acyl-CoA: cholesterol-0-acyl transferase, has been detected in mitochondria and endoplasmic reticulum of liver.

CHAPTER 6   GLYCEROPHOSPHOLIPIDS

INTRODUCTION

Glycerophospholipids are ubiquitous to living organisms, forming an essential structural component of biological membranes (see Chapter 2). Much has been written about the formation and properties of naturally-occurring and non-biological membranes, or lipid bilayers, in which the major lipid component is phospholipids.

NOMENCLATURE AND STRUCTURE

Phospholipids containing glycerol vary widely, but they have several structural features in common. They all contain a phosphate group linked to the C-3 position of *sn*-glycerol via an ester linkage, a hydrocarbon chain attached at the C-1 and/or C-2 positions of *sn*-glycero-3-phosphate via ester, ether, or vinyl ether linkages, and a fourth constituent linked to phosphorus via an ester or P-C bond. The fourth constituent may be hydrogen, or one of several substances the most common of which are choline, ethanolamine, serine, glycerol, and inositol (Figure 6.1 and 6.2).

The stereospecific numbering (*sn*) system of nomenclature described in Chapter 5 for naming neutral acyl derivatives of glycerol is also recommended by the IUPAC-IUB Commission on Biochemical Nomenclature for naming glycerophospholipids [Lipids 12:455 (1976); Chem. Phys. Lipids 21:159 (1978)]. Recommendations have also been made by the commission for the naming of phosphorus-containing compounds of biochemical

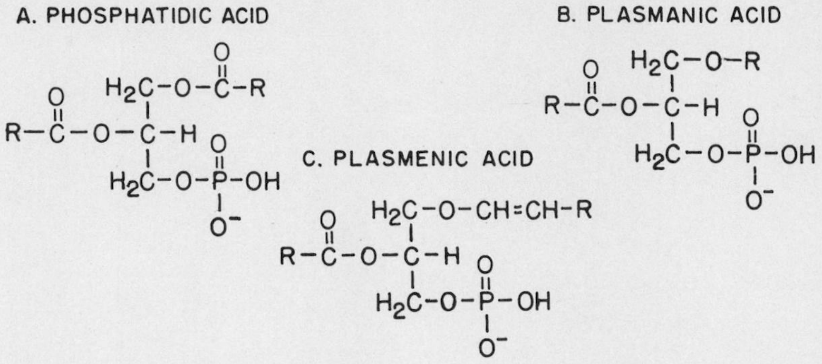

A. PHOSPHATIDIC ACID   B. PLASMANIC ACID

$$
\begin{array}{c}
\text{O} \\
\| \\
\text{H}_2\text{C}-\text{O}-\text{C}-\text{R} \\
| \\
\text{R}-\text{C}-\text{O}-\text{C}-\text{H} \\
\| \qquad\qquad\quad | \\
\text{O} \qquad \text{H}_2\text{C}-\text{O}-\text{P}-\text{OH} \\
\qquad\qquad\qquad | \\
\qquad\qquad\qquad \text{O}^-
\end{array}
$$

C. PLASMENIC ACID

$$
\begin{array}{c}
\text{H}_2\text{C}-\text{O}-\text{CH}=\text{CH}-\text{R} \\
| \\
\text{R}-\text{C}-\text{O}-\text{C}-\text{H} \\
\| \qquad\qquad | \\
\text{O} \qquad \text{H}_2\text{C}-\text{O}-\text{P}-\text{OH} \\
\qquad\qquad\qquad | \\
\qquad\qquad\qquad \text{O}^-
\end{array}
$$

B. PLASMANIC ACID

$$
\begin{array}{c}
\text{H}_2\text{C}-\text{O}-\text{R} \\
| \\
\text{R}-\text{C}-\text{O}-\text{C}-\text{H} \\
\| \qquad\qquad | \\
\text{O} \qquad \text{H}_2\text{C}-\text{O}-\text{P}-\text{OH} \\
\qquad\qquad\qquad | \\
\qquad\qquad\qquad \text{O}^-
\end{array}
$$

Figure 6.1   (A) O-(1-acyl), (B) O-(1-alkyl), and (C) O-(1-alkenyl) glycerophospholipids.

importance [Chem. Phys. Lipids 21:159 (1978)]. Briefly, the carbon atom on top of a Fischer projection with a vertical carbon chain and hydroxyl group at C-2 to the left is designated as C-1. Thus, the systematic name for the simplest glycerophospholipid is 1,2-diacyl-*sn*-glycero-3-phosphate (Figure 6.2). The common name for this lipid is phosphatidic acid. There may be numerous phosphatidic acid 'species' that vary according to the acyl groups at C-1 and C-2 of the glycerol moiety. Partial hydrolysis products of phosphatidic acid may be designated by the prefix 'lyso.' For example, 2-lysophosphatidic acid indicates hydrolysis has occurred at C-2, leaving a hydroxyl group at the second carbon of the glycerol moiety. Glycerophospholipids with substituents other than hydrogen at the 'X' position (Figure 6.2) of phosphatidic acid are named accordingly. For example, the name of the choline derivative of a phosphatidic acid is 1,2-diacyl-*sn*-glycero-3-phosphocholine. A shorthand system of nomenclature for glycerophospholipids is commonly used, and will be employed in this book. The term 'phosphatidyl' is used to denote the phosphatidate portion of the molecule, i.e. 3-*sn* phosphatidylcholine (3-*sn* will be dropped to conserve space, but is recommended by the IUPAC-IUB Commission).

Plasmalogen is a term used to designate glycerophospholipids containing an O-(1-alkenyl) residue and an acyl group at C-2. The systematic name for such a lipid is 2-acyl-1-alkenyl-*sn*-glycero-3-phosphate (for a phosphatidic derivative)

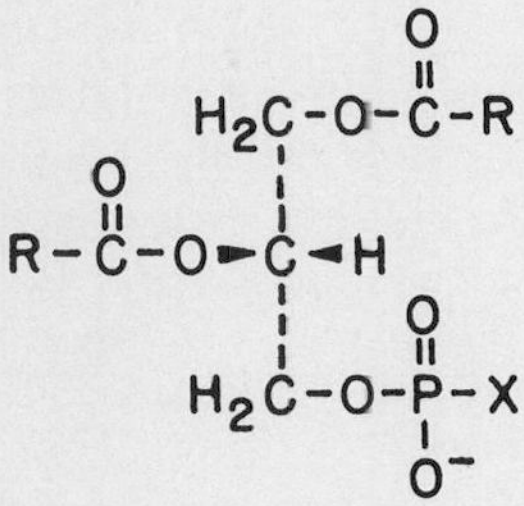

$$
\begin{array}{c}
\quad\quad\quad\quad\quad\quad\quad O \\
\quad\quad\quad\quad\quad\quad\quad \| \\
\quad\quad\quad H_2C-O-C-R \\
\quad O \\
\quad \| \\
R-C-O\blacktriangleright C\blacktriangleleft H \\
\quad\quad\quad\quad\quad O \\
\quad\quad\quad\quad\quad \| \\
\quad\quad\quad H_2C-O-P-X \\
\quad\quad\quad\quad\quad\quad | \\
\quad\quad\quad\quad\quad\quad O^-
\end{array}
$$

I X = OH
1,2-DIACYL-SN-GLYCERO-3-PHOSPHATE
(PHOSPHATIDIC ACID)

II X = OCH$_2$CH$_2$-$^+$N(CH$_3$)$_3$
1,2-DIACYL-SN-GLYCERO-3-PHOSPHOCHOLINE
(PHOSPHATIDYLCHOLINE)

  NH$_2$
III X = - OCH$_2$CHCOOH
1,2-DIACYL-SN-GLYCERO-3-PHOSPHOSERINE
(PHOSPHATIDYLSERINE)

IV X = OCH$_2$CH$_2$NH$_2$
1,2-DIACYL-SN-GLYCERO-3-PHOSPHOETHANOLAMINE
(PHOSPHATIDYLETHANOLAMINE)

V X = -O OH   OH
1,2-DIACYL-SN-GLYCERO-3-PHOSPHO-L-MYO-INOSITOL
^{a}(-4-PHOSPHATE) ^{a,b}(-4,5-BISPHOSPHATE)
(PHOSPHATIDYLINOSITOL-4-PHOSPHATE or
4,5-BISPHOSPHATE)

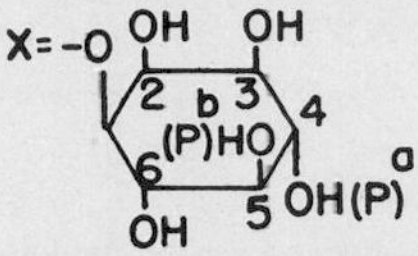

VI X = ^{1}CH$_2$OH^{2}CHOH^{3}CH$_2$O-
1,2-DIACYL-SN-GLYCERO-3-PHOSPHO-3-SN-GLYCEROL
(PHOSPHATIDYLGLYCEROL)

VII X = PHOSPHATIDYLGLYCEROL-
1',3'-(DI-1,2-DIACYL-SN-GLYCERO-3-PHOSPHO-3)-SN-GLYCEROL
(DIPHOSPHATIDYLGLYCEROL, CARDIOLIPIN)

VIII X=-CH$_2$CH$_2$NH$_2$
1,2-DIACYL-SN-GLYCERO-3-PHOSPHONOETHYLAMINE
(PHOSPHATIDYLETHYLAMINE)

Figure 6.2   Common glycerophospholipids.

'Plasmenic acid' is recommended as a shorthand name for
lipids containing one vinyl ether and one acyl group, i.e.
plasmenylcholine. 2-Lysoplasmenylcholine describes a plasmenic acid with a hydroxyl group at C-2. Glycerophospholipids with an alkyl (ether) group at C-1 and acyl group at
C-2 are named 2-acyl-1-alkyl-*sn*-glycero-3-phosphate. The
recommended shorthand name for these lipids is 'plasmanic
acid,' and is used as described above for lipids containing
a vinyl ether linkage, i.e. plasmanylcholine, lysoplasmanyl-
choline.

In addition to phosphodiesters of phosphoric acid,
glycerophospholipids also occur as esters of phosphonic
acid, i.e. 1,2-diacyl-*sn*-glycero-3-phosphono-ethylamine or
phosphatidylethylamine (Figure 6.2).

OCCURRENCE IN FUNGI

Phospholipids of various organisms have been reviewed
by Kates and Wassef (1970), in fungi by Griffin et al. (1974)
and Wassef (1977), and in yeasts by Hunter and Rose (1971),
Mangnall and Getz (1973), and Rattray et al. (1975). Total
lipid extracts of fungi can be easily separated into neutral
and polar lipid fractions. The polar lipid fraction contains
mostly glycerophospholipids which may be accompanied by some
glycolipids. As described in Chapter 2, the polar lipid or
glycerophospholipid *per se* averages about 41% of the total
lipid (Figure 2.5). The phospholipid content of yeast changes
during growth; it is higher in log-phase compared to stationary phase growth (Getz, 1970; Gailey and Lester, 1968).

Yeasts. The glycerophospholipid composition of numerous
yeasts and yeast-like fungi has been reported, particularly
Saccharomyces cerevisiae (Table 6.1). The five principal
glycerophospholipids of fungi are phosphatidylcholine (PC),
phosphatidylethanolamine (PE), phosphatidylinositol (PI),
phosphatidylserine (PS), and diphosphatidylglycerol (cardiolipin, CL). There is a remarkable similarity in the relative
proportions of these major glycerophospholipids among the
yeast and yeast-like species examined, and between the reports from different laboratories where various cultural
conditions and chromatographic techniques have been employed.
With few exceptions, PC is the principal glycerophospholipid
of yeasts, representing 25 to 55% of the total and averaging
40% ± 8. Sometimes PE is predominant, but usually is the

second most abundant glycerophospholipid, representing 13 to 38% of the total averaging 22% ± 7. PI ranges from 7 to 21% of the total (ave. 17% ± 15), PS ranges from 4 to 19% (ave. 11% ± 4), and CL ranges from 1 to 15% (ave. 8% ± 4). Inositol-containing phospholipids have been reviewed by Hawthorne (1960).

TABLE 6.1 Glycerophospholipid Composition of Some Yeast and Yeast-Like Fungi

Fungus	Phospholipid (%)[a]				
	PC	PE	PI	CL	PS
Bulleria alba[b]	54	30	8	–	8
Brettanomyces bruxellensis[b]	41	21	22	–	16
Candida macedoniensis[b]	41	26	16	–	17
C. krusei[c,d]	33	17	20	6	9
C. laurentii[c,d]	30	13	11	8	4
C. mycoderma[c,d]	35	17	17	8	7
C. pulcherrima[c,d]	41	18	13	6	11
C. tropicalis[c,d]	31	19	15	8	14
C. utilis[c,d]	32	27	13	8	12
C. utilis[e]	48	17	12	15	9
Cryptococcus neoformans[c,d]	49	28	7	4	8
Debaryomyces hansenii[c,d]	42	16	9	9	12
D. nilssonii[c,d]	33	20	18	7	12
Endomycopsis selenospora[b]	46	23	17	–	14
Hansenula anomala[c,d]	40	16	11	9	13
H. anomala[b]	53	23	9	–	15
Kluyveromyces polysporus[c,d]	42	15	18	5	7
Lipomyces starkeyi[c,d]	43	25	12	5	11
L. lipoferus[c,d]	38	20	16	3	9
Pichia membranaefaciens[c,d]	38	17	16	7	7
P. farinosa[c,d]	25	18	17	6	13
Rhodotorula glutinis[c,d]	40	29	12	8	5
R. rubra[c,d]	16	29	7	12	16
Saccharomyces carlsbergenesis[c,d]	34	25	16	8	9
S. cerevisiae (YFQ+)[b]	46	15	13	8	10
S. cerevisiae (Guiness 1164)[g]	43	23	21	3	7
S. cerevisiae (CBS 712)[b]	44	24	21	–	11
S. cerevisiae[h]	55	20	15	–	–
S. cerevisiae[i]	45	17	20	–	9
S. cerevisiae[c,d]	42	25	16	9	7

162   CHAPTER 6

TABLE 6.1   Continued

| Fungus | Phospholipid (%)[a] | | | | |
	PC	PE	PI	CL	PS
S. cerevisiae (YE-G)[j]					
aerobically grown	26	38	17	12	-
anaerobically grown	49	20	22	4	-
S. cerevisiae[k]	45	16	16	1	10
S. cerevisiae[b]	46	23	17	-	14
S. cerevisiae[l]	35	26	26	trace	8
S. cerevisiae[m]	39	30	19	-	12
S. fragilis[n,o]	31	24	23	5	9
S. maxianus[b]	42	22	17	-	19
S. occidentalis[e,p]	42	22	28	25	-
S. pastorianus[n,q]	31	24	16	2	8
S. rosei[c,d]	39	22	18	9	7
S. rouxii[c,d]	44	17	15	8	9
Saccharomycodes ludwigii[c,d]	47	14	8	8	19
Schizosaccharomyces pombe[c,d]	51	13	14	6	12
Schwanniomyces occidentalis[c,d]	48	16	10	7	12
Sporobolomyces salmonicolor[c,d]	29	30	12	8	13
Torulopsis colliculosa[c,d]	34	27	17	9	6
T. candida[c,d]	29	16	16	8	14
Trichosporon cutaneum[c,d]	29	15	11	6	12
Trichonopsis variabilis[c,d]	41	18	12	7	12
T. variabilis[b]	51	18	15	-	16

[a]Symbols for phospholipids: phosphatidylcholine, PC; phosphatidylethanolamine, PE; phosphatidylinositol, PI; phosphatidylserine, PS; cardiolipin, CL; phosphatidylglycerol, PG; phosphatidic acid, PA; lysophosphatidylcholine, LPC; lysophosphatidylethanolamine, LPE; monomethylphosphatidylethanolamine MMPE; dimethylphosphatidylethanolamine, DMPE; [b]Graff et al., 1968; [c]Hiroshi et al., 1976; [d]1 to 3.5% PA, 2.7 to 18.7% unidentified phospholipids; [e]Johnson et al., 1972; [f]Getz et al., 1970; 0.6% LPE; 0.2% DMPE; 1.6% PA; 0.9% LPC; 1.6% PG; [g]Letters, 1966; 2.4% DMPE; 1.9% PG; [h]Suomalainen & Nurminen, 1970; <5% LPE; <5% PA; [i]Deierkauf & Booij, 1968; [j]Kellerman & Linnane, 1968; PI + PS; CL includes PG; [k]Mudd & Saltzgaber-Muller, 1978; sphingolipids, 5%; 5% PA; 2.5% monophosphoryl-PI; [l]Ramsay & Douglas, 1979; 5.1% PA; trace LPC; 0.8% DMPE; continuous culture; [m]Cejkova & Jirku, 1978;

TABLE 6.1   Continued

synchronously dividing culture; [n]Hendrix & Rouser, 1976,
[o]1% PA; 8.3% unknown; [p]3.7% PA; [q]1.4% PA; 17.2% unknown.

The most widely studied single fungal species with
respect to glycerophospholipid composition is *S. cerevisiae*.
At least twelve glycerophosphatides have been identified in
lipid extracts of this yeast, and include PC, PE, PI, and PS
as the principal components. Quantitatively minor glycerophospholipids of this yeast include MMPE, DMPE, PA, LPE, PG,
and CL along with two unidentified phosphorus-containing
lipids (see Table 6.1 for abbreviations) (Letters, 1966;
Getz et al., 1970; Steiner and Lester, 1972). The di- and
triphosphate isomers of PI, PI-4-phosphate and PI-4,5-di-
phosphate have also been identified in extracts of *S. cerevisiae* (Lester and Steiner, 1968) and *Kloeckera brevis*
(Prottey et al., 1970).

One of the few exceptions to the generalization that PC
is the principal glycerophospholipid is in *Rhodotorula rubra*
which contains 7 to 16% of this lipid depending on the strain
(Hiroshi et al., 1976). PE is the principal glycerophospholipid of this yeast.

Few molecular species of individual glycerophospholipids
have been identified but 1,2-dipalmitoyl-*sn*-glycero-3-phospho-
choline has been identified in *S. cerevisiae* (Hanahan and
Jayko, 1952), *Candida* sp. (Kates and Baxter, 1962), *Hanseniaspora valbyensis* (Haskell and Snell, 1965), and *S. carlsbergensis* (Shafai and Lewin, 1968).

Plasmalogens (plasmenic acid derivatives) are apparently
not common components of yeasts. Trace quantities have been
detected in *S. cerevisiae* (Jollow et al., 1968), but others
have been unable to detect them (Letters and Snell, 1963).
Trace amounts of these substances have also been detected in
Lipomyces starkeyi (McElroy and Stewart, 1967; Bergelson et
al., 1966). Plasmalogens have been detected in *Pullularia
pullulans*, and are present in both the neutral and glycerophospholipids (1:6 ratio, respectively) (Goni et al., 1978).
Plasmenic acid derivatives of PC and PE (PE > PC) with
traces of PI and CL have been detected in this yeast.

   Filamentous Fungi. Relatively few filamentous fungi
have been analyzed for glycerophospholipid content.   There
appears to be more variability in the relative proportions
of these lipids among filamentous species than yeasts (Table
6.2).   PC (26 to 75%) and PE (14 to 48%) are the principal
glycerophospholipids, but PE is more often greater than or
equal to PC in relative amounts.   PI (3 to 15%), PS (1 to
24%), and CL (2 to 6%) are also present in variable amounts
in most of the filamentous fungi examined.

   The glycerophospholipid content of *Agaricus bisporus*
mycelium, sporophores, and basidiospores varies with the
report and perhaps fungal isolate (Table 6.2).   PC has been
reported in mycelium of this fungus by one group, but not by
another.   Also, small amounts of PS and PI are present in
sporophores of *A. bisporus*, but not the mycelium (Holtz and
Schisler, 1971).   Basidiospores of this fungus contain high
relative proportions of PC and similar amounts of PE, PS, PI,
and PG (O'Sullivan and Losel, 1971).   Polar lipid of both
aeciospores and basidiospores of the rust fungus *Cronartium
fusiforme* contain high levels of polar lipid (ca. 83%),   but
differ considerably in the relative proportions of glycerophospholipids.   Basidiospores contain high amounts of CL and
little or no PC, whereas aeciospores have PC as the most
abundant phospholipid and CL is present in lower relative
amounts (Weete, unpublished).

TABLE 6.2   Glycerophospholipid Composition of Some Filamentous Fungi

Fungus	Phospholipid (%)[a]				
	PC	PE	PI	CL	PS
Agaricus bisporus[b]	39	14	11	–	12
Alternaria dauci[c,d]	20	14	4	–	2
Aspergillus flavus[e]	36	26	14	–	9
Blastocladiella emersonii[f] (zoospores)	55	22	4	–	3
Basidiobolus merestoporus[i]	23	39	–	–	–
Cephalosporium spp.[g]	33	34	5	–	–
C. falciforme[h]	50	15	–	6	9
C. kiliense[h]	50	15	–	6	14
Entomophthora coronata[i]	49	29	–	–	–
Fusarium oxysporum[ee]	27	43	16	–	–

TABLE 6.2   Continued

Fungus	Phospholipid (%)[a]				
	PC	PE	PI	CL	PS
Fusarium solani[g]	34	28	15	–	–
F. solani[c,j]	23	–	–	–	trace
Graphium sp.[g]	30	53	6	2	–
Neurospora crassa[k,l]	44	24	10	5	6
N. crassa[m]	48	24	–	5	5
Phycomyces blakesleeanus[k,l]	31	41	3	3	12
Phytophthora parasitica					
var. *nicotianae*[k,n]	39	20	7	3	1
Polyporus versicolor[k,l]	47	25	3	3	12
Pythium ultimum	38	27	8	2	–
Rhizopus arrhizus[r]	26	48	–	–	–
Schizophyllum commune[k,l]	42	30	4	3	14
Sclerotium rolfsii[c,p]	19	18	9	–	2
Smittium culisetae[s]	75	14	–	–	11
Sporedonema epizoum[t]					
mycelium[u]	35	34	–	–	15
conidia[v]	25	21	–	–	24
Uromyces phaseoli[w]	54	22	–	–	–

GLYCEROPHOSPHOLIPIDS DETECTED IN OTHER FILAMENTOUS FUNGI

Agaricus bisporus
  basidiospores[b]   PC, PE, PS, PA
  mycelium[aa]   PE, PS
  mycelium[bb]   PC, PE
  sporophores[bb]   PC, PE, PS, PI
  sporophores[aa]   PE, PS
Ceratocystis stenoceras[x]   PC, PE, PS, PI, CL, LPE,
                              LPC, MMPE
Clitocybe illudens[y]   PC, PE (or PS)
Coprinus comatus[z]   PC, PE, PS
Humicola grisea var. *thermoidea*[cc] PC, PE, PI, LPC, CL
Neurospora crassa[dd]
  (choline-requiring mutant)   MMPE, DMPE
Sporothrix schenckii[x]   PC, PE, PS, CL, LPE, LPC,
                              MMPE

[a]See Table 6.1 for abbreviations; [b]O'Sullivan & Losel, 1971;

TABLE 6.2   Continued

sporophores; [c]Gunasekaran & Weber, 1972; [d]3.6% LPC; [e]Gupta
et al., 1970; 8.2% LPE; 6.7% polyglycerophosphate; [f]Mills &
Cantino, 1974; 6.3% LPC; 6.3% LPE; 3% PA; [g]Kok & Norris,
1972; 8 to 28% unknown; [h]Sawicki & Pisano, 1977; grown at
28.5 C; 7.2 to 11.5% LPC; sphingomyelin, 3.6 to 13.2%;
[i]DeBievre, 1974; [j]61.5% LPC; [k]Hendrix & Rouser, 1976; [l]2.4-
3.4% PA; 7% unkonwn; [m]Kushwaha et al., 1976; 21.2% PA; [n]1.9%
PG, 2.7% PA, 26.3% unkonwn; [o]Bowman & Mumma, 1967; 5.1% LPC;
4.7% unknown 3; 15.2% PG + LPE; 1.0% PA; [p]Gunasekaran et al.,
1972; 17.9% LPC; 7.6% unknowns (2); [r]48.3% LPC; [s]Patrick et
al., 1973; [t]Lopez & Burgos, 1976; [u]16.1% LPE; [v]30.6% LPE;
[w]Langenbach & Knoche, 1971; relative amount of ^{32}P incorporated into phospholipids; 21% unidentified; 4% CL + PA;
[x]de Bievre & Jourd'huy, 1974; [y]Bentley et al., 1964; [z]Jack,
1966; [aa]Griffin et al., 1970; [bb]Holtz & Schisler, 1971;
[cc]Mumma et al., 1971; [dd]Hall & Nye, 1961; [ee]Barran & de la
Roche, 1979; 3.7% PA; 4.4% OMPE; 6% PG.

Evidence has been reported for a phospholipid from
Neurospora crassa that contains 1-amino-2-methyl-2-propanol
(Ellman and Mitchell, 1954).

Glycerophospholipids of Subcellular Fractions.   The
glycerophospholipid composition of various subcellular fractions of some fungi, particularly *S. cerevisiae*, is given
in Table 6.3.   Since membranes are the structural basis of
most subcellular organeles and are the cellular location of
glycerophospholipids, the relative amounts of these lipids
should generally reflect the overall cellular composition.
This is generally true.   Two reports of *S. cerevisiae* plasmalemma (protoplast membrane) show PE as the predominant glycerophospholipid, while two others show PC as the major lipid
of this type (Table 6.3).   Reports of *S. cerevisiae* mitochondrial phospholipid content are generally consistent, with
PC being the predominant lipid followed by PE in yeast, but
PI in *Claviceps purpurea* (Anderson et al., 1964).   This is
also true for promitochondria of yeast (Paltauf and Schatz,
1969).   Respiratory competence of mitochondria has been correlated with CL content; a threefold increase in respiratory
competence has been associated with a twofold increase in
mitochondrial CL content.   Mitochondria contain 11 to 16% CL,
while promitochondria contain 6 to 9% (Table 6.3).   The lipid
content of the plasmalemma and tonoplast from *S. cerevisiae*
differs considerably (see Chapter 2); in relation to

glycerophospholipids, the tonoplast contains 1.5 times more
PI + PS than the plasmalemma (Kramer et al., 1978).   The
plasmaleema contains 15% PA which is absent in the tonoplast.
The glycerophospholipid content of yeast microsomes is generally similar to membranes of other subcellular fractions,
except for low (12%) levels of PE (Jakovcic et al., 1971).

TABLE 6.3   Glycerophospholipid Composition of Subcellular
   Fractions of *Saccharomyces cerevisiae*, *S. carlsbergensis*,
   and *Claviceps purpurea*

Fungus/ Fraction	Phospholipid (%)[a]						
	PC	PE	PI	CL	PA	PS	PG
Saccharomyces cerevisiae							
cell homogenate[f,g]	29	14	13	7	2	5	3
protoplast membrane[b]	23	33	28[c]	–	11	–[c]	–
protoplast membrane[d]	29	47	24	–	–	–	tr
whole envelopes[e]	45	15	30	<5	5	–[e]	–
plasmalemma[p,q]	34	20	28	–	15	–[q]	–
mitochondria[f,h]	33	22	11	16	1	3	2
mitochondria[i,j]	39	31	8	11	–	4	–
mitochondria[k]	42	35	9	12	–	–	–
promitochondria[i,l]	48	19	13	6	–	10	–
promitochondria[i,m]	34	18	26	9	–	4	–
microsomes[f,n]	41	12	28	2	1	7	–
tonoplast[p]	33	15	43	–	–	–[r]	–
S. carlsbergensis[s]							
mitochondria (+ inositol)							
inner[t]	24	28	10	9	–	8	–
outer[u]	44	16	11	6	–	3	–
mitochondria (- inositol)							
inner[v]	22	15	4	13	–	–	–
outer[w]	78	9	3	1	–	–	–
Claviceps purpurea[o]							
respiratory particles	27	18	28	–	–	13	–

[a]See Table 6.1 for abbreviations; [b]Longley et al., 1968;
NCYC 366; [c]PI + PS; LPC, 2%; [d]Baraud et al., 1970; Trace,
LPE; [e]Suomalainen & Nurminen, 1970; unknown, 5%; PI + PS;

[f]Jakovcic et al., 1971; [g]DMPE, 1.7%; LPC, 10.8%; LPE, 7.9%;
unknown, 8.5%; [h]DMPE, 0.9%; LPE, 3.6%; LPC, 4.2%; [i]Paltauf
& Schatz, 1969; [j]LPE + LPS, 4.5%; LPC, 3.2%; [k]Vignais et al.,
1970; [l]from cells grown in presence of lipids; LPE + LPS,
4.1%; LPC, 0.3%; [m]from cells grown in absence of lipids; LPE
+ LPS, 8.2%; LPC, 0.8%; [n]DMPE, 1.4%; LPC, 2.7%; LPE, 2.0%;
unknown, 3.7%; [o]Anderson et al., 1964; polyglycerophosphate,
6.7%; [p]Kramer et al., 1978; [q]PS + PI reported as single value;
unidentified PL, 3%; [r]LPC, 5%; PS + PI reported as single
value; unidentified PL, 4%; [s]Bednarz-Prashad & Mize, 1978;
phospholipid composition of inner and outer mitochondrial
membranes of cells grown in the presence and absence of
inositol which is required for this yeast; [t]LPC, LPE, LPS,
1 to 8%; unknown, 15%; [u]LPC, 12%; LPE, 3%; LPS, 0.4%; unknown,
5%; [v]LPC, LPE, LPS, 2 to 6%, unknown, 35%; [w]LPC, LPS, 1 to 2%;
unknown, 4%.

   Acyl Groups of Glycerophospholipids. Fatty acids obtained by alkaline hydrolysis of the total glycerophospholipid of fungi generally reflect the total fatty acid (obtained by hydrolysis of total lipid) composition. Some
fungi for which the acyl groups of phospholipids have been
reported are given in Table 6.4.

   The fatty acid composition of individual types of glycerophospholipids of *S. cerevisiae* also reflect the total
fatty acids with $C_{16:1}$ being the predominant acyl group of
PC, PE, PI, PS, and CL followed by $C_{18:1}$ (Trevelyan, 1966;
Suomalainen and Nurminen, 1970). PI from *Schizosaccharomyces
pombe* contains only $C_{18:1}$ (61%) and C_{16} (37.6%) (White and
Hawthorne, 1970).

   Some other fungi for which phospholipid acyl groups
have been determined are *Alternaria oleracea*, *Neurospora
sitophila*, and *Rhizopus nigricans* (Jack and Laredo, 1968),
R. arrhizus (Weete et al., 1970), *S. cerevisiae* (Bertoli et
al., 1971; Trevelyan, 1966; Paltauf and Schatz, 1969; Chavant
et al., 1978), and *Fusarium oxysporum* (Barran and de la Roche,
1979).

   The stereospecific distribution of fatty acids in the
two major phospholipids of *Phycomyces blakesleeanus* mycelium
and sporangiophores has been determined (DeBell and Jack,
1975). Greater than 85% of the fatty acids at the *sn*-1
positions of PE and PC consist of C_{16}, $C_{18:2}$ and $C_{18:3}$. The

sn-2 positions of these phospholipids have about 98% unsaturated fatty acids. Phospholipids from mycelium and sporangiospores are similar with $C_{18:2}$ and $C_{18:3}$ constituting more than 85% of the fatty acids from these lipids.

BIOSYNTHESIS

The biosynthesis of glycerophospholipids has been studied in several plants, animals, and microbes, and the subject has been reviewed by several investigators (Kates and Wassef, 1970; Lennarz, 1970; Mangnall and Getz, 1973). With the exception of S. cerevisiae, and to some extent N. crassa, relatively little is known about glycerophospholipid biosynthesis in fungi, particularly the enzymology.

Phosphatidic Acid. PA is a key intermediate in the formation of triacylglycerides and glycerophospholipids, representing a branching point in the biosynthetic pathway of these two major types of lipid. The biosynthesis of PA has been discussed in Chapter 5 (Figure 5.4). Briefly, in S. cerevisiae sn-glycero-3-phosphate undergoes a sequential double acylation via acyl-CoA (Kuhn and Lynen, 1965). The acylation reactions are catalyzed by two enzymes, sn-glycero-3-phosphate acyltransferase and 1-acyl-sn-glycero-3-phosphate acyltransferase which catalyze the formation of 1-acyl-sn-glycero-3-phosphate and PA, respectively in S. saki (Yamashita et al., 1975). The specific distribution of acyl groups in the C-1 and C-2 positions of sn-glycero-3-phosphate appear to be regulated by substrate specificities of these two enzymes (Morikawa and Yamashita, 1978). The enzyme that catalyzes the first acylation is specific for saturated and monoenoic thioesters, but when only C_{16}-CoA is supplied to the enzyme the product is a LPA. 1-Acyl-sn-glycero-3-phosphate acyltransferase is highly specific for monoenoic acyl-CoA's. This view is not shared by Yamada et al. (1977), who also found a different degree of specificity for acyl-CoA's, but suggest enzyme specificities cannot explain the positional distribution of fatty acids in phospholipid molecules. Different pH optima of the two acylation reactions and dihydroxyacetone phosphate as an initial substrate have been offered previously as possible factors influencing the positional distribution of acyl groups in PA, and hence other phospholipids. These factors cannot be ruled out entirely. Dihydroxyacetone phosphate can serve as the initial substrate for PA synthesis in mitochondrial preparations of S. carlsbergensis (Johnston and Paltauf, 1970). Low concentrations (ca 20 μM) of fatty acids, mainly C_{14}, inhibit both acyl-transferases; the enzyme that catalyzes the second acylation is more sensitive than the one catalyzing the first (Morikawa and Yamashita, 1978). Since fatty acids may accumulate under certain metabolic conditions and there is an inverse relation between lipogenesis and fatty acid concentration, it has been suggested fatty acids may be important regulatory factors in glycerolipid biogenesis (Morikawa and Yamashita, 1978).

The mitochondrial and microsomal cell fractions appear to be the sites of phospholipid synthesis, but the relative contribution of these fractions to the synthesis of individual phospholipids is not well defined. sn-Glycerophosphate and 1-acyl-sn-glycerophosphate transferase activities in yeast have been detected mainly in the microsomal fraction but activities of these enzymes in mitochondria could not be explained by contamination (Yamada et al., 1977). Yeast mitochondrial and microsomal fractions differ in their ability to produce individual phospholipids from 3-sn-$[2-^3H]$ glycero-3-phosphate (Cobon et al., 1974). The principal phospholipids produced by microsomal fractions are PS, PI, and PA while the major phospholipids produced by mitochondria, which have the higher specific activity for glycerolipid synthesis, are PG, PE, PI, PS, and PA. PC synthesis by the methylation and CDP-choline pathways (see below) occurs in the microsomal fraction.

PA can be converted to the various types of glycerophospholipids via two routes (Figure 6.3). The first, and most important in fungi, involves the CDP (cytidine-5'-diphosphate)-diacylglycerol derivative formed in a reaction between PA and CTP catalyzed by CTP; phosphatidic acid cytidylyltransferase (Table 6.4I). This enzyme is present in the particulate fraction of yeast cells (Hutchinson and Cronan, 1968), and has been located in both the inner and outer mitochondrial membranes of yeast (Mangnall and Getz, 1973). It is stimulated by citrate and inhibited by CL, suggesting that this enzyme is important in the biosynthesis of cardiolipin. CDP-diacylglycerol is believed to be a direct or indirect precursor of all the major phospholipids of S. cerevisiae (Steiner and Lester, 1972).

A second way in which PA serves as a precursor of glycerophospholipids is via a diacylglycerol, after dephosphorylation (see below).

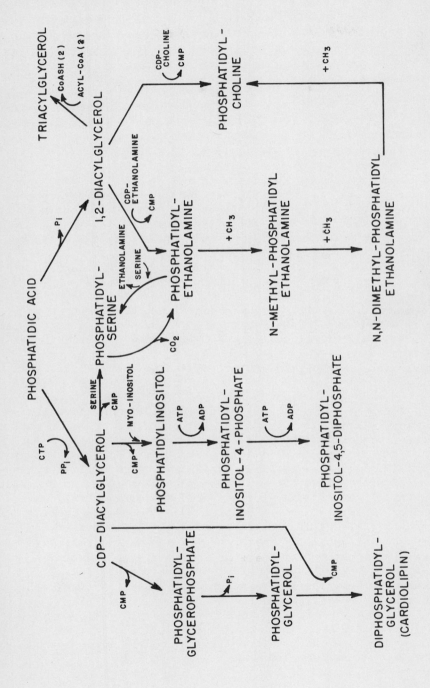

Figure 6.3   Biosynthesis of glycerophospholipids.

TABLE 6.4   Summary of Reactions of Glycerophospholipid
Biosynthesis

   I.   Phosphatidic acid + CTP^a → CDP-diacylglycerol + P-P^b

  II.   CDP-diacylglycerol + Serine → Phosphatidylserine
                                          + CMP^c

 III.   Phosphatidylserine → Phosphatidylethanolamine + CO_2

  IV.   Phosphatidylethanolamine + SAM^d → N-methyl-phospha-
                                              tidylethanolamine
                                              + SAH^e

        MMPE^f + SAM → N,N-dimethyl-phosphatidylethanolamine
                      + SAH

        DMPE^g + SAM → Phosphatidylcholine + SAH

   V.   Choline + ATP → O-Phosphocholine + ADP

        O-Phosphocholine + CTP → CDP-choline + P-P

  VI.   (1) Lysophosphatidylcholine + acyl-CoA → PC + CoASH

        (2) 2 Lysophosphatidylcholine → PC + *sn*-glycero-3-
                                          phosphocholine

 VII.   Ethanolamine + ATP^h → Phosphoethanolamine + ADP^i

        Phosphoethanolamine + CTP → CDP-ethanolamine + P-P

        CDP-ethanolamine + diacylglycerol → Phosphatidyl-
                                              ethanolamine
                                              + CMP

VIII.   Phosphatidylethanolamine + serine → PS + ethanolamine

  IX.   CDP-diacylglyceride + *myo*-inositol → PI + CMP

   X.   CDP-diacylglyceride + GP^j → Phosphatidylglycerophos-
                                        phate

        Phosphatidylglycerophosphate → Phosphatidylglycerol
                                        + Pi

TABLE 6.4   Continued

Phosphatidylglycerol + CDP-diacylglyceride →
Diphosphatidylglycerol + CMP

---

[a]Cytidine triphosphate; [b]Pyrophosphate; [c]Cytidine-5'-mono-
phosphate; [d]S-adenosylmethionine; [e]S-adenosylhomocysteine;
[f]Monomethyl-PE; [g]Dimethyl-PE; [h]Adenosine triphosphate;
[i]Adenosine-5'-diphosphate; [j]sn-glycero-3-phosphate.

Phosphatidylcholine. PC is the principal glycerophospholipid in most fungi, as well as other organisms, and its
synthesis involves several other common phospholipids as
intermediates. There are two principal pathways of PC bio-
sunthesis (Figure 6.3). The first pathway begins with the
formation of PS from CDP-diacylglycerol and serine (Table
6.4II). This reaction is catalyzed by CDP-diacylglycerol:L-
serine-phosphatidyltransferase. This pathway of PS synthesis accurs in S. cerevisiae (Steiner and Lester, 1972) and
N. crassa (Sherr and Byk, 1971), animals (Avidson, 1968),
gram negative bacteria (Kaneshiro and Law, 1964), and algae
(Kates and Volcani, 1966).

The next step in PC formation via this pathway is the
conversion of PS to PE by decarboxylation (Figure 6.3; Table
6.4III). PC is subsequently formed by three sequential
methylations with S-adenosylmethionine (SAM) as the methyl
donor (Table 6.4IV). The first reaction involves the methylation of PE resulting in the formation of N-monomethyl-PE
(MMPE). MMPE is then methylated giving rise to N,N-dimethyl-
PE (DMPE) which is subsequently methylated to give PC. The
N-methyltransferases of N. crassa are microsomal (Scarborough
and Nye, 1967a). Using the microsomal fraction (Scarborough
and Nye, 1967a) and solubilized N-methylase (Scarborough and
Nye, 1967b) of two choline deficient mutants of N. crassa,
evidence has been obtained for the existence of only two
N-methylating enzymes. One catalyzes the methylation of PE
and the second catalyzes the two subsequent methylation
reactions. The two mutants could be distinguished on the
basis of the absence or reduced activity of one of the two
enzymes. The solubililized enzymes exhibit a pH optimum at
8.

The methylation pathway is also present in *S. cerevisiae*.
When the yeast is grown in a medium containing choline, microsomal methylating activity is greatly reduced (Waechter et
al., 1969; Steiner and Lester, 1970). Reduced activity of
microsomal preparations from choline grown cells could be
attributed to less enzyme, suggesting repression of N-methylase synthesis. The effect of choline is reversible (Waechter
and Lester, 1968). A similar regulatory system has been observed in *N. crassa* (Crocken and Nye, 1964) and animals
(Lombardi et al., 1969).

The second, but apparently less important, pathway of
PC synthesis in *S. cerevisiae* and *N. crassa* is the CDP-
choline diglyceride pathway (Figure 6.3). This pathway
occurs when N-methylase synthesis is repressed by the presence
of choline in the growth medium. In animals, choline is activated by phosphorylation involving the enzyme ATP:choline
phosphotransferase (Table 6.4V). CTP:cholinephosphate
cytidylyltransferase catalyzes the transfer of O-phosphocholine to CMP to form CDP-choline. The formation of PC
involves the transfer of O-phosphocholine from CDP-choline
to a diacylglyceride. The enzyme that catalyzes this reaction, CDP-choline diacylglyceride phosphocholine transferase,
has been detected in microsomal and mitochondrial (outer
membrane) membranes of *S. cerevisiae* (Ostrow, 1971). Perhaps under certain conditions both pathways are operative,
the CDP-choline diacylglyceride pathway in mitochondria
furnishing PC solely to these organelles and the methylation
pathway of the microsomal fraction producing PC for other
organelles and the plasmalemma (Mangnall and Getz, 1973).

Two other ways in which PC may be formed involves lyso-
PC. One reaction involves the acylation of LPC, and the
other involves the transfer of an acyl group from one LPC
molecule to another (Table 6.4VI). Both reactions have been
detected in mammalian tissue. The latter reaction has also
been detected in the supernatant fraction of *S. cerevisiae*
(Van den Bosch et al., 1965).

Phosphatidylethanolamine. There are also two known
pathways of PE synthesis (Figure 6.3). One pathway involving
the formation and decarboxylation of PS has been described
above. The second pathway involves CDP-ethanolamine, and is
similar to the CDP-choline diglyceride pathway of PC synthesis (Table 6.4VII). This pathway is not considered important

in fungi, but has been detected in *S. cerevisiae* (Waechter et al., 1969). CTP:ethanolamine cytidylyltransferase, detected in several animal tissues (Kennedy and Weiss, 1956), catalyzes the transfer of O-phosphoethanolamine to a diacylglyceride to form PE.

Phosphatidylserine. See section on PC synthesis above. PS can also be formed via base exchange involving PE and serine (Table 6.4VIII).

Phosphatidylinositol. *Myo*-inositol is a growth factor required by certain yeasts and appears to be involved in polysaccharide biosynthesis in these organisms (Chung and Nickerson, 1954). It is suspected that PI is a storage form of *myo*-inositol. One of the symptoms associated with inositol deficiency in yeast is the accumulation of triacylglycerides (Shafai and Levin, 1968; Johnston and Paltauf, 1970).

The principal pathway of PI formation in mammals is through the enzymatic transfer of 1,2-diacylglyceride from CDP-diacylglyceride to *myo*-inositol with CMP being released (Paulus and Kennedy, 1960). PI may also be formed through the acylation of LPI (Keenan and Hokin, 1964), and by an exchange reaction involving a phospholipid and *myo*-inositol. The inositol residue of PI may be further phosphorylated by enzymes present in the microsomal fraction of animal tissues (Salway et al., 1968; Benjamin and Agranoff, 1969). PI synthesis in *S. cerevisiae* appears to also occur via the CDP-diacylglyceride intermediate (Steiner and Lester, 1972) (Table 6.4IX; Figure 6.3); over 90% of $[^{14}C]myo$-inositol supplied to yeast may be incorporated into PI (Tanner, 1968). There appears to be a rapid PI turnover in yeast (Angus and Lester, 1972).

The CDP-diacylglyceride pathway of PI synthesis could not be detected in *Schizosaccharomyces pombe* (White and Hawthorne, 1970).

In yeast, PI, PI-4,5-diphosphate, and PI-4-phosphate occur in a molar ratio of 140:4:1 (Steiner and Lester, 1972). Experiments with ^{32}P have shown that there is a high rate of turnover of the di- and triphosphoinositides, which are probably formed through successive phosphorylations of PI by the action of kinases as in animals.

Diphosphatidylglycerol (Cardiolipin). Relatively little is known about the biosynthesis of CL. The pathway of CL synthesis in yeast involves CDP-diacylglyceride as an intermediate, as well as phosphatidylglycerophosphate and phosphatidylglycerol (Table 6.4X; Figure 6.3) (Steiner and Lester, 1972).

Plasmanic and Plasmenic Acid Derivatives. Both neutral (see Chapter 5) and phospholipids contain O-alkyl and O-alkenyl groups. The biosynthesis of these lipids has not been studied in fungi. Presumably the immediate precursor of the plasmanic acid is a fatty aldehyde rather than acyl CoA. In animals, the O-alkenyl group is formed by desaturation of an O-alkyl group in a reaction requiring O_2 and NADPH. The enzyme catalyzing this reaction is microsomal and similar to that which catalyzes a similar reaction in fatty acids (see Chapter 4).

DEGRADATION

Enzymes that catalyze the hydrolysis of phospholipids are called phospholipases, and are included in several reviews on phospholipid metabolism (Lennarz, 1970; Van Deenan and de Haas, 1966). These enzymes are widely distributed in nature and have been detected in plants, animals, bacteria, and fungi. Relatively high levels of these enzymes are present in snake venom, extracts of higher plants, and excretions of bacteria.

Although a system of classification has been recommended for these enzymes [Biochemical Journal 105:897 (1967)], the simplified ABCD system will be used here. Phospholipases may be distinguished according to the site of action on the phospholipid molecule. The sites of attack of phospholipases A (A_1 and A_2), B, C, and D are shown in Figure 6.4. Phospholipase A catalyzes the hydrolytic removal of acyl groups from diacylphospholipids, giving rise to a lysophosphatide and a fatty acid. Two enzymes showing phospholipase A activity have been detected, A_1 and A_2, and catalyze the hydrolysis of ester bonds at C-1 and C-2 of the glycerol moiety of glycerophospholipids, respectively. Phospholipase A_1 activity has been detected in a variety of animal tissues and bacteria. Phospholipase A_2 activity has been detected in rattlesnake venom, pancreatic tissue and others. Lysophospholipase catalyzes the further breakdown of the

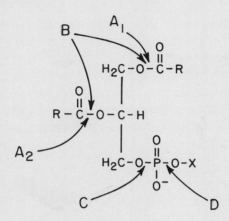

Figure 6.4   Sites of action of phospholipases A_1, A_2, B, C, and D.

lysophosphatide product of phospholipase A action. Lysophospholipase acts on both 1- and 2-acylglycerophosphatides
(Van deen Bosch et al., 1968).

In addition to fungi (see below), phospholipase B has
been detected in several plant species. This hydrolase
reportedly catalyzes the complete deacylation of 1,2-diacyl-
glycerophosphatides to sn-glycero-3-phosphate and two fatty
acids.

Phospholipase C activity has also been detected in
several organisms, including snake venom and bacteria (extracellular). This enzyme catalyzes the hydrolysis of a phosphodiester giving rise to 1,2-diacylglyceride and a phosphorylated base (ca O-phosphocholine). This enzyme has a
rather braod substrate specificity, but the degree of specificity depends on the source.

Phospholipase D activity has been detected only in
plants. This enzyme cleaves the terminal phosphate ester
bond of glycerophospholipids giving rise to PA and a base
(ca choline).

Phospholipases, as described above, are specific for a
particular site on the phospholipid molecule, and are classified accordingly. Phospholipases otherwise show a wide
range of substrate specificity, and vary for the same enzyme
from different sources. The specific configuration of *sn*-
glycero-3-phosphate as a component of the phospholipid molecule and presence of a phosphate group seem to be the principal requirements for phospholipase activity. The rate of
hydrolysis catalyzed by these enzymes is influenced by the
physical state and fatty acid composition of the substrate,
and the charge of the phosphate group. Phospholipase activity
can be stimulated by diethyl ether and certain other organic
solvents (Slotboom et al., 1970; Letters, 1968). This stimulation of phospholipase activity has been attributed to pene-
tration of solvent molecules into the phospholipid micelles
resulting in wider spacing of the substrate molecules at the
lipid-water interface.

Phospholipases have been investigated in few fungi.
Phospholipase A activity has been detected in dried yeast
(Van der Bosch et al., 1967); PC being hydrolyzed to LPC
and *sn*-glycero-3-phosphocholine.

Phospholipase B from *Penicillium notatum* has been more
extensively studied. In one study, only a lysophospholipid
could serve as a substrate for this enzyme (Fairbairn, 1948),
but in another study it has been shown that phospholipase B
activity toward diacylphospholipids could be stimulated by
lipids such as CL or PI. The activation has been attributed
to the lipid imparting a negative charge on the substrate
which is essential for initial attack of the enzyme (Bangham
and Dawson, 1959). Another explanation for the stimulation
of phospholipase B activity by anionic amphipaths is the
dispersion of substrate molecules by the lipid (Kates et al.,
1965); ultrasonification could substitute for the lipid activators (Beare and Kates, 1967). The enhanced activity of the
enzyme by these methods is presumably due to the increased
negative potential on the substrate surface which presumably
assists the enzyme to come into a favorable stereochemical
orientation at the lipid-water interface (Dawson and Hauser,
1967). It has also been suggested that phospholipase B of
P. notatum exists with inhibitory factors that can be dis-
placed electrostatically by anionic activators (Saito and
Sato, 1968). The optimum activity of phospholipase B from
this source is pH 4.0, and there is evidence that S-S
linkages are required for activity (Beare and Kates, 1967).

Phospholipase B activity has also been detected in *Sclerotium rolfsii*, and has optimum activity at a low pH (4.5) (Tseng and Bateman, 1969). This enzyme is extracellular and catalyzes the hydrolysis of soybean PC to palmitic and oleic acids and *sn*-glycero-3-phosphocholine.

Phospholipase C activity has been reported in dried yeast by some investigators (Hoffman-Ostenhoff et al., 1961; Harrison and Trevelyan, 1963), but it could not be detected by others (Van der Bosch et al., 1967). Phospholipase C has been detected in *Erwinia carotovora* (Tseng and Bateman, 1969) and certain phytopathogenic fungi (Tseng and Bateman, 1968).

An unusual enzyme with phospholipase activity has been detected in the lipase preparations of *Rhizopus arrhizus* (see Chapter 5). This enzyme catalyzes hydrolysis at the C-1 and C-3 positions of triacylglycerides (Enbressangler et al., 1966; Semeriva et al., 1967), but also specifically cleaves the acyl moiety from the C-1 position of several glycerophospholipids (Slotboom et al., 1970; de Haas et al., 1965). An enzyme similar to the one from *Rhizopus* has been detected in porcine pancreas (de Haas et al., 1965).

CHAPTER 7   SPHINGOLIPIDS

INTRODUCTION

Sphingolipids are widely distributed in nature, yet
their biological function(s) is not well understood. However, certain groups of these lipids exhibit immunological
activity and they are constituents of cellular membranes.
High amounts of sphingolipids are associated with nerve tissue. Considerable interest in these lipids developed because
they are associated with certain diseases called lipidoses
(sphingolipidoses). These diseases result from inborn errors
in sphingolipid metabolism; the absence or low activity of
certain degradative enzymes resulting in the accumulation of
specific sphingolipids in certain tissues.

STRUCTURE AND NOMENCLATURE

Sphingolipids are a large and complex group of lipids
characterized by long chain 2-amino alcohols (base) as components of their structure. There are four base structures
that occur most frequently in biological materials. Trivial
names of these bases are sphingosine, dihydrosphingosine,
phytosphingosine, and dehydrophytosphingosine (Table 7.1).
The basic structures represented by these 2-amino alcohols
occur in nature as homologous series of normal and sometimes
methyl branched chain isomers. The C_{18}, and often C_{20}, homologues are most abundant. Aliphatic chains of these bases
range from 12 to 20 carbon atoms in chain-length, and the
methyl branch occupies the *iso* position. Sphingosines and
dehydrophytosphingosines have a single *trans* double bond in

TABLE 7.1 Structure and Nomenclature of the Four Common Long Chain 2-Amino Alcohols

STRUCTURE	NOMENCLATURE
$CH_3-(CH_2)_{12}-CH=CH-CH-CH-CH_2-OH$ $\qquad\qquad\qquad\quad OH\ NH_2$	SPHINGOSINE (C_{18}-SPINGOSINE)[a], 4-SPHINGENINE[b], D-ERYTHRO-1,3-DEHYDROXY-2-AMINO-TRANS-4-OCTADECENE[c]
$CH_3-(CH_2)_{14}-CH-CH-CH_2-OH$ $\qquad\qquad\qquad OH\ NH_2$	DIHYDROSPHINGOSINE (C_{18}-DIHYDROSPHINGOSINE), SPHINGANINE, D-ERYTHRO-1,3-DIHYDROXY-2-AMINO-OCTADECANE
$CH_3-(CH_2)_{13}-CH-CH-CH-CH_2-OH$ $\qquad\qquad\quad OH\ OH\ NH_2$	PHYTOSPHINGOSINE (C_{18}-PHYTOSHINGOSINE), 4-D-HYDROXY-SPHINGANINE, D-RIBO-1,3,4-TRIHYDROXY-2-AMINO-OCTADECANE
$CH_3-(CH_2)_8-CH=CH-(CH_2)_3-CH-CH-CH-CH_2-OH$ $\qquad\qquad\qquad\qquad\qquad\quad OH\ OH\ NH_2$	DEHYDROPHYTOSPHINGOSINE (C_{18}-DEHYDROPHYTOSPHINGOSINE), 4-D-HYDROXY-8-SPHINGENINE, D-RIBO-1,3,4-TRIHYDROXY-2-AMINO-TRANS-8-OCTADECENE

[a] Trivial name
[b] IUPAC-IUB nomenclature
[c] Systematic nomenclature

the 4 and 8 positions, respectively, but less common bases
may have one to two additional double bonds (*cis*). Bases
with two adjacent asymmetric carbons always have the *erythro*
configuration and those with three always have the *ribo* configuration. Over 30 individual naturally occurring sphingolipid bases have been identified in biological materials.
They vary according to the base homologue, number of hydroxy
groups, degree and location of unsaturation, and presence or
absence of methyl branching. The usual rules of organic
nomenclature may be applied to sphingolipids. For example,
sphingosine would be D-*erythro*-1,3-dihydroxy-2-amino-*trans*-
4-octadecene. The IUPAC-IUB recommends a semi-systematic
system of nomenclature based on the saturated base dihydrosphingosine, proposing the name sphinganine to mean D-*erythro*-
1,3-dihydroxy-2-amino-octadecane. 4-Sphingenine is the term
for the corresponding unsaturated isomer. Similarly, phytosphingosine and dehydrosphingosine are named 4-D-hydroxy-
spinganine and 4-D-hydroxy-8-sphingenine, respectively. The
trivial names are often used to refer to the entire homologous series of 2-amino alcohols, or the chain length may be
specified, i.e. C_{20}-4-sphingenine or *erythro*-eicosa-*trans*-4-
sphingenine. The trivial names are used in this chapter.

The term sphingolipid refers to a complex group of substances with a large number of molecular species. They may
vary not only according to the type of long chain base, but
also according to the type, number, and location of substituents on the base. Unsubstituted bases do not occur in
nature; instead they are linked to fatty acids, phosphate
and/or simple sugars or oligosaccharides. Most sphingolipid
bases occur as N-acyl derivatives called ceramides (Figure
7.1). Ceramides occur naturally, but generally in low concentrations. Ceramides occur more frequently with various
substituents linked via ester or glycoside bonds at the 1-
position of the base. Depending on the substituent, ceramides are classifed as (1) phosphosphingolipids, (2) glycosphingolipids, and (3) glycosphosphingolipids.

Phosphosphingolipids are one of the most widely occurring types of sphingolipids in animals. The most common
example of this type are the sphingomyelins which are ceramides linked to phosphocholine via a phosphoester bond at
carbon number one of the base (Figure 7.1).

Glycophosphosphingolipids are the largest group of
sphingolipids that include cerebrosides, ceramide

oligosaccharides, globosides, and gangliosides. Cerebrosides
are the simplest of this group and are ceramide monosaccharides with the sugar attached at the 1-position of the base
through a β-glycoside linkage (Figure 7.1). Glucose and
galactose are the principal sugars of cerebrosides. The
sulfuric acid ester of galactosyl ceramide (sulfatides or
sulfocerebrosides) also occur in animals with the sulfate
attached to carbon three of the sugar. Sulfocerebrosides
occur as potassium salts.

Ceramide oligosaccharides constitute a more complex
group of glycosphingolipids that are classified as cytosides
(contain sugars of only carbon and hydrogen), globosides
(cytosides containing amino sugars) and gangliosides (globosides containing sialic acid, N-acetyl neuramic acid) (Figure
7.1).

Glycophosphosphingolipids are ceramides linked to oligosaccharides via a phosphodiester bond. Phytoglycolipid is
an example of this type of lipid containing inositol, glucuronic acid, glucosamine, mannose and other sugars (Carter
et al., 1969) (Figure 7.1).

Fatty acids obtained upon hydrolysis of sphingolipids
range from 12 to 26 carbons in chain length with long chain
acids predominant. α-Hydroxy fatty acids are also abundant
in sphingolipids with 2-hydroxy-stearic (2-OH C_{18}) and 2-
hydroxyhexaeicosanoic (2-OH C_{26}) acids being common.

OCCURRENCE

Bacteria, Plants, and Animals. Sphingolipids were first
thought to be restricted to higher animals, but they are now
known to be widely distributed in nature. Sphingolipid occurrence and synthesis has been included in several reviews
(Svennerholm, 1964; Carter et al., 1965; Olsen, 1966;
Sharpiro, 1967; Kates and Wassef, 1970; Karlsson, 1970a;
1970b). These lipids appear to be rare in bacteria, but a
ceramide phosphoethanolamine and ceramide phosphoglycerol were detected in *Bacteriodes melaninogenicus* (Labach
and White, 1969). Bacterial sphingolipids contain normal
and branched saturated C_{17} to C_{19} long chain bases and the
N-acyl groups are primarily C_{14}, br C_{15}, C_{17}, C_{18}, and
$C_{18:1}$.

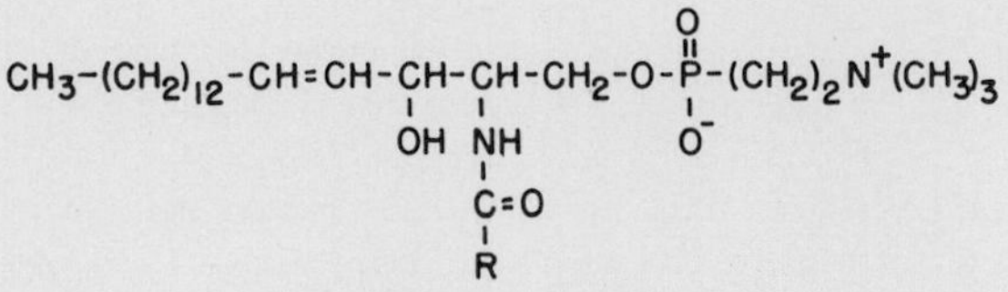

$$CH_3-(CH_2)_{12}-CH=CH-CH-CH-CH_2-O-\overset{\overset{O}{\|}}{\underset{\underset{O^-}{|}}{P}}-(CH_2)_2N^+(CH_3)_3$$

with OH, NH and C=O / R substituents

SPHINGOMYELIN
(N-ACYL-4-SPHINGENYL-I-O-PHOSPHORYLCHOLINE)

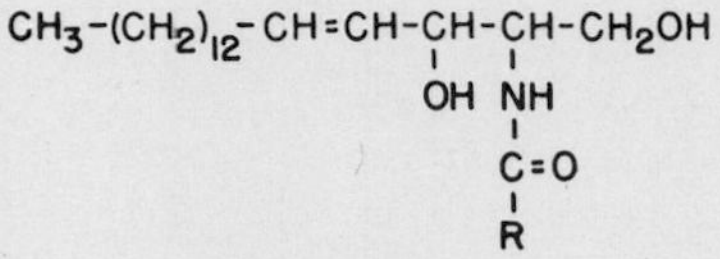

CERAMIDE
[R = -(CH_2)_N CH_3]

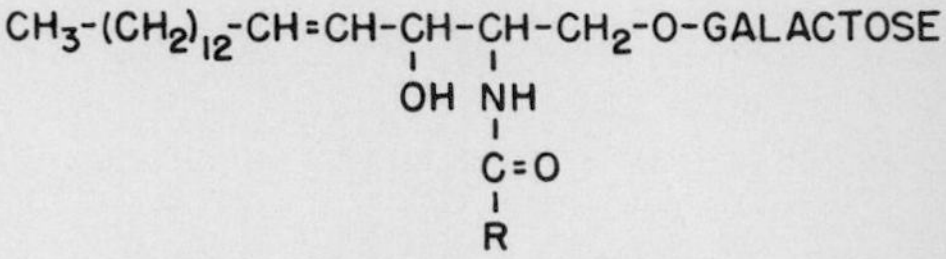

CEREBROSIDE
(N-ACYL-I-O-D-GALACTOSYL-4-SPHINGENINE)

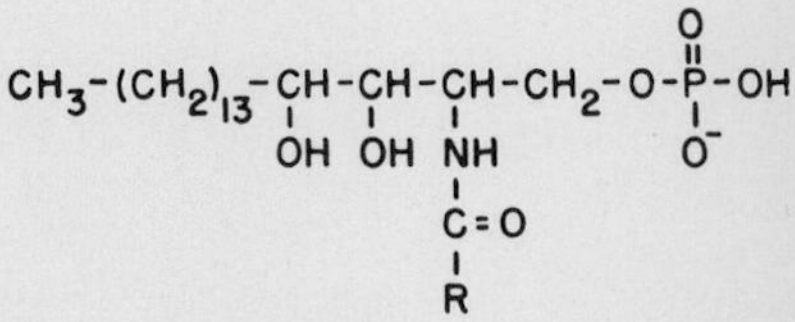

CEREBRIN PHOSPHATE
(N-ACYL-4-D-HYDROXYSPHINGANINYL-
I-O-PHOSPHATE)

FUCOSE
GALACTOSE
ARABINOSE

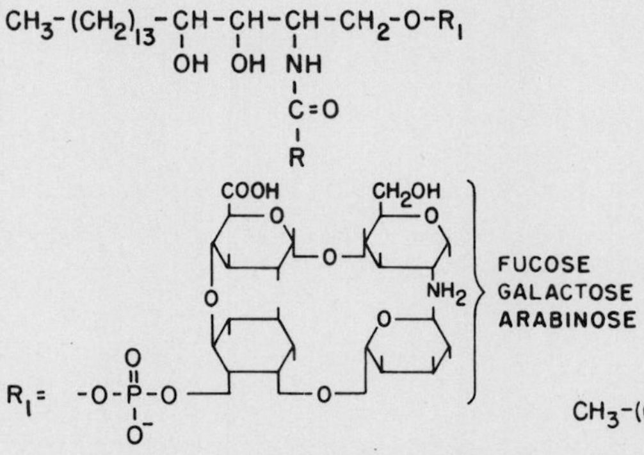

PLANT PHYTOGLYCOLIPID

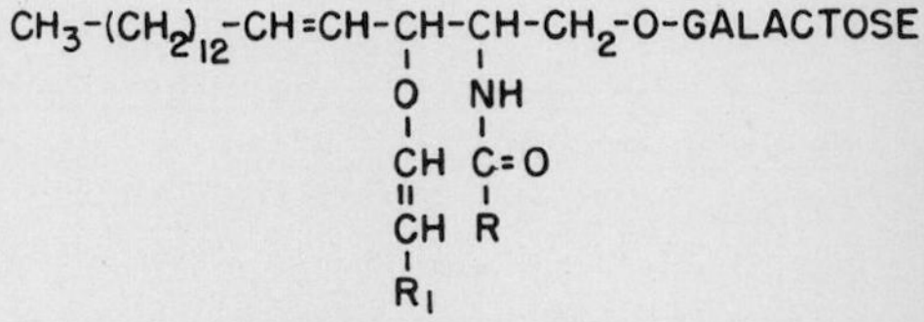

SPHINGOPLASMALOGEN
(3-O-ALKENYL-N-ACYL-I-O-D
GALACTOSYL-4-SPHINGENINE)

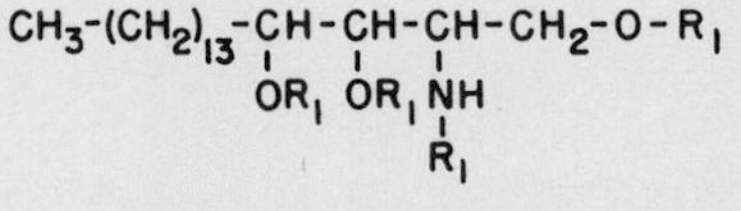

TETRA ACETYLPHYTOSPINGOSINE
R_1 = (COCH_3)

Figure 7.1   Sphingolipid structures.

Sphingolipids of the cerebroside and ceramide oligogly-
coside (glycophosphosphingolipids) types are predominant in
plants with phytosphingosine as the principal long chain
base. Other dihydroxy and trihydroxy bases are found in
plant sphingolipids, and evidence for a tetrahydroxy base
has been reported (Carter and Koob, 1969; Sastry and Kates,
1964). N-acyl groups of plant sphingolipids range from 16
to 26 carbon atoms in chain length with 2-hydroxy derivatives
predominant. Plant ceramide oligoglycosides are structurally
composed of phytosphingosine, phosphate, fatty acids, hexose
sugars, an amino sugar, inositol, and glucuronic acid as
shown in Figure 7.1 (Carter and Koob, 1969). Structures of
the phytoglycolipids have been only partially characterized
and are discussed in more detail by Carter et al. (1965).

Sphingomyelin, cerebrosides, and gangliosides are the
major sphingolipids of animals, with dihydroxy bases being
predominant. Animal cerebrosides generally contain galactose
as the principal sugar constituent, and plant cerebrosides
most often contain glucose.

Fungi. Cerebroside-like substances have been detected
in mushrooms in the early 1900's (Bamberger and Landsiedl,
1905; Zellner, 1911; Rosenthal, 1922; Hartmann and Zellner,
1928; Froeschl and Zellner, 1928), but the first detailed
structural studies on sphingolipids were reported in 1940
by Reindel et al. using the yeasts *Saccharomyces cerevisiae*
and *Hansenula ciferrii*. They identified a C_{20}, 1,3,5-tri-
hydroxy,2-amino compound (and called it dihydrosphingosine)
as the base component of yeast cerebrin. The term cerebrin
(for cerebrine) has been used by early investigators to
refer to a ceramide phosphate isolated from yeast and other
fungi. Prostenik and Stanacev (1958) confirmed that this
base contains 20 carbon atoms, but the hydroxyl groups are
located at the 1, 3, and 4 positions. This compound is a
homologue of the C_{18}-phytosphingosine identified in higher
plant tissues by Carter et al. (1965). The corresponding
C_{20} base is also found as a component of cerebrin from
mycelium of *Aspergillus sydowii* (Bohonos and Peterson, 1943).
C_{18}-Phytosphingosine has been identified as the major base
of cerebrin from *Penicillium notatum* with lower relative
amounts of the C_{20} homologue (Oda, 1952). Oda and Kamiya
(1958) have reported that the C_{18}- and C_{20}-phytosphingosines
are present in yeast cerebrin phosphate in a 2:1 ratio. The
yeast *Torulopsis (Candida) utilis* contains a mixture of at
least three types of sphingolipids with base constituents

C_{18}-phytosphingosine, C_{18}-dihydrosphingosine, and C_{20}-phyto-
sphingosine in a ratio of 1.0:0.19:0.03, respectively (Stanacev and Kates, 1963). Similar relative proportions of these
bases have been found in Baker's yeast (Nurminen and Suomalainen, 1971; Reindel, 1930), while Karlsson and Holm (1965)
detected only the two phytosphingosine homologues in *T.
utilis*. The two homologues are present in another isolate
of Baker's yeast (Duphar) in a ratio of 1:2 (C_{18} and C_{20}),
respectively (Kisic and Prostenik, 1960). The base of a
novel inositol-phosphorus containing sphingolipid of *S.
cerevisiae* is hydroxysphinganine (see below) (Smith and
Lester, 1974). The total purified sphingolipid fraction
from yeast is only 0.04% of the dry cell weight (Trevelyan,
1966).

In 1960, Wickerham and Stodola have reported for the
first time an extracellular sphingolipid produced by *H.
ciferrii* (strain NRRL Y-1031). This substance has been
identified as tetraacetylphytosphingosine (Stodola and Wickerham, 1960; Stodola et al., 1962) (Figure 7.1), which is
accompanied by a triacetyldihydrosphingosine (Maister et al.,
1962). Green et al. (1965) have found only low amounts of
sphingolipids in a *Saccharomyces* species, but like Stodola
and Wickerham (1960), they found the fully and partially
acetylated sphingolipids as *H. ciferrii* products. Glycosyl-
ceramides have also been identified as extracellular products
of this yeast (Kaufman et al., 1971). Pentadecanoic acid
(0.1%) added to the culture medium of *H. ciferrii* causes
marked changes in the base composition of extracellular
sphingolipids (Kulmacz and Schroepfer, 1978). C_{17}-Phyto-
sphingosine (39%), C_{19}-phytosphingosine (15%), and C_{18}-
phytosphingosine (46%) were the bases characterized by
GLC-MS. The odd-numbered carbon bases are not normal constituents of yeast.

The most thorough studies of fungal sphingolipids have
been with the filamentous fungi *Phycomyces blakesleeanus*,
Fusarium lini, *Agaricus bisporus*, *Amanita muscaria* and *A.
rubescens* (Weiss and Stiller, 1972; Weiss et al., 1973).
Ceramides and cerebrosides of these fungi represent between
0.2 and 0.7% of the mycelial dry weight. Generally, the
long chain base constituents of sphingolipids from these
species are similar, each consisting of a complex mixture of
normal, branched, saturated, and unsaturated phytosphingosine
homologues from C_{17} to C_{22} in chain length (Table 7.2). All
branched chain bases are of the *iso* type. Relative proportions of each base vary between the ceramide and cerebroside

TABLE 7.2   Long-Chain Base Constituents of Ceramides and
Cerebrosides from *Agaricus bisporus*, *Amanita muscaria*,
A. rubescens[a]

Phytosphingosine Base	Chain Length
4-Hydroxyheptadecasphinganine[b]	nC_{17}
16-Methyl-4-hydroxyheptadecasphinganine	iC_{18}
4-Hydroxyoctadecasphinganine	nC_{18}
17-Methyl-4-hydroxyoctadecasphinganine	iC_{19}
18-Methyl-4-hydroxynonadecasphinganine[c]	iC_{20}
19-Methyl-4-hydroxyeicosasphinganine	iC_{21}
20-Methyl-4-hydroxyheneicosasphinganine[c]	iC_{22}
20-Methyl-4-hydroxyheneicosa-X_1-sphingenine[c,d]	$iC_{22:X_1}$
20-Methyl-4-hydroxyheneicosa-X_2-sphingenine[c]	$iC_{22:X_2}$
4-Hydroxydocosasphinganine[c]	nC_{22}
4-Hydroxydocosa-X_1-sphingenine[c]	$nC_{22:X_1}$
4-Hydroxydocosa-X_2-sphingenine[c]	$nC_{22:X_2}$

[a] Taken from Weiss and Stiller (1972) and Weiss et al. (1973).
[b] Not found in *P. blakesleeanus* and *F. lini*.
[c] Not previously reported.
[d] X_1 and X_2 indicate unknown positions, geometry, and nature
of unsaturation.

fractions and among the species studied. The ceramides, and
in some cases the cerebrosides, contain C_{18}-phytosphingosine
as the predominant base. Cerebrosides and ceramides of the
basidiomycete *Lactarius delicious* contain mainly phytosphingo-
sines with lesser amounts of sphingosines and dihydrosphingo-
sines (Ondrusek and Prostenik, 1978).

An unusual type of phosphosphingolipid has been isolated
from the pythiacious fungus *Pythium prolatum* (Wassef and Hendrix, 1977). Ceramide aminoethylphosphonate has been separated from phosphoacylglycerols of the aquatic phycomycete;
sphingosine represents 98% of the long chain bases, and individual molecular species differ mainly according to the N-
acyl group.

Hydrolysis of most fungal sphingolipids yields a homologous series of fatty acids ranging from 12 to 26 carbon atoms
in chain length. Saturated and unsaturated, normal and methyl-
branched (*iso*) chain isomers are generally components of these

lipids. As mentioned above, α-hydroxy acids are abundant in fungal sphingolipids. The first studies by Reindel (1930) show that the base of yeast cerebrin is linked to an α-hydroxy acid believed to be 2-hydroxy-n-hexacosanoic acid (α-OH C_{26}). This acid and α-OH C_{24} have been identified in yeast and other fungi (Kaufman et al., 1971; Chibnall et al., 1953; Sweeley, 1959; Sweeley and Moscatelli, 1959; Wagner and Zofcsik, 1966a; 1966b). The "mycoglycolipid" fraction from yeast (see below) yields 2-OH C_{24}, 2-OH C_{22}, C_{26}, C_{24}, C_{22} and C_{20} acids on hydrolysis. Less than 1% of the sphingo-lipid fatty acids of *T. utilis* have 18 or less carbon atoms, while α-OH C_{26} represents 70% (Stanacev and Kates, 1963). Similar results have been found with the yeast whole cell en-velope where 68.3% of the fatty acids of sphingolipids are α-OH C_{26} and 21.1% C_{26} (Nurminen and Suomalainen, 1971). The α-hydroxy acids of sphingolipids have the D-configuration (Karlsson, 1966). Nine ceramides of *L. delicious* vary con-siderably according to the fatty acid composition, namely the presence or absence of hydroxy fatty acids, degree of unsatu-ration, as well as the relative proportions of acids (Ondrusek and Prostenik, 1978). α-Hydroxy C_{16} (63.4%) is the principal fatty acid of mixed cerebrosides of this fungus. No α-OH C_{26} has been detected in either the ceramide or cerebroside frac-tions. Other fatty acids of fungal sphingolipids are gener-ally those typical of fungal lipids, C_{16}, C_{18}, $C_{18:1}$, and $C_{18:2}$. It is interesting to note that hexadecenoic acid ($C_{16:1}$) does not tend to accumulate in sphingolipids from true yeasts (i.e. *S. cerevisiae*) as in the total lipids from these fungi. See Table 7.3 for the fatty acid composition of sphingolipids from representative fungi.

A group of inositol-containing phosphorylceramides have been identified from *S. cerevisiae*. The major component of these sphingolipids contain equimolar amounts of hydroxy-sphinganine, phosphorus, inositol, α-OH C_{26}, and sodium (inositol phosphorylceramide) (Smith and Lester, 1974). Sev-eral other related lipids have also been detected as minor components of the sphingolipid fraction, differing mainly by their base and fatty acid composition. Manosyl-(inositolphos-phoryl)$_2$-ceramide (Steiner et al., 1969) and mannosylinositol phosphorylceramide (Smith and Lester, 1974) have also been identified in yeast extracts. A mannosylinositol-containing sphingolipid has also been detected in yeast cell wall and plasma membrane preparations (Nurminen and Suomalainen, 1971). Approximately 39% of the inositol-containing lipid phosphorus is found in the sphingolipids.

Fungal cerebrosides are mainly D-glucosyl ceramides (Kaufman et al., 1971; Weiss and Stiller, 1972; Weiss et al., 1973; Ondrusek and Prostenik, 1978). A cerebroside sulfate has been detected in yeast preparations (Nurminen and Suomalainen, 1971).

SPHINGOLIPID METABOLISM

There has been considerable interest in sphingolipid metabolism; in animals because the accumulation of some types of these lipids are associated with certain human disorders, and in certain yeasts because they produce relatively large quantities of sphingolipids and are easily cultured in the laboratory. Sphingolipid metabolism has been reviewed by Stoffel (1971; 1974).

Biosynthesis. Early studies with animal systems suggested that the two principal long chain bases, sphingosine and dihydrosphingosine, are formed by the condensation of serine and palmitic acid or palmitaldehyde. A microsomal fraction from the yeast *H. ciferrii* catalyzes the formation of *erythro*-sphingosine and *erythro*-dihydrosphingosine from palmityl-CoA (or palmitic acid, ATP, and CoASH) and serine (Braun and Snell, 1967; 1968). In the absence of NADPH, the corresponding 3-keto intermediates accumulate (Braun and Snell, 1968; Braun et al., 1968; Brady et al., 1971). Pyridoxal phosphate is required for the condensation reaction which, along with the loss of CO_2 from serine, is believed to occur in a conserted fashion (Morell and Braun, 1972). The preferred acyl-CoA substrates are: $C_{16} > C_{18} > C_{14} > C_{10} > C_{12}$. Under appropriate conditions, *trans*-2-hexadecenoyl-CoA is a good substrate for base synthesis (equal to palmityl-CoA) and, like palmityl-CoA, leads to a mixture of sphingosine and dihydrosphingosine as products. α-Hydroxypalmityl-CoA is not a substrate for the condensing enzyme. Sphingosine and dihydrosphingosine have two asymmetric carbon atoms, one at C-2 which has the 2S configuration, and a second at C-3. The second optically active center is formed with the stereospecific reduction of the 3-keto group by the hydride ion from the B-side of NADPH. This gives rise to the 3R configuration and D-*erythro* stereoisomer (Braun and Snell, 1968; Snell et al., 1970).

Based on the following evidence, DiMari et al. (1971) have proposed that in *H. ciferrii* the pathways of saturated and unsaturated long chain base synthesis diverge at the acyl-CoA

level: (1) Chemically synthesized $[3\text{-}^{14}C]$ 3-ketodihydro-
sphingosine is converted by a cell-free extract to dihydrosphingosine exclusively, (2) palmityl-CoA and *trans*-2-hexa-
decenoyl-CoA are almost equally effective substrates for
the condensing enzyme, and (3) the enzyme preparation converts palmitic acid to *trans*-2-hexadecenoic acid.   In brain
tissue, on the other hand, dihydrosphingosine is an immediate precursor to sphingosine and the reaction involves a *cis*
hydrogen elimination, first the 4R- and then 5S-hydrogen,
leading to the formation of the 4-*trans* double bond of
sphingosine (Stoffel, 1974).

In addition to the dihydroxy bases, the trihydroxy base
phytosphingosine (4-hydroxy-sphinganine) is also produced by
H. ciferrii, but how it is synthesized remains uncertain.
$[1\text{-}^{14}C, 3\text{-}^{3}H]$ Dihydrosphingosine is converted to phytosphingosine with the loss of a tritium atom at C-3 and the
4R hydrogen (Stoffel et al., 1968; Weiss and Stiller, 1967).   It
has been subsequently proposed that the dihydroxy base is converted to the corresponding 3-keto intermediate, hydroxylated
at C-4, and then reduced to the trihydroxy base which is
known to have the 2S, 3S, 4R configuration.   Neither molecular oxygen nor water appears to provide the oxygen atom of
the 4-hydroxy of phytosphingosine (Thorpe and Sweeley, 1967).

Synthesis of the ceramide and cerebroside forms of
sphingolipids has been reviewed by Morell and Braun (1972).
There are two pathways leading to the formation of cerebrosides: (1) the addition of a monosaccharide to a ceramide
and (2) the addition of a monosaccharide to a long chain
base followed by N-acylation.   Ceramides are formed by
N-acylation of a long chain base.   There is some preference
for certain fatty acids in the formation of ceramides that
varies according to the developmental stage and age of the
tissue and organism.   Acyl-CoA:LCB (long chain base) acyl
transferase from mouse brain shows a chain length specificity of $C_{18} > C_{24} > C_{16} > C_{18:1}$.   In some organisms there
is a preference for 2-hydroxy fatty acids, particularly
2-OH C_{26}.   Microsomal fractions show little difference
in the *in vitro* utilization of *erythro* and *threo*-sphingosines
in the formation of ceramides and there is no discrimination
between sphingosine and dihydrosphingosine.   Glycosylation of
a long chain base gives rise to a substance called psychosine
which is subsequently acylated to form a cerebroside.   Galactose and glucose are the sugars most often present in cerebrosides and both are incorporated into cerebrosides

experimentally by transfer from the UDP-glycose (Figure 7.2) (Burton et al., 1958; Basu et al., 1968).

An unknown sulfur-containing sphingolipid has been detected in yeast (Nurminen and Suomalainen, 1971), but the biosynthesis of this compound has not been investigated. In animals, cerebroside sulfatides are formed by the transfer of sulfate from 2'-phosphoadenosine-5'-phosphosulfate. For a review of sulfolipids see Goldberg (1961).

More complex sphingolipids include the ceramide oligoglycosides. Among the known most widely distributed complex sphingolipids are the phytoglycolipids in higher plants, mycoglycolipids in fungi, and gangliosides in animals. Synthesis of ceramide oligoglycosides appears to occur through the sequential transfer of sugar moieties to a cerebroside or glycolipids with carbohydrate residues of increasing chain lengths. The biosynthesis of gangliosides by brain preparations has received the greatest attention and was reviewed by Morell and Braun (1972) and Burton (1974).

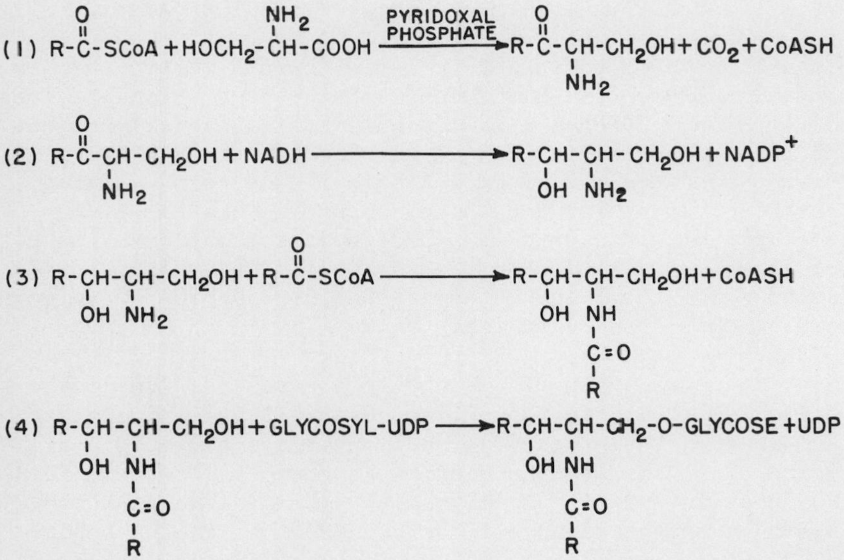

$$(1) \quad R\text{-}\overset{O}{\overset{\|}{C}}\text{-}SCoA + HOCH_2\text{-}\overset{NH_2}{\underset{|}{C}H}\text{-}COOH \xrightarrow{\text{PYRIDOXAL PHOSPHATE}} R\text{-}\overset{O}{\overset{\|}{C}}\text{-}\underset{\underset{NH_2}{|}}{C}H\text{-}CH_2OH + CO_2 + CoASH$$

$$(2) \quad R\text{-}\overset{O}{\overset{\|}{C}}\text{-}\underset{\underset{NH_2}{|}}{C}H\text{-}CH_2OH + NADH \longrightarrow R\text{-}\underset{\underset{OH}{|}}{C}H\text{-}\underset{\underset{NH_2}{|}}{C}H\text{-}CH_2OH + NADP^+$$

$$(3) \quad R\text{-}\underset{\underset{OH}{|}}{C}H\text{-}\underset{\underset{NH_2}{|}}{C}H\text{-}CH_2OH + R\text{-}\overset{O}{\overset{\|}{C}}\text{-}SCoA \longrightarrow R\text{-}\underset{\underset{OH}{|}}{C}H\text{-}\underset{\underset{NH}{|}}{C}H\text{-}CH_2OH + CoASH$$
$$\underset{\underset{R}{|}}{\underset{\underset{C=O}{|}}{}}$$

$$(4) \quad R\text{-}\underset{\underset{OH}{|}}{C}H\text{-}\underset{\underset{NH}{|}}{C}H\text{-}CH_2OH + GLYCOSYL\text{-}UDP \longrightarrow R\text{-}\underset{\underset{OH}{|}}{C}H\text{-}\underset{\underset{NH}{|}}{C}H\text{-}CH_2\text{-}O\text{-}GLYCOSE + UDP$$
$$\underset{\underset{R}{|}}{\underset{\underset{C=O}{|}}{}}$$

Figure 7.2 Reactions in the synthesis of sphingolipid bases, ceramides, and cerebrosides.

Sphingomyelin is one of the most important sphingolipids in animal tissues, but is not otherwise widely distributed in nature. See Morell and Braun (1972) for a review of the bio-synthesis of this sphingolipid.

Degradation. Sphingolipids first became of interest because of their association with several human disorders such as sphingomyelinosis, Tay-Sachs disease, Gaucher's disease, and others. These diseases are generally charac-terized by the accumulation of certain sphingolipids caused by defects in the catabolic pathways. These defects are usually caused by the absence of an enzyme required for the breakdown of a particular sphingolipid. For this reason, sphingolipid catabolism has received more attention than synthesis. Sphingolipid breakdown has not been studied in fungi and will be only outlined here.

Beginning with the complex ceramide oligoglycosides, breakdown starts with the sequential hydrolysis of the oligo-saccharide portion of the molecule. Enzymes which catalyze the breakdown of certain complex sphingolipids have been isolated and purified from several animal tissues. These enzymes, most of which are specific for a certain type of glycoside bond, generally cleave a terminal sugar giving rise to a sphingolipid containing one less sugar residue. Gluco-sidases and galactosidases specific for the respective gly-cosyl ceramides have also been isolated from animal tissues. For example, glycosyl ceramide galactosidase catalyzes the breakdown of ceramide-glu-gal to ceramide-glu plus galactose. Enzymes specific for lipid sulfatides have been isolated, and phospholipases C and D (see Chapter 6) catalyze the breakdown of sphingomyelin. A ceramidase that seems to be specific for ceramides containing medium chain fatty acids has been isolated, and catalyzes the degradation of ceramides to long chain bases and fatty acids.

The degradation of sphingolipid bases has been reviewed by Stoffel (1971) and Morell and Braun (1972). The first step in long chain base degradation is the ATP-dependent kinase catalyzed phosphorylation at the C-1 position (Keenan and Haegelin, 1969). A microsomal pyridoxal phosphate-dependent enzyme catalyzes the breakdown of the phosphory-lated long chain base to phosphorylethanolamine (C-1 and C-2 of the base) and palmitaldehyde, hexadec-trans-2-enal, or 2-hydroxy-palmitaldehyde for the substrates dihydrosphin-gosine, sphingosine, or phytosphingosine, respectively

(Stoffel, 1970). Sphingolipid bases are degraded in a similar manner by the protozoan *Tetrahymena pyriformis* and the
kinase requires Mg^{+2} or Mn^{+2} and CTP, ATP, or UTP. Hexadec-
trans-2-enal is reduced to palmitaldehyde by a NADPH-
dependent reductase. Palmitaldehyde may be oxidized to the
corresponding acid which can be subsequently oxidized to CO_2
or incorporated into acyl lipids. Palmitaldehyde, or its
alcohol reduction product, may also be incorporated into
plasmalogens. Phosphorylethanolamine is incorporated into
phosphoglycerides. The degradation of ceramide oligoglycosides is outlined in Figure 7.3.

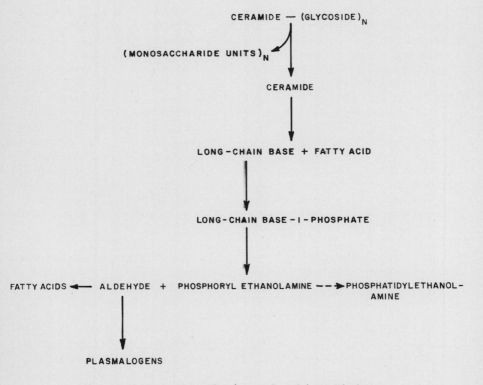

Figure 7.3   Degradation of sphingolipids.

TABLE 7.3   Fatty Acids of Fungal Sphingolipids

Fatty Acids	Percentage						
	Torulopsis utilis[a]	Baker's yeast[b,c]	Phycomyces blakesleeanus[d,e]	Fusarium linid[,e]	Amanita muscariae[,f]	Agaricus bisporus[f]	Pythium prolactum[g,h]
C12	–	–	0.2	0.2	0.6	1.7	–
OH-C12	–	–	–	trace	–	–	–
C13	–	–	1.4	0.2	0.7	1.7	–
OH-C13	–	–	–	trace	–	–	–
C14	0.1	–	1.5	0.2	0.7	–	0.3
iC14	–	–	0.6	0.1	0.6	–	–
OH-C14	–	–	–	0.1	8.1	3.4	–
C15	0.5	–	2.3	0.2	2.5	1.7	–
iC15	–	–	–	–	trace	–	–
OH-C15	0.2	–	1.3	0.4	15.9	8.4	–
C16	13.0	3.4	17.9	0.2	9.1	5.0	19.6
iC16	–	–	–	0.1	–	–	–
C16:1	–	3.7	0.6	–	1.3	2.1	0.3
OH-C16	40.0	–	41.5	2.3	21.3	42.7	–
C17	0.2	–	–	0.1	0.9	2.8	–
iC17	–	–	–	–	0.3	–	–
OH-C17	1.5	–	–	–	–	1.7	–
C18	1.0	1.9	7.6	0.4	8.7	1.7	1.6
C18:1	42.0	8.4	15.0	0.4	15.7	6.4	21.2
C18:2	–	–	7.1	0.5	11.5	5.9	21.6
C18:3	–	–	0.4	–	–	–	4.7

OH-C_{18}	1.2	–	1.0	94.2	0.4	13.4	–
C_{19}	–	–	–	0.3	0.3	–	–
$C_{19:1}$	–	0.2	–	–	–	–	5.9
C_{20}	0.2	–	1.2	0.1	0.9	–	–
iC_{20}	–	–	0.4	–	0.4	–	–
$iC_{21:1}$	–	–	–	–	0.1	–	–

[a] Karlsson (1966). Another isolate of *T. utilis* contained ca 70% 2-OH C_{26} and 7% 2-OH C_{18} (Stanacev and Kates (1963)).
[b] Nurminen and Suomalainen (1971).
[c] Fatty acids of sphingolipids isolated from the whole cell envelope (cell wall + plasmalemma); principal fatty acids are 2-OH C_{26} (68.3%), $C_{18:1}$ (8.4%), 2-OH C_{24} (4.0%).
[d] Weiss et al. (1973).
[e] Fatty acids of the cerebroside fraction. Ceramide fractions generally were qualitatively the same.
[f] Weiss and Stiller (1972).
[g] Wassef and Hendrix (1977).
[h] Unknown fatty acid, 24.9%.

CHAPTER 8   ALIPHATIC HYDROCARBONS

INTRODUCTION

The surfaces of most terrestrial organisms are coated
with a complex mixture of lipophilic non-glyceride substances
collectively referred to as wax. The composition of wax
varies quantitatively and qualitatively depending on the
source, but may include aliphatic hydrocarbons, primary and
secondary alcohols, acids, ketones, aldehydes, wax esters,
and others. Each class of lipid may be comprised of a homologous series of compounds with a relatively high molecular
weight range. Hydrocarbons are present in wax of most organisms examined and consist mainly of n-alkanes, but may include n-alkenes, singly and doubly branched saturated and
unsaturated hydrocarbons, cyclic alkanes, and isoprenoid
hydrocarbons. The composition and biosynthesis of hydrocarbons in higher plants has been reviewed by Eglinton and
Hamilton (1967), and Kolattukudy (1968; 1970a; 1970b; 1976);
in algae by Weete (1976); and in bacteria by Albro and
Dittmer (1970) and Albro (1976).

The first report of alkanes from a biological source
was in 1929 when nonacosane (C_{29}) and hentriacontane (C_{31})
were identified as components of plants (Cannon and Chibnall,
1929; Clenshaw and MacLean, 1929; Chibnall et al., 1934).
Because of the complex nature of wax from biological sources,
progress on the chemistry of the component substances was
delayed until the development of analytical techniques such
as thin-layer and gas-liquid chromatography, mass spectrometry, and nuclear magnetic resonance.

The composition and biosynthesis of hydrocarbons in a
variety of plant, animal, and microbial organisms have been
reviewed in a recent book edited by Kolattukudy (1976) and
their occurrence is summarized in Table 8.1. Hydrocarbons
may comprise up to 90% of the surface wax depending on the
source. In higher plants n-alkanes are generally the most
abundant type of hydrocarbon, which typically range from
C_{15} to C_{36} in chain length. The odd-numbered carbon chain-
lengths characteristically predominate with C_{31}, C_{29}, and
C_{33} being the principal alkanes in decreasing order of importance. Alkanes with even-numbered carbons are present in
low relative proportions. Exceptions to the odd over even
chain-length predominance seem to apply to non-epidermal
intracellular alkanes. For example, the alkanes of tobacco
callus tissue range from C_{16} to C_{28} with the odd/even ratio
equal to ca 1 and C_{22}, C_{23}, and C_{24} are the principal homologues (Weete et al., 1971). Similar results have been obtained with chloroplasts (Gulz, 1968) and leaf tissues
(Kaneda, 1969; Herbin and Robins, 1969).

Hydrocarbons of blue-green algae range from C_{15} to C_{18}
in chain length with C_{17} representing 32 to 96% of the total
alkanes. Several species such as *Nostoc muscorum* and *Ana-
cystis nidulans* produce a mixture of 7- and 8-methylhepta-
decanes (Fehler and Light, 1970; Gelpi et al., 1970). Small
amounts of 4- and 6-methylheptadecane have been detected in
some species (McCarthy et al., 1968; Han et al., 1968; Gelpi
et al., 1970). The GLC chromatograms of hydrocarbons from *A.
montana* and many taxonomically more advanced species show a
bimodal distribution of chain-lengths (C_{13} to C_{33}) with
principal alkanes being C_{15} or C_{17} and C_{27} or C_{29}. n-Alkanes
comprise 0.05 to 0.12% of the dry weight of blue-green algae,
but only trace amounts have been detected in the advanced
algae. The alkenes $C_{15:1}$, $C_{16:1}$, $C_{17:1}$, and $C_{19:1}$, are
present in some phytoplankton (Winters et al., 1969), but
the predominant hydrocarbons of most marine algae are polyunsaturated (tetra, penta, and hexa) C_{19} to C_{21} alkenes with
all *cis*- 3,6,9,12,15,18-heneicosahexaene representing 80 to
90% of total hydrocarbons in photosynthetic diatoms, dino-
flagalates, cryptomonads, phaeophytes, chrysophytes, and
lesser amounts in some green algae (Halsall and Hills, 1971;
Youngblood et al., 1971). The isoprenoid hydrocarbons pristane and phytane have been detected in some algae (Gelpi
et al., 1970).

TABLE 8.1   Hydrocarbons of Plants, Animals, and Microbes

Hydrocarbons	Comments
I. n-Alkanes	Almost universally distributed; C_{16} to C_{36}; odd chain predominance; shorter or longer chain-lengths occur but are of limited distribution.

II. Branched Alkanes
  A. Single Branched
    1. 2-Methyl (*iso*)
	Present in plants, animals, and microbes but occurrence is generally limited to certain species or genera; odd chain predominance; C_{29} to C_{31} in insects; C_{29} to C_{33} in plants (tobacco).

2. 3-Methyl (*anteiso*)	Occurrence same as above except even chains predominate; C_{24}, C_{26}, C_{28}, C_{30}, C_{32} in insects; C_{30} and C_{32} in plants (tobacco).
3. Internal	Blue-green algae, 4-, 6-, 7-, 8-methyl-heptadecane. Insects, 11-, 13-, or 15-methyl branching, odd carbon chains predominate.
B. Multiple Branched 1. Dimethyl	In bacteria, C_{23} to C_{30} mono-unsaturated (*cis*) with *iso-iso*, *iso-anteiso*, *anteiso-anteiso* structures. In insects, C_{25} to C_{55} with methyl branches having isoprenoid spacing on odd-carbon atoms. Odd chains predominate.

TABLE 8.1   Continued

Hydrocarbons	Comments
2. Trimethyl Alkanes	
a. Branches near end of chain: 3, 7, 11 and 4, 8, 12	Limited occurrence, only in insect genus *Atta*. Isoprenoid spacing of methyl groups (C_{34} to C_{39}).
b. Internal	Limited occurrence, only two insect species. Isoprenoid spacing of methyl groups (C_{22} to C_{44}).
3. Isoprenoids	Universal occurrence of squalene (C_{30}) but low abundance except under low O_2 tension. Major lipid of shark liver oils (30%). Dihydrosqualene in bacteria. Pristane (C_{19}) and phytane (C_{20}) in algae, bacteria, copepods, and zooplankton.
III. Alkenes	Limited occurrence in plants; flower petals, C_{19} to C_{33} *cis*-5-alkenes (rose petal wax), odd chain predominance; 1-alkenes with even chain predominance in sugar cane wax; some conjugated dienes. In insects, C_{23} to C_{43} with *cis* double bonds, odd chain predominance. Monounsaturated, normal and branched alkenes in bacteria. Not detected in fungi except hexadecatriene A in *Candida utilis*. Predominant olefin in marine phytoplankton is $C_{21:6}$ and in diatoms may account for up to 10% of total lipids.

Hydrocarbon compositions may vary depending on the algal growth form and stage of growth. For example, green actively growing colonies of the green alga *Botryococcus braunii* produces hydrocarbons (C_{17} to C_{33}) constituting 0.1 to 0.3% of the cellular dry weight with 98% being saturated, mono-, di-, and triunsaturated C_{27}, C_{29}, and C_{31} (Gelpi et al., 1968). However, 76% of the dry weight and 90% of the hydrocarbons from the brown resting stage of this alga is composed of two isomeric hydrocarbons (9:1 mixture), botryococcene and isobotryococcene (Maxwell et al., 1968; Brown et al., 1969). Structures of the botryococcenes have not been fully characterized; but they are highly branched and unsaturated molecules with empirical formula of $C_{34}H_{58}$. A third growth stage with large green cells contain relatively few hydrocarbons.

Several types of animals have been analyzed for their hydrocarbon content, but most attention has been given to the insects. Hydrocarbons may represent 3 to 90% of insect cuticular wax and are often a complex mixture of normal and methyl branched alkanes. Types of branching exhibited by insect hydrocarbons include *iso-* and *anteiso-*, internal monomethyl, di-, and trimethyl branching with isoprenoid spacing of methyl groups.

Alkanes are generally present in mammalian skin lipids, but in low amounts, and they are generally considered contaminants. Internal organs contain very low levels of alkanes ranging from C_{18} to C_{34} in chain length with no odd over even chain predominance.

Hydrocarbons may represent up to 20% of the cellular dry weight of some bacteria. Bacteria of the family Micrococcaceae have been studied more thoroughly than others. Complex mixtures (26 to 36 individual molecular species) of normal and methyl branched alkanes and monounsaturated alkenes ranging from C_{16} to C_{29} in chain length have been detected in these bacteria. Methyl branching includes *iso*, *anteiso*, *iso-iso*, *iso-anteiso*, and *anteiso-anteiso* structures. Olefins have the double bond in the center of the hydrocarbon chain with the *cis*-configuration (Tornabene et al., 1967; Albro and Dittmer, 1969; Tornabene and Oro, 1967; Tornabene and Markey, 1971; Albro, 1971). The type and chain length range of hydrocarbons in other bacteria varies considerably. Most photosynthetic bacteria contain hydrocarbons in the C_{14} to C_{20} chain length range, which is

similar to blue-green algae. Heptadecane comprises 42 to
50% of total hydrocarbons in *Clostridium acidivrici*, *Rhodo-*
pseudomonas spheroides, and *Chlorobrium* sp. Pristane
(46.5%) is the principal hydrocarbon in *Pseudomonas sher-*
manii.

                        OCCURRENCE IN FUNGI

   The composition and occurrence of aliphatic hydrocarbons
in fungi have been previously reviewed by Weete (1971; 1976).
Hydrocarbons were first reported in fungi by Baker and
Strobel (1965) who identified by GLC a homologous series of
n-alkanes ranging from C_{22} to C_{31} in chain length in a hexane
extract of intact uredospores of the rust fungus *Puccinia*
striiformis. Their identification has been later confirmed
by mass spectrometry and they comprise 13% of the spore wall
extract (Jackson et al., 1973). The alkane content of
fungal spores is relatively low, ranging from 40 to 150 ppm.
Fungal spore alkanes are similar to those of higher plant
epicuticular wax, ranging in chain length from C_{14} to C_{37}
with odd-numbered carbon chains predominant. Principal
alkanes are typically C_{27}, C_{29}, and C_{31}. *Iso-* and *anteiso-*
methyl branched alkanes and alkenes are of limited occurrence
in fungi. Typical GLC separation patterns of alkanes from
rust and smut spores are shown in Figures 8.1, 8.2, and 8.3.

   The presence of paraffinic hydrocarbons in extracts of
intact fungal spores suggests that possibly a wax coating
analogous to that on the exterior surfaces of higher plants
and insects may be present. Scanning and transmission (using
the freeze-etch technique) electron microscopy has failed to
detect such a coating on the surfaces of spores of pathogenic
fungi collected from their hosts and spores of saprophytic
fungi grown in culture (Laseter et al., 1968; Schwinn, 1969)
(Figures 8.4-8.8). However, using electrophoretic mobility
of spores as evidence for the presence of surface wax, it
appears that wax is present on the surface of spores of some
species but not on others (Fisher et al., 1972). *Alternia*
tenius, *Botrytis fabae* and *Neurospora crassa* conidia and
sporangiospores of *Rhizopus stolonifer* have surface wax
whereas conidia of *Erysiphe cichoracearum*, *E. graminis*,
Nectria galligena, *Penicillium expansum*, and *Verticillium*
albo-atrum, and sporangiospores of *Mucor rouxii* do not. When
present, surface wax represents 0.15 to 0.2% of the spore dry
weight. The surface wax and spore wall fatty acid composition

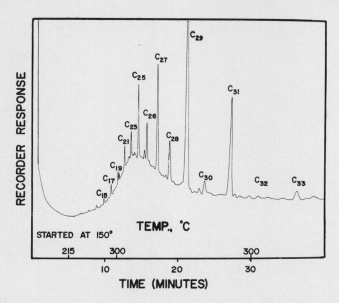

Figure 8.1   Alkanes of *Puccinia graminis* uredospores from wheat.

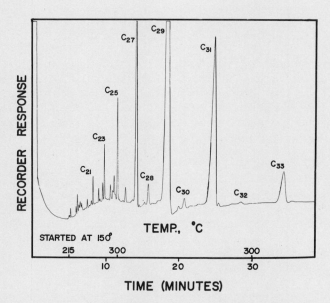

Figure 8.2   Alkanes of *Urocystis agropyri* chalmydospores (Laseter et al., 1969).

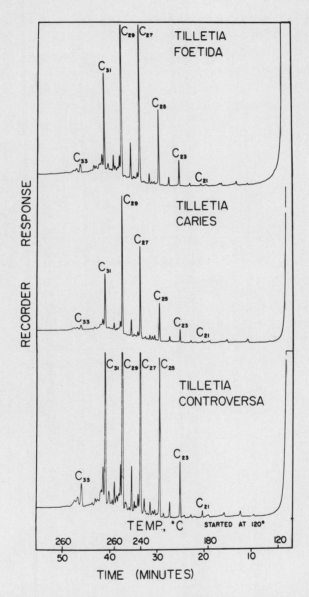

Figure 8.3   Alkanes of teliospores of three *Tilletia* species from wheat kernels (Weete et al., 1969).

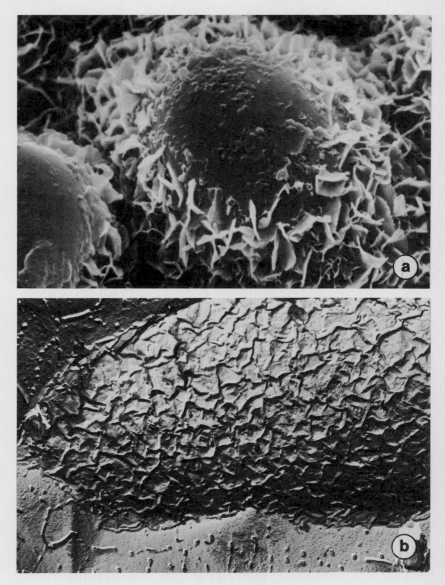

Figure 8.4   (a) Scanning electron micrograph of *Salicornia virginica*, a salt tolerant plant, showing the wax crystal pattern. 6000x. (Courtesy of J. V. Allen.) (b) Freeze-etch replica of an *Erysiphe graminis* conidium showing the waxlike nature of the surface. 6000x. (Courtesy of W. M. Hess.)

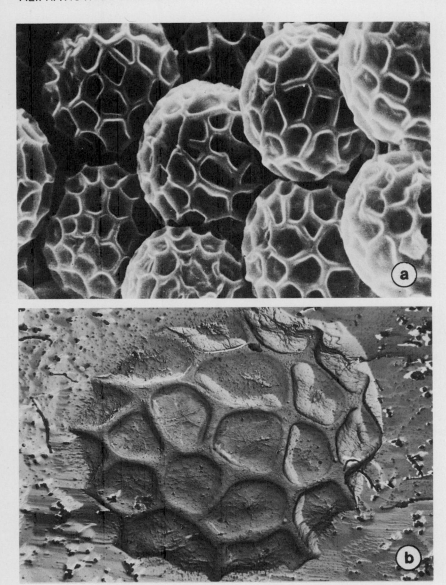

Figure 8.5   (a) Scanning electron micrograph of *Tilletia*
controversa teliospores showing the reticulated surface
patterns. 2400x. (Courtesy of J. V. Allen.)   (b) Freeze-
etch replica of a single *Tilletia caries* teliospore showing the netlike pattern on the surface. 6000x. (Courtesy
of W. M. Hess.)

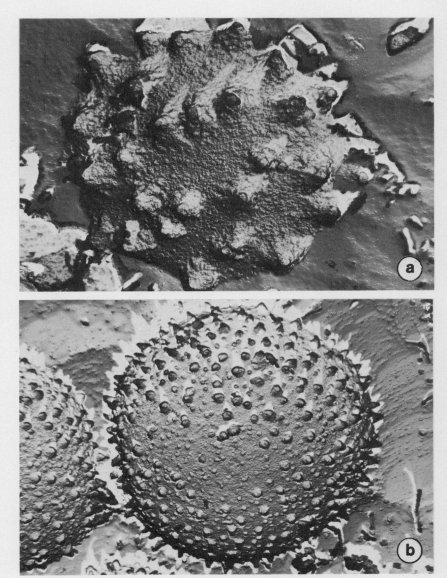

Figure 8.6   (a) Freeze-etch replica of a *Ustilago maydis*
teliospore showing the surface protrusions. 12000x.
(Courtesy of W. M. Hess.) (b) Freeze-etch replica of a
Sphacelotheca reiliana teliospore showing the fine
echinulations present on the surface. 11000x. (Courtesy of W. M. Hess.)

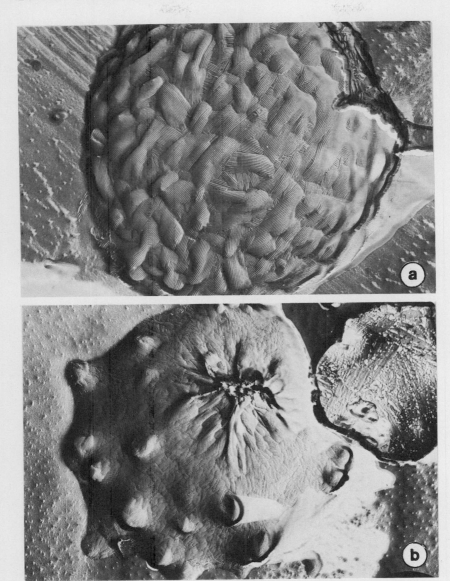

Figure 8.7   (a) Freeze-etch replica of an *Aspergillus
melleus* conidium showing the pattern of the surface "rod-
lets." 34000x. (Courtesy of W. M. Hess.) (b) Freeze-
etch replica of an *Aspergillus niger* conidium showing the
"rodlet" pattern and the characteristic surface protrusions. 20000x. (Courtesy of W. M. Hess.)

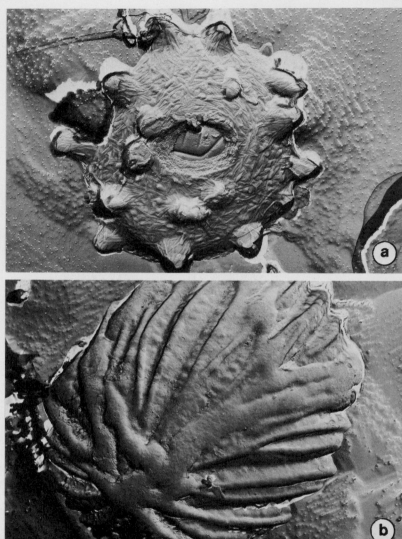

Figure 8.8   (a) Freeze-etch replica of a *Lycoperdon perla-*
tum teliospore showing the "rodlet" pattern on the surface
and a portion of the exposed membrane where the cell wall
has been fractured away. 23000x. (Courtesy of W. M.
Hess.)   (b) Freeze-etch replica of a *Rhizopus arrhizus*
sporangiospore showing the ridges and grooves which make
up the surface of the dormant spore. 12000x. (Courtesy
of W. M. Hess.)

varies according to the species, but C_{16} (29 to 47%) and C_{18}
(11 to 18%) are the principal components with low relative
proportions of monoenes, and no polyunsaturated fatty acids.
The high degree of saturation appears to be characteristic
of spore surface and wall wax (see below), although fatty
acids from the two locations differ quantitatively and qualitatively. Spore walls of *B. fabae* contain C_{23} (8.6%) and
$C_{23:1}$ (16.3%) which are not known in other fungi. Spore surface and wall wax contain a highly complex mixture of hydrocarbons, many of which have not been identified, with no odd
over even numbered carbon chain preference among n-alkanes.
Spore hydrocarbon content also differs quantitatively and
qualitatively, with C_{20}, C_{21}, C_{22}, C_{23}, C_{24} generally being
the predominant homologues. Spores of the powdery mildew
fungus *Sphaerotheca fuliginea* contain a complex mixture of
wax substances, 50% of which are wax esters with C_{20} as the
predominant alcohol and C_{22} and C_{24} as the principal acids
(Clark and Watkins, 1978). Also, *trans* Δ^2 acids are components of the wax esters, mainly $\Delta^{2t}C_{22}$; minor components of
the spore surface wax are Δ^{2t} methyl esters, diol esters
(mainly 1,12-dodecanediol), and free fatty acids.

Analyses of spores for hydrocarbons is limited to a
relatively few species, mostly rust and smut fungi. Based
on both qualitative and quantitative differences in the alkane
content of chlamydospores of three smut fungi, Oro et al.
(1966) have suggested that hydrocarbon patterns of fungal
spores may be sufficiently distinct to be chemotaxonomically
useful. Moreover, Laseter and Valle (1971) have reported quite
different alkane GLC patterns between the uredospores of *P.
graminis* var. *avenae* and *P. graminis* var. *tritici*. Three
closely related *Tilletia* species (*caries*, *foetida*, and *controversa*) could not be distinguished on the basis of hydrocarbon content (see Figure 8.3) (Laseter et al., 1968). GLC
separation patterns of alkanes from fungal spores may prove
useful in supporting the identification of certain species
such as *U. bullata* with C_{23} and C_{25} as the predominant hydrocarbons (Gunasekaran et al., 1972) and *P. graminis* var.
avenae with C_{23} being the most abundant alkane (Laseter and
Valle, 1971), but because of the general similarities in
spore hydrocarbon content it appears to be an unlikely chemotaxonomic tool for general use. However, the value of fungal
spore hydrocarbon content as a chemotaxonomic tool has not
been sufficiently tested. The hydrocarbon content of spores
from several rust and smut fungi is shown in Table 8.2.

TABLE 8.2   Relative Hydrocarbon Distribution in the Spores of Certain Phytopathogenic Fungi

Species	Hydrocarbon Chain Length								
	C_{19}	C_{20}	C_{21}	C_{22}	C_{23}	C_{24}	C_{25}	iC_{25}	C_{26}
Tilletia foetida[c]	---	---	0.3	0.3	3.2	1.0	11.9	---	1.3
Tilletia caries[c]	---	---	0.1	0.2	5.4	0.9	9.7	---	1.0
Tilletia controversa[c]	---	---	0.4	0.2	3.9	0.9	11.4	---	0.4
Puccinia graminis tritici	---	---	---	---	3.1	7.2	7.3	---	8.1
Puccinia striiformis	---	---	---	---	3.9	3.2	7.9	---	2.2
Spacelotheca reiliana[d]	---	---	---	---	2.1	---	3.2	1.8	1.1
Ustilago maydis[d]	1.1	1.5	---	---	1.9	1.3	11.1	---	2.1
Ustilago maydis[a,e]	1.6	1.6	1.3	1.3	3.1	1.5	11.0	---	3.0
Urocystis agropyri[d]	1.1	---	1.5	---	2.0	1.2	3.3	1.6	1.0
Ustilago bullata[b,e]	10.3	3.8	11.3	4.3	21.8	---	23.5	---	---

TABLE 8.2   Continued

Species	Hydrocarbon Chain Length									
	C_{27}	iC_{27}	C_{28}	C_{29}	iC_{29}	C_{30}	C_{31}	iC_{31}	C_{32}	C_{33}
Tilletia foetida[c]	24.7	0.9	5.7	28.8	1.6	1.9	15.2	1.3	---	1.7
Tilletia caries[c]	23.5	1.0	4.3	34.8	1.3	1.6	13.7	1.1	---	1.5
Tilletia controversa[c]	24.7	0.8	2.7	25.6	1.9	1.3	21.5	1.7	---	2.7
Puccinia graminis tritici	17.6	---	7.3	29.4	---	1.3	16.2	---	---	2.0
Puccinia striiformis[d]	20.5	---	2.5	23.9	---	---	23.9	---	---	3.4
Spacelotheca reiliana[d]	9.1	2.8	3.8	34.2	8.3	13.1	16.2	5.1	2.7	2.5
Ustilago maydis[d]	33.4	8.4	7.7	13.8	6.9	3.8	---	3.1	1.7	3.6
Ustilago maydis[a,e]	36.0	---	7.0	13.0	---	5.0	4.0	---	1.0	---
Urocystis agropyri[d]	12.2	---	0.9	46.3	---	1.2	23.7	---	---	---
Ustilago bullata[b,e]	1.1	---	---	0.4	---	---	---	---	---	---

[a] C_{14} to C_{18} (9.65%).
[b] C_9 to C_{18} (23.13%).
[c] Data from Laseter et al. (1968).
[d] Data from Weete et al. (1969).
[e] Data from Gunasekaran et al. (1972).

In studies on the hydrocarbon content of fungal spores, material for analysis has been collected from host tissue. Although care has been taken to remove plant debris prior to extraction, the close similarity of the hydrocarbons from plant and fungal sources raises the question of whether the spore alkanes are indeed fungal products and makes it difficult to determine the origin of alkanes/in the spores. This is difficult to test since axenic cultures of rust and smut fungi generally do not sporulate or produce sufficient quantities of the spore-types required for study. However, there is some evidence to suggest that the fungal spore alkanes are truly fungal products. Oro et al. (1966) and Laseter and Weber (1966) have pointed out that hydrocarbons of *Ustilago maydis* chlamydospores differ from that of infected and noninfected host (corn) tissue. The alkane content of spores from the same batch as those from the above studies determined after eight years storage is similar to that previously reported and qualitatively similar to but quantitatively different from that of 15 day old corn leaves (Weete, 1976). There are close similarities in hydrocarbon content between teliospores of three *Tilletia* species (Laseter et al., 1968) and *P. striiformis* uredospores (Jackson et al., 1973). Only trace quantities of hydrocarbons have been detected in aeciospores and basidiospores of *Cronartium fusiforme* collected from their natural hosts (*Pinus taeda* and *Quercus rubra*, respectively), and they are atypical of rust spore hydrocarbons with the absence of an odd over even chain predominance (Carmack et al., 1976; Weete and Kelley, 1977).

Mycelial hydrocarbons have been reported for the three soil fungi: *Penicillium* sp., *Aspergillus* sp., and *Trichoderma viride* grown in aerated non-defined medium (Jones, 1969). Alkanes (C_{15} to C_{36}) identified only by GLC have been detected in each species, but no odd over even chain predominance could be found for the *Penicillium* and *Aspergillus* species, and C_{22} and C_{24} and an unknown compound are predominant in *T. viride*. Only the isoprenoid hydrocarbon squalene has been detected in the hydrocarbon fraction of mycelial extracts of *Rhizopus arrhizus* (Weete et al., 1970). Sclerotia from two fungal species also do not have the hydrocarbon composition typical of the rust and smut spores. Sclerotia of *Sclerotinia sclerotiorum* collected from the natural host contain a single unidentified component in the hydrocarbon fraction of the lipid extract. This compound is absent from sclerotia taken from the fungus grown on a synthetic medium containing different carbon sources. Alkanes are also absent from laboratory grown sclerotia of *Sclerotium rolfsii* (unpublished data).

The hydrocarbon content of yeasts has been estimated to
be 2 to 20% of the total lipid as determined by gravimetric
measurements of the non-saponifiable lipid fraction (Barron
and Hanahan, 1961; Kovac et al., 1967; Baraud et al., 1967).
Squalene probably accounts for the greatest portion of this
fraction in *Saccharomyces cerevisiae* (Jollow et al., 1968)
and *S. carlsbergenesis* (Shafai and Lewin, 1968). Noniso-
prenoid hydrocarbons have been reported for several yeasts,
but in most cases identifications have not been confirmed
by mass spectrometry and consideration has not been given to
the possibility of contamination from the media. *S. oviformis*
and *S. ludgigii* contain over 40 normal and branched-chain
hydrocarbons with chain lengths ranging from C_{10} to C_{31}, C_{10}
to C_{19} being predominant (Baraud et al., 1967). Another *Saccharomyces* species contains hydrocarbons ranging in chain
length from C_{15} to C_{34} (Jones, 1969). Alkanes from C_{16} to
C_{39} have been reported for *Debaryomyces hansenii* (Merdinger
and Devine, 1965). *Pullularia pullulans* contain a hydrocarbon
pattern similar to that of bacteria; a complex mixture of
saturated and unsaturated, normal and branched-chain isomers
ranging in chain length from C_{16} to C_{28} with predominant
homologues being nC_{19} to nC_{22} (Merdinger et al., 1968).
Over 60% of the hydrocarbons from an unknown yeast have
molecular weights greater than 282 (C_{20}) and there is no
odd over even chain length predominance among C_{15} to C_{20}
alkanes (Hans et al., 1968). Eicosane (C_{20}) is the principal alkane of this species identified by GLC-MS.

Hydrocarbons of *Candida utilis* grown anaerobically on a
defined medium have been more thoroughly studied than those
of other yeasts (Fabre-Joneau et al., 1969). Normal alkanes
from C_{14} to C_{29} and monounsaturated alkenes from C_{16} to C_{23}
have been identified by GLC-MS. Cytoplasmic membranes, cell
walls, and cellular contents from 9 grams of the yeast material contain 15, 0.5 and 0.4 mg of hydrocarbons, respectively. Odd-chain alkanes, particularly C_{27}, were predomi
nant, but the odd over even predominance is not pronounced
as in hydrocarbons of the rust and smut spores. In addition,
squalene represents over 50% of the hydrocarbons from the
cytoplasmic membranes and a hexadecatriene represents 82%
of the cell wall hydrocarbons.

With the exception of spores from the few rust and smut
species cited above, few Basidiomycete fungi have been examined for their hydrocarbon content. The wood-rotting fungus
Fomes igniarius, collected from its natural habitat (*Populus
tremuloides*) contains a homologous series of n-alkanes from

C_{15} to C_{27}, with C_{23} (30.3%) and C_{25} (26.8%) as the predominant homologous (Epstein et al., 1966).

Suggested roles for the fungal spore alkanes are similar to those for the epicuticular wax of higher plants: (1) prevention of desiccation and insulation from extreme temperatures, (2) provide resistence to microbial attack, (3) inhibition of germination, and (4) playing a role in the infection process. The fact that membranes of *C. utilus* contain over 20 times the hydrocarbon content of other cellular components suggests a role in membrane structure.

HYDROCARBON BIOSYNTHESIS

Very little is known about alkane biosynthesis in fungi. $[U-^{14}C]$Acetate is incorporated into alkanes by a soluble fraction (105,000 xg) of a *S. cerevisiae* homogenate (Cassagne and Larrougere-Regnier, 1972). ATP, coenzyme A (CoA), Mg^{+2}, NADPH, $KHCO_3$, and pyridoxal phosphate are required. Fatty acids are also produced by this preparation.

Hydrocarbon synthesis has been studied extensively in higher plants and has been recently reviewed by Kolattukudy et al. (1976). With the identification of nonacosane (C_{29}) and hentriacontane (C_{31}) as the principal alkanes in wax of most higher plant species, it has been proposed that these substances are produced by the head-to-head condensation of two fatty acids of equal chain-length with the subsequent loss of the carboxyl group of one fatty acid as CO_2. For example, condensation of two C_{15} fatty acids and decarboxylation would result in a C_{29} alkane. Intermediates with internal oxo, hydroxy, and double bond functions would be expected in order of their occurrence in the head-to-head condensation pathway. Alkanes become labeled when leaves are given $[^{14}C]$ labeled acetate or fatty acids. Although this pathway was initially attractive, no direct evidence for its presence in higher plants has been obtained. Evidence against the head-to-head condensation pathway includes the following: (1) pentadecanoic acid occurs in low relative proportions in higher plants, (2) although the expected oxo and hydroxy intermediates were detected in the first plants studied (cabbage, *Brassica oleracea*), no unsaturated intermediate has been found, (3) no precursor-product relationship has been established between the oxo and hydroxy intermediates and alkanes of corresponding chain-length, (4) the carboxyl group of precursor fatty acids is incorporated intact into alkanes rather than being lost as required in the head-to-head condensation pathway.

Since supporting evidence for the head-to-head condensation pathway has not been obtained, an alternative pathway of alkane synthesis has been proposed by Kolattukudy (1966; 1967), the elongation-decarboxylation pathway (Figure 8.9). According to this hypothesis, the principal product of the fatty acid synthetase (C_{16}) is elongated via malonyl CoA to very long-chain fatty acids (C_{18} to C_{36}), and hydrocarbons are formed following decarboxylation of these acids. De novo fatty acid synthesis, chain-elongation, and decarboxylation occurs in the leaf epidermis which, except for guard cells, contain no chloroplasts. There is considerable indirect evidence supporting this pathway, such as the precursor-product relationship between very long-chain fatty acids and alkanes and the incorporation of intact fatty acids into alkanes; but the most convincing evidence is the direct conversion of radio-labeled exogenous very long-chain fatty acids to the alkanes containing one less carbon atom. Based on the study by Han et al. (1969), who has shown that $[18-^{14}C]$ stearic acid is incorporated exclusively into heptadecane (C_{17}) by Nostoc muscorum, it is generally believed that alkane synthesis also occurs via elongation-decarboxylation in blue-green algae.

Although of limited occurrence in fungi and higher plants, branch-chain alkanes with the iso and anteiso structures are produced by some species and are the most frequently encountered methyl branched alkanes in these organisms. Methyl branches appear to originate from very short-chain acids, derived from appropriately branched amino acids by deamination and decarboxylation, which serve as the "primer" molecule for the fatty acid synthetase instead of acetate (see Chapter 3). Iso- and anteiso-branched chain products of the fatty acid synthetase are elongated and decarboxylated as described above for the elongation-decarboxylation pathway of n-alkane synthesis. Amino acid precursors for iso- and anteiso-branched chain fatty acids and, hence, the corresponding branched-chain alkanes, are valine, leucine, and isoleucine. Experimental evidence supporting this pathway in tobacco leaves is that labeled valine and isobutryic acid, which can be derived from valine and can serve as the "primer" molecule in fatty acid synthesis, are incorporated into even-chained iso-branched fatty acids and odd-chained iso-branched alkanes (Kaneda, 1967; Kolattukudy, 1968). Odd-chained fatty acids and even-chained alkanes with the anteiso structure became labeled when $[^{14}C]$ isoleucine was supplied to the leaves.

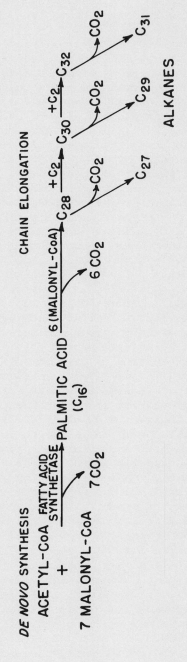

Figure 8.9   Elongation-decarboxylation pathway of alkane biosynthesis in higher plants.

Studies on hydrocarbon biosynthesis in bacteria are restricted to a very few species, particularly the isoleucine-
requiring mutant *Sarcina lutea* strain FD-SCC-533 isolated
from soil. These studies have been reviewed by Albro and
Dittmer (1970) and Albro (1976). *S. lutea* produces a complex
mixture of singly and doubly *iso-* and *anteiso*-branched mono-
olefins ranging in chain-length from C_{23} to C_{29}. Synthesis
of these hydrocarbons appears to occur via a head-to-head
condensation pathway, but one that is different from and
more complicated than that originally proposed for higher
plants. The presence or absence of acetate in the culture
medium can alter the route by which fatty acids are incorporated into hydrocarbons by *S. lutea*. In a medium containing
acetate, palmitic acid is incorporated intact into hydrocarbons, whereas in acetate-free medium most of the C_{16} acid
is decarboxylated prior to being incorporated. Similarly,
the form of the fatty acid substrate determines whether the
molecule is incorporated intact into hydrocarbons. The carbonyl carbon of oxygen and thioesters (CoA) is not incorporated into hydrocarbons, but that of the free acid (C_{16})
is incorporated as described above. Hydrocarbon synthesis
from labeled acetic acid and $[1-^{14}C]$ or $[16-^{14}C]$ palmitic
acid by a dialyzed cell-free extract of *S. lutea* requires
ATP, CoA, Mg^{+2}, NADPH, and either pyridoxal or pyridoxamine
phosphate (Albro and Dittmer, 1969c). Deletion of pyridoxal
phosphate from the incubation medium results in decreased
incorporation of $[16-^{14}C]$ palmityl CoA into hydrocarbons by
95%, but the free acid by only 44%. Chemical degradation
of the principal hydrocarbon product of *S. lutea*, dianteiso
C_{29} alkene (Albro and Dittmer, 1968a) produced from radiolabeled *anteiso*-pentadecanoic acid, yields two fatty acids,
C_{15} and C_{14}, each containing half the radioactivity originally
present in the alkene. Degradation studies have also shown
that the double bond is located between the C-1 and C-2 of
the fatty acid incorporated intact into the alkene. This
suggests that hydrocarbon synthesis in *S. lutea* occurs by
head-to-head condensation and fatty acid precursors may enter
the pathway by more than one route, one of which involves the
CoA derivative (Figure 8.10).

Presumably fatty acid methyl esters, triglycerides, and/
or the coenzyme A derivatives serve as the intermediate(s)
through which the decarboxylated acyl moiety is incorporated
into hydrocarbons. The pathway of entry for intact fatty acid
chains into hydrocarbons appears to involve vinylic ethers.
The vinylic ether portion of either 1-hexadec-1-enyl-glycerol
or a neutral plasmalogen labeled in either the C-1 or C-16 of

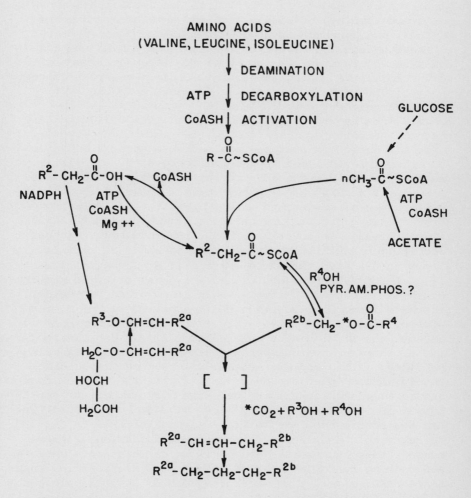

Figure 8.10   Hydrocarbon biosynthesis in *Sarcina lutea* FD-SCC-533 (redrawn from Albro, 1976).

the hexadecenyl moiety is incorporated into hydrocarbons more
rapidly than label from [16-^{14}C] palmitic acid. Alkanes are
produced by reduction of alkenes (Albro and Dittmer, 1969b).

The absence of long-chain branched alkanes and presence
of medium-chain hydrocarbons such as C_{17} in *Chlorobrium*,
suggest that there may be other pathways of hydrocarbon
synthesis in these bacteria. Perhaps the decarboxylation
of medium-chain fatty acids, as seems to be the case with
blue-green algae, also occurs in certain bacteria.

OXIDATION OF HYDROCARBONS

Aliphatic hydrocarbons are secondary metabolites not
readily recycled through intermediary metabolism. Although
most organisms possess the enzymes for degrading hydrocarbons, or the ability to produce such enzymes, alkanes are
not often accessible to the enzymes because of their deposi-
tion on the cuticular surface, i.e., higher plants and
insects. Some organisms, particularly certain yeasts and
bacteria, readily oxidize aliphatic hydrocarbons and in some
cases, can use them as sole sources of carbon. The first
step in the degradation of alkanes is the enzymatic hydroxylation of a terminal methyl carbon atom, ω-hydroxylation.
Further oxidative reactions lead to the production of carboxylic acids of corresponding chain-length that may be
metabolized according to particular cellular requirements,
i.e., energy or carbon skeletons. Alkane oxidation was
first recognized in bacteria and yeasts, but it has also
been observed in mammalian and plant systems. Hydrocarbon
oxidative processes in bacteria have been extensively reviewed (Davis and Updegraff, 1954; Davis, 1967; Foster, 1962;
Van der Lindon and Thysse, 1965; McKenna and Kallio, 1965;
Jurtshuk and Cardini, 1972; Albro, 1976). Using cell-free
preparations, the mechanisms of alkane oxidation in mammals,
yeast and certain bacteria have been worked out and are outlined below.

Corynebacterium 7EIC. This bacterium grows well on
n-alkanes ranging from C_3 to C_{18}, with n-octane being the
best substrate for oxidation. Alkenes (C_{12} to C_{18}) and certain 1-halogenated alkanes are also oxidized, but aromatic
hydrocarbons serve as poor substrates for hydroxylation.
This bacterium is capable of diterminal oxidation of C_{10} to
C_{14} alkanes (Kester and Foster, 1963). The hydroxylation
reaction is carried out by a mixed function oxygenase, i.e.,

one atom of molecular oxygen is transferred to the substrate
and the other is reduced to water by a pyridine nucleotide
coenzyme. Enzyme systems of this type are widespread in
plant, animal, and microbial organisms and catalyze the oxidation of a variety of substances such as alkanes, fatty
acids, camphor, steroids, drugs, and others. In addition to
the hydroxylase, electron carriers such as cytochromes are
involved in the transfer of electrons from the primary donor
(NADH or NADPH) to oxygen in the formation of water.

The *Corynebacterium 7EIC* oxidation system can be separated into two, possibly three, components which include a
flavoprotein, possibly a nonheme iron component, and a hemo-
protein cytochrome P-450 (Jurtshuk and Cardini, 1972). The
combined fractions carry out the ω-hydroxylation of n-octane
as shown in Figure 8.11a. This enzyme system is specific
for NADH. It is postulated that electrons from the donor
NADH are sequentially transferred to the NADH-flavoprotein,
possibly through several nonheme components, and then to a
cytochrome P-450 reductase. Incorporation of oxygen into
the n-alkane probably occurs via cytochrome P-450. The
reaction is inhibited by electron acceptors such as cytochromes and dichloroindophenol (DCIP). The ω-hydroxylation
system is inducible. Cell-free preparations of *Corynebacterium 7EIC* grown on a n-alkane substrate contain membrane-
bound cytochrome P-450 (opposed to P-420) and NADH-dependent
flavoprotein, but not when the bacterium is grown on glucose
or acetate.

Pseudomonas olevorans. The most extensive studies on
n-alkane hydroxylation have been conducted using soluble,
inducible enzymes of *P. olevorans* (Gholson et al., 1963;
Baptist et al., 1963; McKenna and Coon, 1970; Peterson et
al., 1969; Kusunose et al., 1967a; 1967b; 1968). This bacterium oxidizes substrates from C_6 to C_{16}. This enzyme system has been separated into three fractions: (1) a rubre-
doxin-like, nonheme iron protein, (2) a NADH-rubredoxin
reductase, and (3) an ω-hydroxylase (Peterson et al., 1966).
The flow of electrons from NADH to the electron acceptor
proceeds in a manner similar to that described for the
coryneform bacterium (Figure 8.11b). Unlike the coryneform
system, *P. olevorans* contains no cytochrome P-450 and its
hydroxylase is insensitive to carbon monoxide.

Mammals. The hydroxylating enzyme complex of mammalian
liver appears to more closely resemble that of the *corynebacterium* than that of *P. olevorans*. The mammalian system

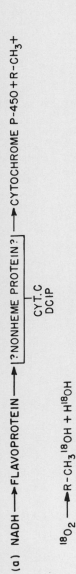
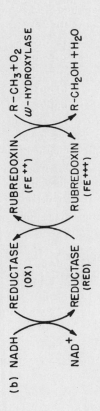
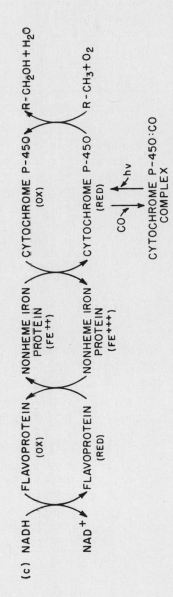

Figure 8.11   Mechanisms of n-alkane ω–hydroxylation in (a) Cornyebacterium 7EIC, (b) *Pseudomonas olevorans*, and (c) mammalian hepatic microsomes.

is composed of a flavoprotein, a nonheme iron protein, and a carbon monoxide-sensitive cytochrome P-450 component (Figure 8.11c). The system is activated by phosphatidyl-choline. NADPH appears to be the preferred primary electron donor, but NADH can substitute for it.

Yeasts. The oxidation of alkanes by yeasts is well-documented (Tulloch et al., 1962; Tizuka et al., 1966; Wagner et al., 1967; Klug and Markovetz, 1967; Lebeault et al., 1970; Lowery et al., 1968), but little is known about the hydroxylation mechanism in these organisms (Lebeault et al., 1971). An inducible and soluble mixed-function oxy-genase similar to that of the mammalian hepatic system has been isolated from *Candida tropicalis* grown on tetradecane. This system oxidizes fatty acids to their corresponding ω-hydroxy derivatives and n-alkanes to primary alcohols, and can be separated into two fractions. The yeast system contains cytochrome P-450, requires NADPH, and is inhibited by carbon monoxide but not by cyanide as in *P. olevorans*. Molecular oxygen is also required. Medium chain-length n-alkanes are preferred substrates while short chain ($<C_9$) hydrocarbons are less readily oxidized (Lebeault et al., 1970; Lowery et al., 1968). A cell-free preparation from yeast oxidizes [^{14}C] hexadecane to the corresponding primary alcohol and acid.

The conversion of oleate to 17-hydroxyoleate by a *Torulopsis* species is sensitive to carbon monoxide (Heinz et al., 1970). The "450-CO" pigment has been detected in *S. cerevisiae* (Lindenmayer and Smith, 1965) and the P-450 hemo-protein is present in a particulate fraction of disrupted yeast cells (Ishidate et al., 1969). However, Peterson (1970) has failed to detect cytochrome P-450 in *Candida lipo-lytica* grown on hexadecane. As in other systems, primary alcohol products are converted to corresponding aldehydes and acids as shown for the oxidation of n-octane in Figure 8.12.

Higher Plants. Although plants have the ability to oxidize hexadecane at the terminal methyl carbon atom to the corresponding acid (Kolattukudy, 1969), little is known about hydrocarbon oxidation in higher plants.

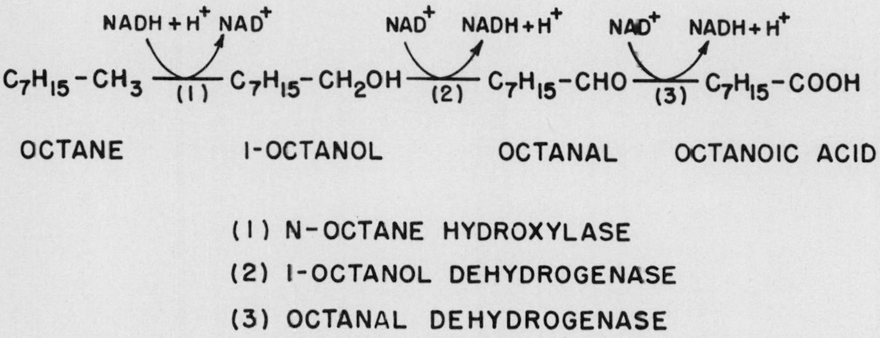

(1) N-OCTANE HYDROXYLASE

(2) I-OCTANOL DEHYDROGENASE

(3) OCTANAL DEHYDROGENASE

Figure 8.12   Oxidation of n-octane to octanoic acid.

CHAPTER 9   STEROLS, CAROTENOIDS, AND POLYPRENOLS

INTRODUCTION

Isoprenoid substances comprise a large, complex, and important group of natural products, most of which are produced by plants. The basic building block of these substances is isoprene (2-methyl-1,3-butadiene):

$$CH_2 = \overset{\overset{\displaystyle CH_3}{|}}{C} - CH = CH_2$$

Condensation of the pyrophosphorylated form of isoprene, isopentenyl-pyrophosphate (IPP), with its isomer dimethyl-allyl-pyrophosphate (DMPP) followed by subsequent additions of IPP to the growing chain, and in some cases cyclization, leads to the formation of the various classes of compounds called isoprenoids or terpenoids: monoterpenes (C_{10}), sesquiterpenes (C_{15}), diterpenes (C_{20}), and C_{30} to C_{85} compounds that includes triterpenes (sterols), carotenes, xanthophylls, penta- and hexacyclic alkaloids, rubber, and the polyprenols. This chapter is concerned only with the occurrence of sterols, carotenoids, and polyprenols in fungi.

STEROLS

In spite of the wide occurrence of sterols throughout the plant and animal kingdoms, insects and members of the fungal family Pythiaceae apparently do not produce sterols,

and require them for reproduction (Hendrix, 1970). Although
some prokaryotic organisms apparently cannot produce them,
sterols have been identified in extracts of certain blue-
green algae and bacteria (see below).

Early studies established that cholesterol is the predominant sterol in animals and sitosterol is the major sterol
of higher plants. Ergosterol has been considered the principal sterol of fungi (Foster, 1949; Cochrane, 1958) since
first being identified by Tanret (1889) in extracts of ergot,
and later in other fungi by Gerard (1892; 1895a; 1895b; 1898)
This is supported by recent studies of fungal sterols where
the identifications of ergosterol and other fungal sterols
have been confirmed using gas chromatography, mass spectrometry, and nuclear magnetic resonance. Sterols occur as
free alcohols, esters, glycosides, and a "water-soluble"
form. They are associated with membranous cell fractions
where they are believed to regulate permeability by altering
the internal viscosity and molecular motion of other lipids
(Demel and DeKruyff, 1976). A sterol, along with oleic acid,
is required for yeast growth under anaerobic conditions.
This requirement is for planarity rather than a specific
sterol (Nes, 1974). Sterols are also required for optimum activity of certain enzymes, such as ATPase (Cobon and
Haslam, 1973). Sterols are also precursors to steroids in
higher plants (Heftmann, 1974) and hormones such as antheridiol in fungi (Barksdale, 1969).

STRUCTURE AND NOMENCLATURE

The structure common to all steroids is 1,2-cyclopentano-
perhydrophenanthrene,

The steric configuration of the ring nucleus is *trans* at the junction of rings B and C, and is almost always *trans* at the junction of rings C and D. The configuration at the junction of rings A and B may be either *cis* (5β) or *trans* (5α) and are called cholestane and coprostane, respectively, when based on the C_{27} hydrocarbon skeleton (Figure 9.1b and c). Additions of various substituents to the ring nucleus give rise to individual steroids. Discussion in this section is restricted to sterols having a 3β hydroxy group, methyl groups at C-10 and C-13, and a side-chain containing 8 to 10 carbon atoms at C-17 (Figure 9.1a). Also, cyclic intermediates in the synthesis of 4-desmethyl sterols are included. Individual sterols with this basic structure differ by the presence or absence of methyl groups at C-4 and C-14 (intermediates in the synthesis of 4-desmethyl sterols), stereochemistry at asymmetric carbon atoms, substitution at C-24 (C_1 or C_2), and the position and number of double bonds in the molecule.

There is a considerable lack of consistency in sterol nomenclature in the literature. The most common sterols are generally referred to by trivial names that give little or no clue as to their structure. The systematic naming of sterols is based on the C_{27} hydrocarbon skeleton cholestane (Figure 9.1a). For example, the systematic name for cholesterol is cholest-5-en-3β-ol (hereafter -enol will be used in place of en-3β-ol). The RS system is used to designate configuration at sites of unsaturation and at asymmetric carbon atoms (Cahn et al., 1966). Hence, the systematic name for ergosterol is (22E,24R)-24-methyl-cholesta-5,7,22-tetraenol. The naming of sterols with methyl or ethyl substituents at C-24 of cholestane may be based on the hydrocarbon skeleton created by the substitution; ergostane for methyl substitution at C-24 and stigmastane for ethyl substitution. Thus, (22E,24R)-ergosta-5,7,22-trienol would be the name for ergosterol and (24R)-stigmast-5-enol for sitosterol (Figure 9.1a).

There are eight asymmetric carbon atoms in cholesterol (C-3, C-8, C-9, C-10, C-13, C-14, C-17, and C-20). The hydroxyl group at C-3, methyl groups at C-10 and C-13, hydrogen at C-8, side chain at C-17, and C-20 methyl group (R configuration) are indicated above the plane of the paper with a solid line, and have the β-configuration (Figure 9.1a). Hydrogens at C-3, C-9, C-14, and C-17 have the α-configuration and are indicated below the plane of the paper with a dotted line. Asymmetric carbons may be eliminated during

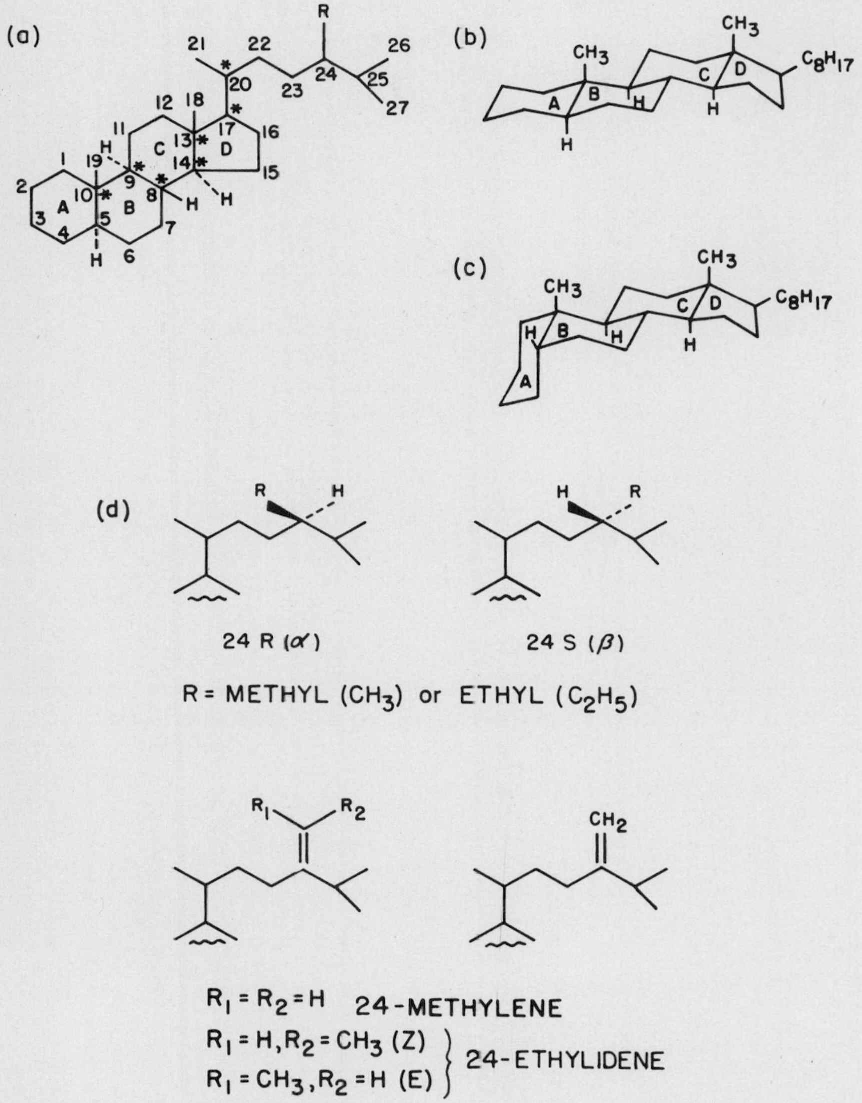

Figure 9.1 a) Parent hydrocarbon structures of sterols: Cholestane, R = H; Ergostane, R = CH_3; Stigmastane, R = C_2H_5. Astericks indicate asymmetric carbons of cholesterol (cholest-5-enol), b) Stereo-configuration of cholestane (5α), c) Stereo-configuration of coprostane (5β), d) Most common sterol side chain structures.

sterol biosynthesis by introducing a double bond at the appropriate carbon; they may be introduced by substitution or, in
some cases, reduction of a double bond. As noted above, substitution at C-24 introduces chirality at that carbon atom
and both stereoisomers occur in nature. 24R is equivalent
to the α-configuration and 24S to the β-configuration if the
sterol side-chain is saturated. However, if a double bond
is introduced into the side-chain, for example at C-22 as in
ergosterol and stigmasterol, priority assignments change and
the 24S isomer becomes 24R or 24R becomes 24S, respectively.
A double bond at C-22 or C-24(28) for C_{29} sterols introduces
geometrical isomerism into the side-chain and is designated
as Z (*cis*) or E (*trans*). The C-22 double bond of naturally
occurring sterols have the E-configuration, and both isomers
of 24-ethylidene sterols have been identified.

Over eighty sterols identified from various sources are
listed in Table 9.1 by their systematic and trivial names.
Nomenclature of the respective authors cited is used in this
chapter.

TABLE 9.1   Systematic and Trivial Names of Sterols

Systematic Name	Trivial Name
4,4,14α-TRIMETHYL STEROLS	
1 Lanosta-8,24-dienol[a]	Lanosterol (Cryptosterol)
2 Lanost-8-enol	24-Dihydrolanosterol
3 24-Methyl-lanost-8-enol	24-Methyl-dihydrolano-sterol
4 9,19-Cyclo-5α,9β-lanost-24-enol	Cycloartenol
5 24-Methylene-9,19-cyclo-5α,9β-lanosterol	24-Methylenecycloartenol
6 24(S)-24-Methyl-9,19-cyclo-5α,9β-lanost-25-enol	Cyclolaudenol(24β)
7 24-Methylene-lanost-8-enol	24-Methylene-24(25)-dihydrolanosterol
4,4-DIMETHYL STEROLS	
8 4,4-Dimethyl-cholesta-8,24-dienol	14-Desmethyl lanosterol, 4,4-dimethylzymosterol
9 4,4-Dimethyl-cholesta-8,14,24-trienol	-

TABLE 9.1   Continued

Systematic Name	Trivial Name
10 4,4-Dimethyl-5α-ergosta-8, 24(28)dienol	–
4,14α-DIMETHYL STEROLS	
11 4,14-Dimethyl-cholesta-8, 24-dienol	4,14-Dimethylzymosterol
12 4α-14α-Dimethyl-ergosta-8, 24(28)-dienol	Obtusifoliol
13 4α,14α-Dimethyl-9,19-cyclo-5α,9β-ergost-24(28)-enol	Cycloeucalenol
14 4α,14α-Dimethyl-ergost-8(9)-enol	24-Dehydro-obtusifoliol
15 4α,14α-Dimethyl-5α-(24S)-stigmast-8-enol	–
16 4α,14α-Dimethyl-9,19-cyclo-5α,9β-ergost-24-enol	31-Norcycloartenol
4α-METHYL STEROLS	
17 4α-Methyl-cholesta-8,24-dienol	4-Methylzymosterol
18 4α-Methyl-5α-cholesta-8,22-dienol	–
19 4α-Methyl-5α-cholesta-8,22-dienol	–
20 4α-Methyl-cholesta-8,14,24-trienol	–
21 4α-Methyl-cholest-7-enol	Lophenol
22 4α-Methyl-5α-ergosta-8,22-dienol	–
23 4α-Methyl-ergosta-8,24(28)-dienol	–
24 4α-Methyl-5α-ergost-8(9)-enol	–
25 4α-Methyl-ergosta-7,24(28)-dienol	24-Methylenelophenol, gramisterol
26 4α-Methyl-5α-(24S)-stigmast-8(9)-enol	–
27 4α-Methyl-5α-(24S)-stigmasta-8,14-dienol	–

TABLE 9.1   Continued

Systematic Name	Trivial Name
28 4α-Methyl-stigmasta-7, 24(28)-dienol	24-Ethylidenelophenol, Citrostadienol
14-METHYL STEROLS	
29 14-Methyl-ergosta-8,24(28)- dienol	14-Methylfecosterol
30 14α-Methyl-9,19-cyclo-5α,9β- ergost-24(28)-enol	24-Methylene pollinas- tanol
31 14α-Methyl-9,19-cyclo-5α,9β- ergostanol	24-Methyl pollinastanol
32 14α-Methyl-9,19-cyclo-5α,9β- cholestanol	Pollinastonal
33 14α-Methyl-5α-ergost-8-enol	–
34 14α-Methyl-5α-ergosta-8, 24(28)-dienol	–
35 14α-Methyl-5α-(24S)-stigmast- 8-enol	–
4-DESMETHYL C_{27} STEROLS	
36 Cholestanol	–
37 Cholest-5-enol	Cholesterol
38 Cholest-7-enol	Lathosterol
39 Cholesta-5,7-dienol	7-Dehydrocholesterol
40 Cholesta-5,24-dienol	Desmosterol
41 Cholesta-7,24-dienol	–
42 Cholesta-8,24-dienol	Zymosterol
43 Cholesta-8,22,24-trienol	–
44 Cholesta-5,22-dienol	22-Dihydrocholesterol
45 Cholesta-5,8,24-trienol	–
46 Cholesta-7,22,24-trienol	
47 Cholesta-5,7,22-trienol	–
48 Cholesta-5,7,24-trienol	
49 Cholesta-5,8,22,24-tetraenol	–
4-DESMETHYL C_{28} STEROLS	
50 Ergost-5-enol	24α-Methyl cholesterol, campesterol(24α), 22-dihydrobrassica- sterol(24β)

TABLE 9.1   Continued

Systematic Name	Trivial Name
51 Ergost-7-enol	Fungisterol(24β), 24β-methyllathosterol, Δ^7-ergostenol, Epifungisterol(24α-methyllathosterol)
52 5α-Ergost-8(14)-enol	-
53 Ergost-8(9)-enol	-
54 Ergosta-8,24(28)-dienol	Fecosterol, 24-methylenezymosterol
55 Ergosta-8,23-dienol	Ascosterol
56 Ergosta-8(9),22-dienol	-
57 5α-Ergosta-8(14),22-dienol	-
58 5α-Ergosta-8,14-dienol	Ignosterol
59 Ergosta-7,22-dienol	5-Dihydroergosterol(24β), 22-Dehydrofungisterol
60 Ergosta-7,24(28)-dienol	Episterol, 24-Methylenelathosterol
61 (24S)-Ergosta-7,16-dienol	-
62 (22E)-Ergosta-5,22-dienol	Brassicasterol(24β), Diatomasterol(24α)
63 Ergosta-5,8(9)-dienol	-
64 (24S)-Ergosta-5,25-dienol	Codisterol(24β)
65 Ergosta-5,7-dienol	22-Dihydroergosterol(24β), 7-Dehydrocampesterol(24α)
66 Ergosta-5,24(28)-dienol	24-Methylene cholesterol, Chalinasterol
67 Ergosta-8,22,24(28)-trienol	-
68 5α-Ergosta-8,14,22-trienol	-
69 Ergosta-8,14,24(28)-trienol	-
70 Ergosta-7,22,24(28)-trienol	-
71 Ergosta-5,7,24(28)-trienol	5-Dehydroepisterol
72 Ergosta-5,7,14-trienol	-
73 (22E)-Ergosta-5,7,22-trienol	Ergosterol(24β), 24-Epiergosterol(24α)
74 Ergosta-5,8,22-trienol	Lichesterol(24β)
75 Ergosta-5,7,24(28)-trienol	-
76 Ergosta-5,8,22,24(28)-tetraenol	-
77 Ergosta-5,7,22,24(28)-tetraenol	24(28)-Dehydroergosterol

TABLE 9.1   Continued

Systematic Name	Trivial Name
78 Ergosta-5,7,14,22-tetraenol	14-Dehydroergosterol
79 Ergosta-5,7,9(11),22- tetraenol	–

4-DESMETHYL C_{29} STEROLS

80 (24S)-Stigmast-5-enol	Clionasterol, 24β-ethyl- cholesterol
81 (24R)-Stigmast-5-enol	Sitosterol, 24α-Ethyl- cholesterol
82 5α-(24S)-Stigmast-7-enol	Δ^7-Chondrillasterol, Schottenol, 24α-Ethyl- lathosterol, Δ^7-Stigma- stenol
83 5α-(24S)-Stigmast-8(9)-enol	–
84 5α-(24S)-Stigmasta-8,14- dienol	–
85 Stigmasta-7,24(28)-dienol	Δ^7-Avenasterol($\Delta^{24(28)}$t)
86 5α-(24S)-Stigmasta-7,25- dienol	–
87 5α-(22E,24R)-Stigmasta-7,22- dienol	Chondrillasterol(24β)
88 (24S)-Stigmasta-5,7-dienol	7-Dehydroclionasterol(24β) $\Delta^{5,7}$-Chondrillasterol
89 Stigmasta-5,24(28)-dienol	Fucosterol(24E), 24-*cis*- ethylidene cholesterol, 28-*trans*-isofucosterol (24Z)
90 (22E)-Stigmasta-5,22-dienol	Stigmasterol(24α,Δ^{22}t)
91 (22E,24S)-Stigmasta-5,22- dienol	Poriferasterol(24β,Δ^{22}t)
92 (24S)-Stigmasta-5,25-dienol	Clerosterol, [$\Delta^{25(27)}$- dehydroclionasterol] (24β)
93 (22E,24S)-Stigmasta-5,7,22- trienol(24β)	Corbisterol, 7-Dehydro- poriferasterol
94 (24S)-5α-Stigmasta-7,22- dienol	Spinasterol(24α,Δ^{22}t)

[a]Lanostane = 4,4,14-trimethylcholestane.

OCCURRENCE IN FUNGI

Yeasts appear to have the greatest potential for sterol
production among fungi. Sterols generally comprise 0.1 to
2.0% of the yeast cellular dry weight (Bills et al., 1930),
but may represent up to 10% (Dulaney et al., 1954). The
early reports considered ergosterol as the sole sterol representing the total sterol content. These studies showed
differences in sterol content according to species that
could be manipulated by changing the growth medium composition (MacLean and Hoffert, 1928; Preuss et al., 1931; 1932;
1934; Bernhauer and Patzelt, 1935; Wenck et al., 1935a;
1935b; Vanghelovici and Serban, 1940; 1941; Cavallito, 1944;
Ellis, 1945; Angelitti and Tappi, 1947; Dulaney et al.,
1954). The sterol content of mycelium is generally lower
than that of yeast cells, ranging from 0 to 1.2% of the dry
weight (Heiduschka and Lindner, 1929; McCorkindale et al.,
1969). Free sterols plus sterol esters comprise 4.6% (4.0 +
0.6) of the dry mycelial weight of *Neurospora crassa*
(Kushwaha et al., 1976).

The combined works of Bergmann (1954) and Fieser and
Fieser (1959) summarize the early reports of sterols in fungi
In his review of plant sterols, Bergmann (1954) has listed
over 60 fungal species for which ergosterol is one of the
principal sterols. Weete (1973; 1976), Brennan et al. (1974)
and Goodwin (1973) have reviewed the more recent studies
concerning the occurrence of fungal sterols. Some species
representative of the major fungal taxa are given in Table
9.2. Identifications made without the aid of modern instru-
mentation should be viewed with caution; however, some of the
early reports of sterol identifications have been confirmed
using GLC-MS. Although the sterol composition of numerous
species is given in Table 9.2, the discussion in this chapter
emphasizes identifications made by GLC or GLC-MS.

Phycomycetes. The most comprehensive single study of
sterols from different fungal species is that conducted by
McCorkindale et al. (1969), who examined 25 Phycomycete
species using GLC and in some cases mass spectrometry. Fungi
belonging to the orders Mucorales (chitin cell walls), Sapro-
legnales (cellulose cell walls), and Leptomitales (cellulose
cell walls) were included in their study. Cholesterol is the
only sterol common to each of the taxonomic groups. Species
of the cellulose containing fungi also contain desmosterol,
24-methylcholesterol, and fucosterol. Cholesterol,

24-ethylcholesterol, 22-dehydrocholesterol, 24-methylcho-
lesterol, and 24-ethyl-22-dehydrocholesterol are produced by
the aquatic fungi (cellulose cell walls) *Allomyces macrogynus*,
Rhizidiomyces apophysatus, *Rhizophlyctis rosea*, and *Hypochy-
trium catenoides* (Bean et al., 1973). There are considerable
quantitative and qualitative differences in sterols among
these fungi. Ergosterol and 22-dihydroergosterol are the
only sterols detected in the Mucorales fungi analyzed by
McCorkindale et al. (1969) (Table 9.2).

In addition to ergosterol, sterols of other Mucorales
fungi include episterol, ergosta-5,7,24(28)-trienol, lanosterol, and 22-dihydroergosterol for *Phycomyces blakesleeanus*
(Jaureuiberry et al., 1965; Akhtar et al., 1969; Lederer,
1969; Lenfant et al., 1969; Barton et al., 1970); and fungisterol, 5-dihydroergosterol, and ergosta-5,7,14(15)-trienol
for *Rhizopus arrhizus* (Weete et al., 1973). Ergosterol represents 90% of the sterols (0.69% of mycelial dry weight)
of *Mucor rouxii* and is accompanied by ergosta-5,7,9(11),22-
tetraenol, episterol, ergosta-5,7,24(28)-trienol, and ergosta-
5,22,24(28)-trienol (Safe, 1973). Bound sterols represent
0.2% of the total sterols in this fungus and the individual
sterols are not randomly distributed between the free and
bound sterol fractions.

Spores of relatively few fungal species have been analyzed for sterol content. Sterols of *R. arrhizus* sporangiospores are similar to those of the mycelium (see above)
(Weete et al., 1973). Spore sterols of *Linderina pennispora*
are atypical of the order Mucorales; fungisterol is the major
sterol and ergosterol is absent (Weete and Laseter, 1974).
The primitive fungus *Plasmodiophora brassicae* is the only
other Phycomycete for which spores have been analyzed for
sterols, and none considered to be fungal products could be
detected (Knights, 1970).

Pythiacious fungi take up and metabolize sterols (Elliott
and Knights, 1974), but as noted above they do not produce
them (see Chapter 10).

Ascomycetes and Deuteromycetes (Fungi Imperfecti).
Saccharomyces cerevisiae is the most extensively studied
fungus for sterol composition, including numerous strains
grown under a variety of conditions. Ergosterol is the
major sterol of most strains, but over 14 other sterols have
been identified as yeast products. The discovery of many

yeast sterols has been facilitated through the use of various
sterol synthesis inhibitors that promote the accumulation of
sterol synthesis intermediates otherwise present below detectable levels, and the use of mutants with metabolic
blocks at specific sites in ergosterol synthesis (see Chapter
10). Lanosterol generally represents a minor component of
yeast sterol extracts. This sterol is the first sterol produced in the pathway of ergosterol formation. It was first
identified as a fungal (yeast) product by Stanley and Wieland
(1931) and called cryptosterol. Zymosterol is another sterol
often accompanying ergosterol in yeast. Ergosterol is the
predominant sterol of *Candida albicans*, but when the yeast is
grown in the presence of miconazole nitrate 14-methylfeco-
sterol, obtusifoliol, lanosterol, and 24-methylene-dihydro-
lanosterol accumulate (Van Den Bossche et al., 1978). Several yeast strains with their sterol composition are listed
in Table 9.2.

Ergosterol is the predominant sterol of fungi belonging
to the orders Euascomycete and Deuteromycete and is often
accompanied by several other sterols in low relative proportions. *Aspergillus fumigatus* has been analyzed in two
laboratories and, in addition to ergosterol, this fungus
produces ergosta-5,7-dienol, 24-methylene lanosterol, 4,4-
dimethyl-ergosta-8,24(28)-dienol (Sherald and Sisler, 1975),
lanosterol, and 24-methylene lophenol (Goulston et al., 1967).
In the presence of triforine and triarimol, this fungus also
produces obtusifoliol and 14α-methyl-ergosta-8,24(28)-dienol
(Sherald and Sisler, 1975). Ergosterol is produced by the
imperfect marine fungi *Zalerion maritima* and *Pyrenochaeta*
sp., but not *Culcitalna achraspora*, *Flagellospora* sp., and
Clavatospora stellatacula (Kirk and Catalfomo, 1970). Ergosterol, 22-dihydroergosterol, and 7-dehydroclionasterol are
products of *Fusarium roseum* (Tillman and Bean, 1970). Ergosterol is the major sterol of *Monilinia fructigena*, but when
grown in the presence of the fungicide S-1358 the ergosterol
content is reduced and obtusifoliol and 24-methylene dihydro-
lanosterol accumulates (Kato et al., 1975). Ergosterol
represents 81.4% of the sterols of *N. crassa* mycelium, and
is accompanied by 13 other sterols (Renaud et al., 1978).
Ergosterol is the major sterol of *Penicillium expansum*, but
when the fungus is grown in the presence of the fungicide
Imazalil 24-methylene-dihydrolanosterol, obtusifoliol, and
14α-methyl-ergosta-8,24(28)-dienol accumulate (Leroux and
Gredt, 1978).

Sterols of *Spicaria elegans* and *A. niger* conidia are similar to those of related fungi, with ergosterol as the major sterol and an unidentified diunsaturated C_{28} sterol as a relatively minor component (Weete and Laseter, 1974). Conidia of these two species also contain a C_{28} tetraene which is uncommon in fungi. Ergosta-5,7,14,22-tetraenol has been reported in extracts of *A. niger* mycelium (Barton and Bruun, 1951). Conidia of *Penicillium claviforme* differ from those of other imperfecti fungi in that fungisterol is the predominant sterol; this sterol is accompanied by ergosterol and several unidentified sterols. Spore sterols seem to be different from those of the mycelium of this species. This also appears to be true for the imperfect stage of *Neurospora crassa* (see above). Cholesterol is present in conidia and young mycelium of this fungus but apparently is not present in mature mycelium (Elliott et al., 1974). Sterols of conidia are bound, while those of rapidly growing mycelium are readily extractable; the proportion of bound sterols increases with senescence.

Basidiomycetes. The sterol composition of few mushroom, or Homobasidiomycete, fungi has been reported, but ergosterol appears to be most abundant in these fungi (Milazzo, 1965; Bentley et al., 1964; Angeletti and Tappi, 1947). A crystalline material from *Daedalea quercina* was called neosterol initially (Wieland et al., 1941; Wieland, 1929), but has been identified as a mixture of ergosterol (75.5%) and 5-dihydroergosterol (24.5%) (Tanahashi and Takahashi, 1966). *Agaricus campestris* basidiocarps contain ergosterol, fungisterol, 22-dihydroergosterol (Holtz and Schisler, 1972) as well as several 4,4-dimethyl and 4α-methyl sterols (Goulston et al., 1973, cited in Goodwin, 1973). *Fomes applanatus* of the family Polyporaceae produces 5α-dihydroergosterol and the corresponding 3-keto compound, but not ergosterol (Pettit and Knight, 1962). Both of these substances are also produced by *Coriolus sanguineus* Fr. of the same family (Cambie and LeQuesne, 1966). A different isolate of *F. applanatus* produces fungisterol and 24-methyl-5-α-cholesta-7,16-dienol but not ergosterol (Strigma et al., 1971).

Ergosterol is the major sterol among seven others in the sporidia of the corn smut fungus *Ustilago maydis* (Heterobasidiomycete, Ustilaginales) (Table 9.2). When treated with the sterol synthesis inhibitor triarimol, over 95% of the sterol fraction is composed of 24-methylene-dihydrolanosterol, obtusifoliol, and 14-α-methylergosta-8,24(28)-dienol.

Ergosterol is also the principal sterol of *U. maydis* and *U. nuda* teliospores, collected from their hosts, along with fungisterol and cholesterol (*U. nuda*) as minor components (Weete and Laseter, 1974).

Sterols of rust fungi (Heterobasidiomycete, Uredinales) differ considerably from those of other fungi. The most striking difference is the predominance of C_{29} sterols and the absence of ergosterol. *Puccinia graminis* (wheat stem rust) uredospores have been analyzed in several laboratories and contain fungisterol, stigmast-7-enol, cholesterol, and an unknown sterol determined only by GLC retention times (Hougen et al., 1958; Nowak et al., 1972). The C_{29} sterol and fungisterol have been confirmed by mass spectrometry, and are accompanied by a C_{29} diene (Weete and Laseter, 1974). *P. striiformis* uredospores have a similar sterol content. Uredospores of *Melampsora lini* (flax rust) contain stigmast-7-enol and stigmasta-7,24(28)-dienol as major sterols and stigmasta-5,7-dienol as a minor component (Jackson and Frear, 1968). Uredospores of the bean rust fungus *Uromyces phaseoli* contain (24Z)-stigmasta-7,24(28)-dienol as the major sterol; it is accompanied by the corresponding Δ^7 monoene (Lin et al., 1972). Sterols of different spore forms of the fusiform rust fungus *Cronartium fusiforme*, a pathogen of slash and loblolly pine in the southeastern United States, have been compared. Like the bean rust uredospores, *C. fusiforme* aeciospores contain stigmasta-7,24(28)-dienol (98%) as the principal sterol (Carmack et al., 1976). The sterols of *C. fusiforme* basidiospores, on the other hand, differ completely from aeciospores of the same fungus and resemble those of uredospores of *P. graminis* and *M. lini*. Stigmast-7-enol (57.6%) is the predominant sterol of these spores; it is accompanied by stigmasta-5,7-dienol (22.7%), fungisterol (16.0%) and an unidentified C_{28} diene (4%) (Weete and Kelley, 1977).

The rust and smut spores analyzed for sterol content have been collected from their natural hosts and, as with hydrocarbons (see Chapter 8), the question of their origin arises. Although stigmast-7-enol (Jeong et al., 1974) and stigmasta-7,24(28)-dienol (Knights and Laurie, 1967) are minor sterols of some higher plants, the fact that the "typical" higher plant sterols cholesterol, campesterol, stigmasterol, and sitosterol are absent from the spores suggests that rust and smut sterols are true fungal products. Furthermore, germinating bean rust uredospores synthesize the Δ^7 and $\Delta^{7,24(28)}$ C_{29} sterols from radiolabeled substrates (Lin and Knoche,

1974). Sterols of *C. fusiforme* mycelium grown in axenic
culture on semi-synthetic medium are similar to those of
aeciospores of the same species collected from galls on pine
trees, stigmast-7-enol, fungisterol, and an unidentified
sterol (Weete and Kelley, 1979).

The above presentation of fungal sterols has been restricted to 3β-hydroxy sterols that vary according to the
degree and location of double bonds and substitution at C-4,
C-14, and C-24. However, closely related substances are
often present in sterol preparations from fungi. One of
these substances is ergosta-4,6,8(14),22-tetraen-3-one (ETO)
which has been identified as a product of several *Aspergillus*
and *Penicillium* species (Seitz and Paukstelis, 1977; Price
and Worth, 1974; White and Taylor, 1970; White et al., 1970;
Brown and Jacobs, 1975), *Fomes officinalis* (Schulte et al.,
1968), *Lampteromyces japonicus* (Endo et al., 1970), *Balansia
epichloe* (Porter et al., 1975), and *Candida utilis* (Morimoto
et al., 1967).

Another compound commonly present in sterol extracts of
fungi is ergosterol peroxide (5α,8α-epidioxy-ergosta-6,22-
dienol) which has been idientified in extracts of *Alternaria
kikuchiana* (Starratt, 1976), *Scleroderma aurantium* (Vrkoc et
al., 1976), and other species (Breivak et al., 1954; Tanahashi and Takahashi, 1966; Clark and McKenzie, 1967). There
is some question as to this compound being a natural product
since it is only detected in extracts of *Piploporus betulinus*
and *Daedalea quercina* sporophores exposed to light for several
days and not in fresh extracts. Ergosterol peroxide is also
present in extracts of *Rhizoctonia repens* and its concentration increases with time. Arditti et al. (1972) suggest that
ergosterol peroxide may be formed by the attack on ergosterol
by singlet oxygen.

Other oxygenated 3β-hydroxy substances may be present
in sterol extracts of fungi. A $C_{28}H_{46}O_3$ compound accompanies
ergosterol and ergosterol peroxide in extracts of *Aspergillus
niger* that is believed to be cervesterol (ergosta-7,22-dien-
3β,5,6-triol), an artefact of the extraction procedure
(Vacheron and Michel, 1968). 9(11)-Dehydroergosterol peroxide, lanosta-8,23-dien-3β,25-diol, and lanosta-8,24-dien-
3β,23-diol have been identified in extracts of *Scleroderma
aurantium* (Vrkoc et al., 1976).

TABLE 9.2   Sterols of Some Fungi

Fungi		Sterols[a]
Class	Species (Reference)	

PHYCOMYCETES

	Saprolegnia fera (McCorkindale et al., 1969)	37,40,66,89
	S. megasperma (McCorkindale et al., 1969)	37,40,66,89
	Leptolegnia caudata (McCorkindale et al., 1969)	37,40,66,89
	Aplanopsis terrestris (McCorkindale et al., 1969)	37,66,89
	Achlya caroliniana (McCorkindale et al., 1969)	37,66,89
	Phythiopsis cymosa (McCorkindale et al., 1969)	37,66,89
	Apodachyla minima (McCorkindale et al., 1969)	37,40,66,89
	A. brachynema (McCorkindale et al., 1969)	37,40,66,89
	Apodachylella completa (McCorkindale et al., 1969)	37,66,89
	Mucor hiemalis (McCorkindale et al., 1969)	37,65,73
	M. dispersus (McCorkindale et al., 1969)	65,73
	M. rouxii (Safe, 1973)	60,65,70,71,73,79
	M. azygospora (Wieland, 1941)	73
	M. inaequisporus (Wieland, 1941)	73
	R. stolonifer (McCorkindale et al., 1969)	65,73
	R. arrhizus (Weete et al., 1973)	51,59,72,73
	R. japonicus (Wieland, 1941)	51
	Phycomyces blakesleeanus (Goodwin, 1973; McCorkindale et al., 1969; Goulston et al., 1967; Goulston and Mercer, 1969)	1,60,65,71,73
	Absidia glauca (McCorkindale et al., 1969)	65,73
	Blakeslea trispora (Goad et al., 1966)	73
	Linderina pennispora (Weete and Laseter, 1974)	51,82

TABLE 9.2   Continued

Fungi		Sterols
Class	Species (Reference)	

ASCOMYCETES

	Saccharomyces cerevisiae D587-4B wild-type (Trocha et al., 1974)	1,51,53,54,56,59, 60,67,73
	S. cerevisiae (Nysl or Erg 2) (Trocha et al., 1974)	1,11,12,29,73
	S. cerevisiae (Nysl or Erg 2) (Tyoringa et al., 1974)	1,37,42,60,73
	S. cerevisiae (Ole 2,3,4) (Bard et el., 1974)	42,54,60,73
	S. cerevisiae (Longley et al., 1968)	42,73
	S. cerevisiae (Wieland, 1941)	54,55,60
	S. cerevisiae (Kato and Kawase, 1976)	1,10,17,42,60,73
	S. cerevisiae (Hays et al., 1977)	58
	S. cerevisiae (Erg 2) (Pierce et al., 1979) .	1,11,17,42,43,45, 49,53,54,56,67, 69,74,76
	S. cerevisiae (Erg 3) (Pierce et al., 1979)	1,11,17,41,42,43, 46,51,53,54,56, 59,60,67,70
	S. cerevisiae (Erg 5) (Pierce et al., 1979)	1,11,17,41,42,48, 51,52,54,58,60, 65,69,71
	S. cerevisiae (Pierce et al., 1978)	1,9,11,17,20,41, 42,54,60,67,70, 73,77
	Yeast (Ponsinet and Ourisson, 1965)	1,10,17
	Pullularia pullulans (Merdinger et al., 1968)	73,90
	Candida utilis (Morimoto et al., 1967)	10,17,51,73,81
	C. albicans (Hamilton and Miller, 1972)	73
	C. albicans (Van Den Bossche et al., 1978)	1,3,12,29,73
	Claviceps purpurea (Wieland and Benend, 1943; Barton and Cox, 1948; Tanret, 1908; Wieland and Coutelle, 1941)	51,59

TABLE 9.2   Continued

Fungi		Sterols
Class	Species (Reference)	
	Penicillium funiculosum (Chen and Haskens, 1963)	37
	P. expansum (Leroux and Gredt, 1978)	7,12,34,73
	P. claviforme (Weete and Laseter, 1974)	51,73
	Aspergillus fennelliae (Kim and Kwon-Chung, 1974)	37,73
	A. flavus (Rambo and Bean, 1974)	37,65,73
	A. fumigatus (Goodwin, 1973)	1,21,73
	A. fumigatus (Sherald and Sisler, 1975)	7,10,12,34,65,73
	A. parasiticus (Rambo and Bean, 1974)	37,65,73
	A. niger (Weete and Laseter, 1974; Barton and Bruun, 1951)	73,77
	Neurospora crassa (Morris et al., 1974)	54,75
	N. crassa (Elliott et al., 1974; Kushuaha et al., 1976; Renaud et al., 1976)	37,60,73
	N. crassa (Renaud et al., 1978)	1,3,7,10,22,51, 54,56,59,60,67, 70,73,74
	Spicaria elegans (Weete and Laseter, 1974)	73
FUNGI IMPERFECTI		
	Fusarium roseum (Tillman and Bean, 1970)	65,73,88
	Monilinia fructigena (Kato et al., 1975)	7,12,73
	Alternaria kikuchiana (Starratt, 1976)	1,7,8,21,73
	A. alternata (Seitz and Paukstelis, 1977)	73
BASIDIOMYCETES		
	Lenzites trabea (Villanveva, 1971)	7,51
	Scleroderma aurantium (Vrkoc et al., 1976)	73

TABLE 9.2   Continued

Fungi		Sterols
Class	Species (Reference)	
	Daedalea quercina (Tanahashi and Takahashi, 1966)	59,73
	Agaricus species (Holtz and Schisler, 1972; Goulston et al., 1972)	7,8,21,51,65,73
	Fomes applanatus (Pettit and Knight, 1962; Strigina et al., 1971)	51,61,65
	Polyporus paragamenus (Budzikiewicz et al., 1964)	65
	Coriolus sanguineus Fr. (*Trametes cinnabarina*, *Polyporus cinnabarina*) (Cambie and LeQuesne, 1966)	59,73
	Ustilago maydis (Sporidia) (Ragsdale, 1975)	7,12,29,51,59, 65,73
	U. maydis (teliospores) (Weete and Laseter, 1974)	37,51,73
	U. nuda teliospores (Weete and Laseter, 1974)	51,73,82
	Puccinia graminis (uredospores) (Hougan et al., 1958; Nowak et al., 1972)	37,51 or 82,90
	P. graminis (uredospores) (Weete and Laseter, 1974)	51,82
	P. striiformis (uredospores) (Weete and Laseter, 1974)	51,82
	Melampsora lini (uredospores) (Jackson and Frear, 1968)	82,85,88
	Cronartium fusiforme (aeciospores) (Carmack et al., 1976)	85
	C. fusiforme (basidiospores) (Weete and Kelley, 1977)	51,82,88
	Uromyces phaseoli (uredospores) (Lin et al., 1972)	82,85

[a]See Table 9.1 for sterol identifications.

STEROLS OF PHOTOSYNTHETIC PLANTS AND BACTERIA

Reviews of the occurrence of sterols illustrate the diversity of structure encountered in photosynthetic plants

(Goad, 1967; Bean, 1973; Weete, 1976; Patterson, 1971; Goodwin, 1973; Nes, 1977). Sitosterol, stigmasterol, and campesterol are generally the major higher plant sterols, and may be accompanied by cholesterol and others that vary with the species.

The occurrence of sterols in algae has been reviewed by Patterson (1971), Goodwin (1973), and Weete (1976). Until relatively recently, it was believed that prokaroytic organisms are incapable of producing sterols. However, sterols have been detected in some bacteria (Schubert et al., 1968) and blue-green algae (Reitz and Hamilton, 1968; deSouza and Nes, 1968). Major sterols of the blue-green algae analyzed are cholesterol and 24-ethyl sterols with Δ^7, $\Delta^{5,7}$, and/or Δ^{22} double bonds. Although detected in some species, sterols may be of limited occurrence in prokarotic organisms but there has been relatively little interest in them.

Eukaroytic algae produce a wide variety of sterols. Red algae (Rhodophyta) may produce cholesterol, 24-ethylcholesterol, or desmosterol as the predominant sterol depending on the species, but trans-22-dehydrocholesterol is the major sterol of Porphyridium cruentum. Fucosterol appears to be the principal sterol produced by members of the Chrysophyta (golden brown algae); although poriferosterol, the C-24 epimer of stigmasterol, and ergosterol are produced by Ochromonas malhamensis. Brassicasterol is the major sterol of some diatoms (Bacillariophyta). Ergosterol, chondrillasterol, and ergost-7-enol are the major sterols among thirteen others of Euglena gracilis (Eugenophyta). Fucosterol is also the major sterol of brown algae (Phaeophyta) and may be accompanied by low relative proportions of Δ^5 C_{28} and C_{29} sterols, some of which have a trans C-22 double bond. Clionasterol, the C-24 epimer of sitosterol, and 28-isofucosterol are the major sterols of the few Charophyta (stoneworts) algae studied.

The great diversity of sterols occurring in photosynthetic plants is illustrated in Table 9.3.

INTRODUCTION TO CAROTENOIDS

Carotenoids are pigmented tetraterpenes (C_{40}) consisting of eight isoprene units. Hydrocarbon carotenoids are called carotenes and those containing oxygen are called xanthophylls.

TABLE 9.3   Sterols of Some Photosynthetic Plants

Species (Reference)	Sterols[a] (%)															
	1	2	3	4	5	6	7	8	9	10	11	12	13	14	15	16
ALGAE																
Chlorella emersonii (Patterson et al., 1974)	–	–	–	–	–	–	–	–	–	–	0.3	–	0.8	0.2	0.9	0.4
C. ellipsoidea (Patterson et al., 1974)	–	–	–	–	–	–	–	–	–	–	–	–	–	–	–	–
Thallasiosira pseudonana (Orcutt and Patterson, 1975)	–	–	–	1.7	–	–	–	–	–	–	–	–	–	–	–	–
Chlamydomonas rheinhardi (Patterson, 1974)	–	–	–	–	–	–	–	–	–	–	–	–	–	–	–	–
MOSSES																
Brachythecium rivulare BSG (Catalano et al., 1976)	57.7	–	5.9	–	–	–	–	–	8.7	–	–	–	–	–	–	–
Scleropodium touretii (Catalano et al., 1976)	54.2	–	25.0	4.1	–	–	–	11.8	–	–	–	–	–	–	–	–
HIGHER PLANTS																
Phaseolus vulgaris (Brandt and Benveniste, 1972)	57	34	6	1	2	–	–	–	–	–	–	–	–	–	–	–
Pinus elliotti (Laseter et al., 1973)	48	9	7	3	–	2	1	16	5	3	1	tr	–	–	–	–

Species (Reference)	Sterols[a] (%)															
	17	18	19	20	21	22	23	24	25	26	27	28	29	30	31	32
ALGAE																
Chlorella emersonii (Patterson et al., 1974)	0.8	16.3	0.5	8.0	70.8	–	–	–	–	–	–	–	–	–	–	–
C. ellipsoidea (Patterson et al., 1974)	–	–	–	–	–	5.6	21.9	68	65.6	–	–	–	–	–	–	–
Thallasiosira pseudonana (Orcutt and Patterson, 1975)	–	–	36.0	–	–	–	–	–	–	39.4	14.6	5.7	–	–	–	–
Chlamydomonas rheinhardi (Patterson, 1974)	–	–	–	–	–	–	–	–	–	–	–	–	56	44	–	–
MOSSES																
Brachythecium rivulare BSG (Catalano et al., 1976)	–	–	–	–	–	–	–	–	–	–	–	–	–	27.6	0.1	–
Scleropodium touretii (Catalano et al., 1976)	–	–	–	–	–	–	–	–	–	–	–	–	–	–	–	0.7
HIGHER PLANTS																
Phaseolus vulgaris (Brandt and Benveniste, 1972)	–	–	–	–	–	–	–	–	–	–	–	–	–	–	–	–
Pinus elliotti (Laseter et al., 1973)	–	–	–	–	–	–	–	–	–	–	–	–	–	–	–	–

TABLE 9.3   Continued

a

1.	Sitosterol
2.	Stigmasterol
3.	Campesterol
4.	Cholesterol
5.	Isofucosterol
6.	Desmosterol
7.	Lophenol
8.	24-Methylene-lophenol
9.	Cycloeucalenol
10.	Cycloartenol
11.	24-Methylene cycloartenol
12.	24-Ethylidene-lophenol
13.	24-Dihydroobtusifoliol
14.	Obtusifoliol
15.	14α-Methyl-5α-ergost-8-enol
16.	14α-Methyl-5α-(24S)-stigmast-8-enol
17.	14α-Methyl-5α-ergosta-8,24(28)-dienol
18.	5α-Ergost-7-enol
19.	5α-Ergosta-7,22-dienol
20.	5α-Stigmast-7-enol
21.	Chondrillasterol
22.	Brassicasterol
23.	Ergost-5-enol
24.	Clionasterol
25.	Poriferasterol
26.	5α-Ergosta-7,22-dienol
27.	Ergosta-7,24(28)-dienol
28.	Fucosterol
29.	7-Dehydroporiferasterol
30.	Ergosterol
31.	31-Norcyclolaudenol
32.	Cyclolaudenol

There are over 300 known naturally occurring carotenoids that
are widely distributed in higher plants, algae, bacteria,
animals, and fungi. β-Carotene and lutein are the principal
carotenoids of leaves, but other plant parts such as flowers
and fruits may have different major carotenoid pigments.
The distribution of carotenoids among algae appears to follow
taxonomic lines.

The precise functions of carotenoids are not well-
defined but they are implicated in photosynthesis, photo-
protection, phototropism and "blue-light effects" (photo-
induction of carotenogenesis in blue light). For example,
the conjugated double bonds of carotenoids can prevent lethal
photosensitizations by quenching excited states of singlet
oxygen.

STRUCTURE AND NOMENCLATURE

Carotenes include acyclic and cyclic, highly unsaturated
hydrocarbons. There are several acyclic polyenes that are
early intermediates in carotenoid biosynthesis, some of
which are non-pigmented. These polyenes include phytoene,
phytofluene, ξ-carotene, neurosporene, and lycopene (Figure
9.2). Cyclic carotenes may have a single terminal ring
structure or a ring at both ends of the molecule. Cyclization at either end of the polyene molecule laeds to the
formation of an α- or β-ionone ring. α-Carotene is a common
carotenoid containing both ring structures (Figure 9.3a).
The conjugated olefinic bonds constitute the chromophoric
group that gives rise to the orange, yellow, and red colors.
Numbering priority follows the β > α > open end (ψ) sequence
with plain numerals as shown for phytoene (Figure 9.2) and
α-carotene (Figure 9.3a). Most carotenoids are assigned
trivial names that give little clue as to their structure.
A semi-systematic nomenclature often used is based on the
portion of the carotenoid molecule from C-7,8 to C-7',8'
having a conjugated double bond system as shown in Figure
9.3b, called "carotene." Naming of a carotenoid would indicate the terminal groups by α, β, or ψ, the degree of saturation, and other substituents as in xanthophylls. For example,
a semi-systematic name for lycopene would be ψ,ψ-carotene;
the name for β-zeacarotene would be 7,8-dihydro-β,ψ-carotene,
and the name for cryptoxanthin would be β,β-carotene-3-ol
(Figure 9.2).

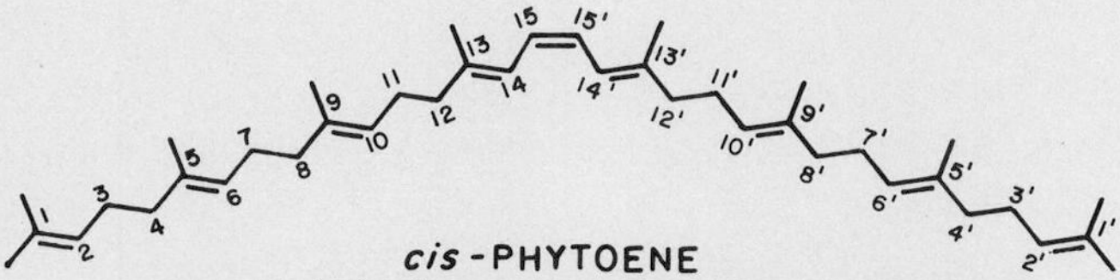

cis-PHYTOENE
(7,8,11,12,7',8',11',12'-OCTAHYDRO-ψ, ψ-CAROTENE)

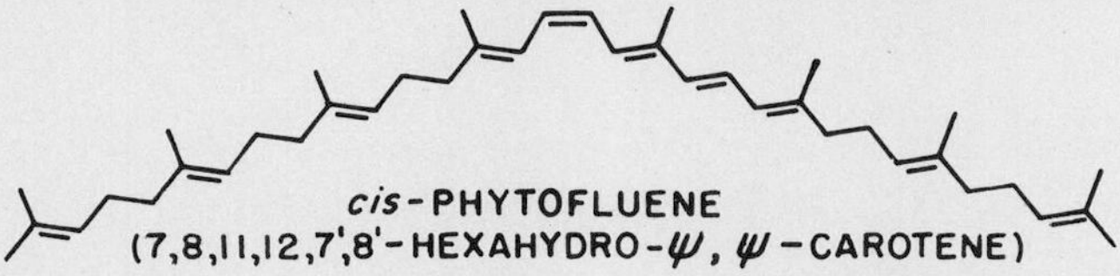

cis-PHYTOFLUENE
(7,8,11,12,7',8'-HEXAHYDRO-ψ, ψ-CAROTENE)

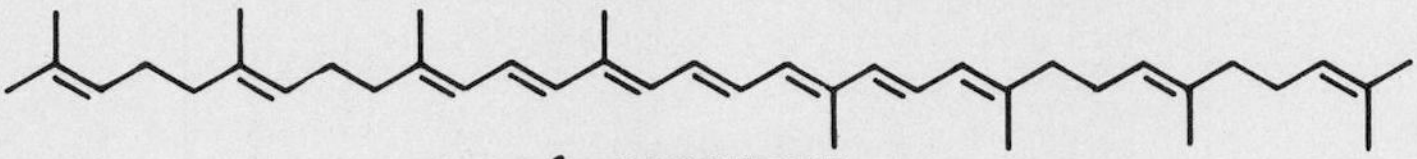

ζ-CAROTENE
(7,8,7',8'-TETRAHYDRO-ψ, ψ-CAROTENE)

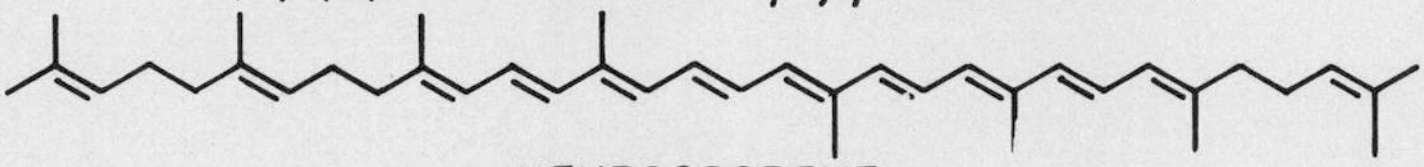

NEUROSPORENE
(7',8'-DIHYDRO-ψ,ψ-CAROTENE)

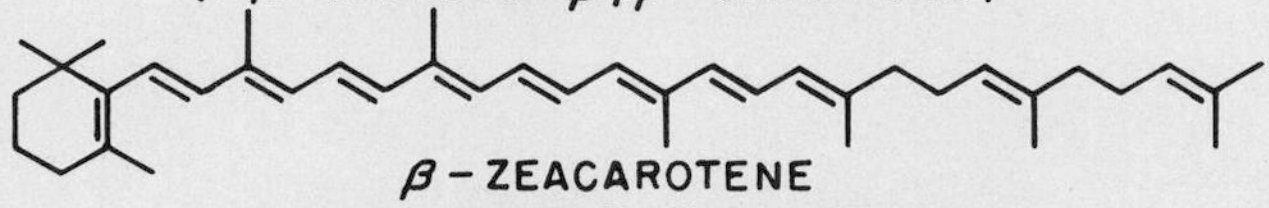

β-ZEACAROTENE
(7,8-DIHYDRO-β,ψ-CAROTENE)

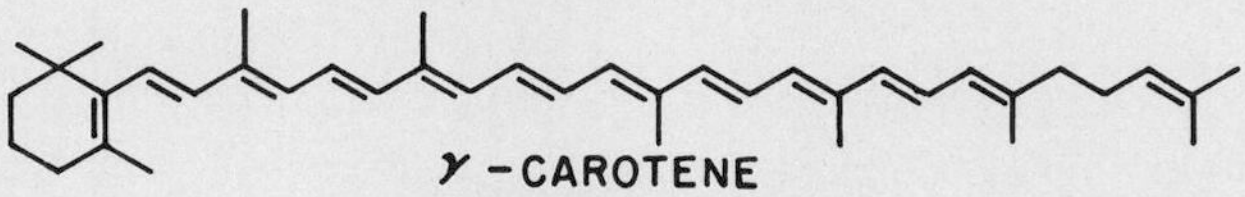

γ-CAROTENE
(β,ψ-CAROTENE)

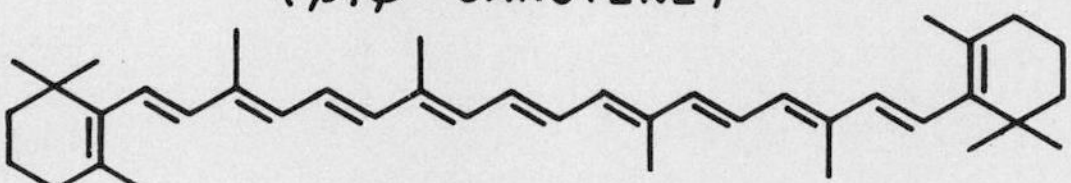

β-CAROTENE
(β,β-CAROTENE)

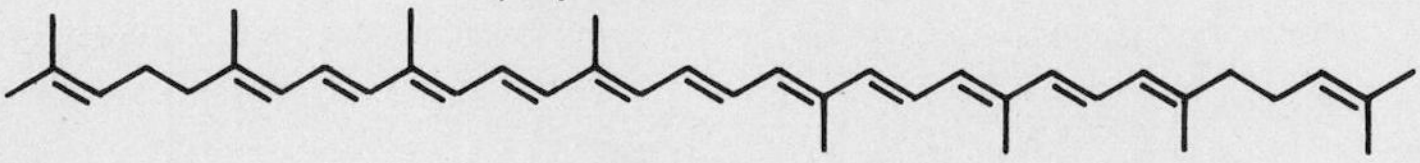

LYCOPENE
(ψ,ψ-CAROTENE)

Figure 9.2 Structures of some fungal carotenoids.

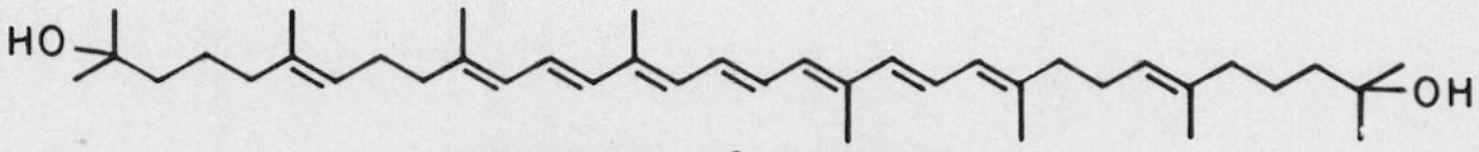

DIHYDROXY-ζ-CAROTENE
(ψ,ψ-CAROTENE-1,1'-DIOL)

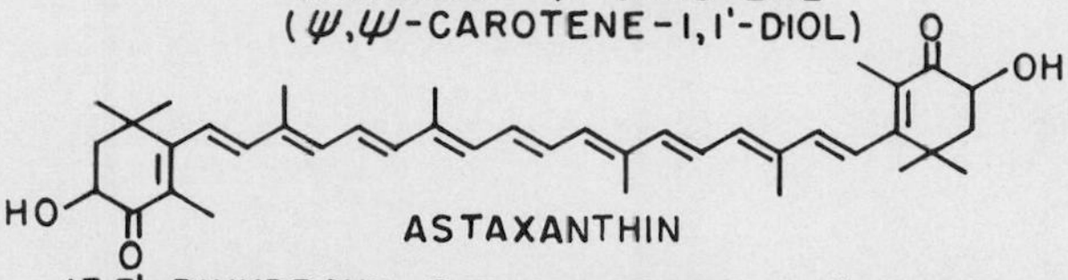

ASTAXANTHIN
(3,3'-DIHYDROXY-β,β-CAROTENE-4,4'-DIONE)

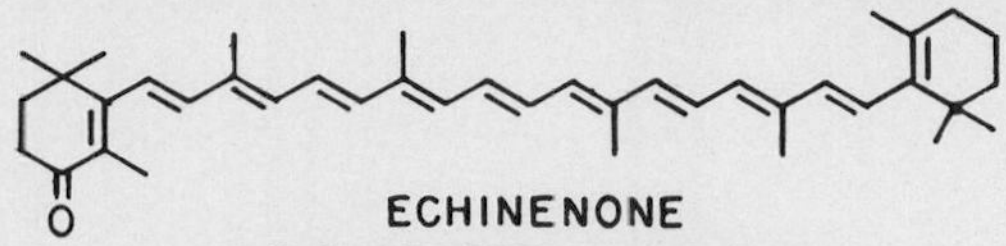

ECHINENONE
(β,β-CAROTENE-4-ONE)

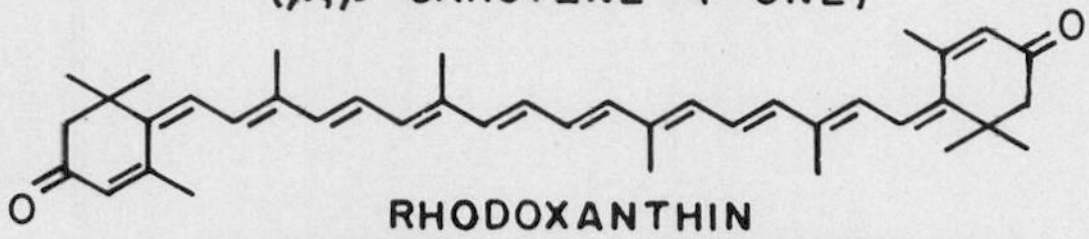

RHODOXANTHIN
(α,α-CAROTENE-3,3'-DIONE)

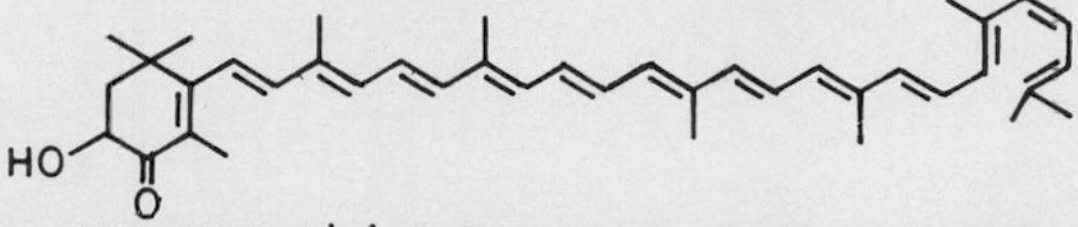

(3-HYDROXY-3',4'-DIDEHYDRO-β,ψ-CAROTENE-4-ONE)

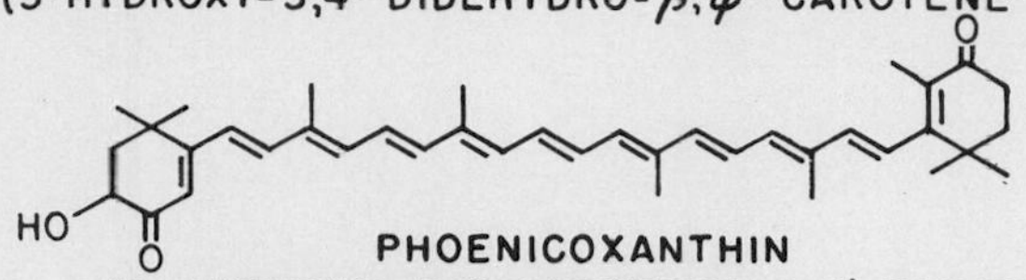

PHOENICOXANTHIN
(3-HYDROXY-β,β-CAROTENE-4,4'-DIONE)

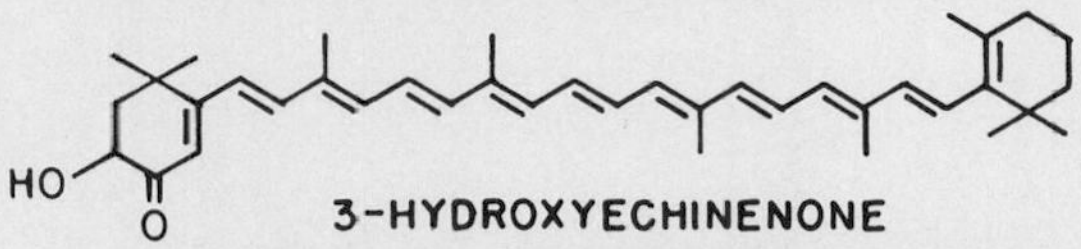

3-HYDROXYECHINENONE
(3-HYDROXY-β,β-CAROTENE-4-ONE)

Figure 9.2 Continued

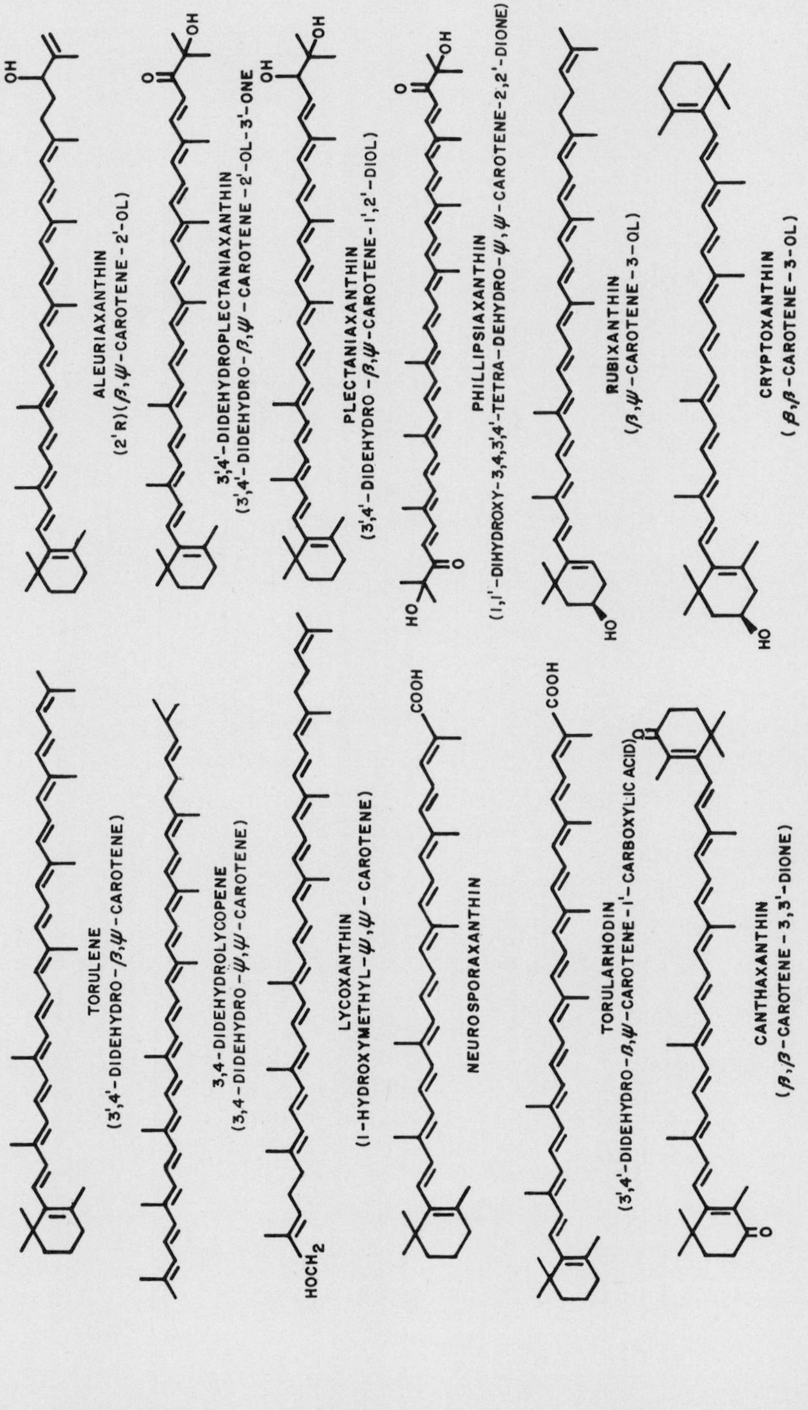

ALEURIAXANTHIN
(2'R)(β,ψ-CAROTENE-2'-OL)

3',4'-DIDEHYDROPLECTANIAXANTHIN
(3,4'-DIDEHYDRO-β,ψ-CAROTENE-2'-OL-3'-ONE)

PLECTANIAXANTHIN
(3,4'-DIDEHYDRO-β,ψ-CAROTENE-1',2'-DIOL)

PHILLIPSIAXANTHIN
(1,1'-DIHYDROXY-3,4,3',4'-TETRA-DEHYDRO-ψ,ψ-CAROTENE-2,2'-DIONE)

RUBIXANTHIN
(β,ψ-CAROTENE-3-OL)

CRYPTOXANTHIN
(β,β-CAROTENE-3-OL)

TORULENE
(3',4'-DIDEHYDRO-β,ψ-CAROTENE)

3,4-DIDEHYDROLYCOPENE
(3,4-DIDEHYDRO-ψ,ψ-CAROTENE)

LYCOXANTHIN
(1-HYDROXYMETHYL-ψ,ψ-CAROTENE)

NEUROSPORAXANTHIN

TORULARHODIN
(3',4'-DIDEHYDRO-β,ψ-CAROTENE-1'-CARBOXYLIC ACID)

CANTHAXANTHIN
(β,β-CAROTENE-3,3'-DIONE)

Figure 9.2   Continued

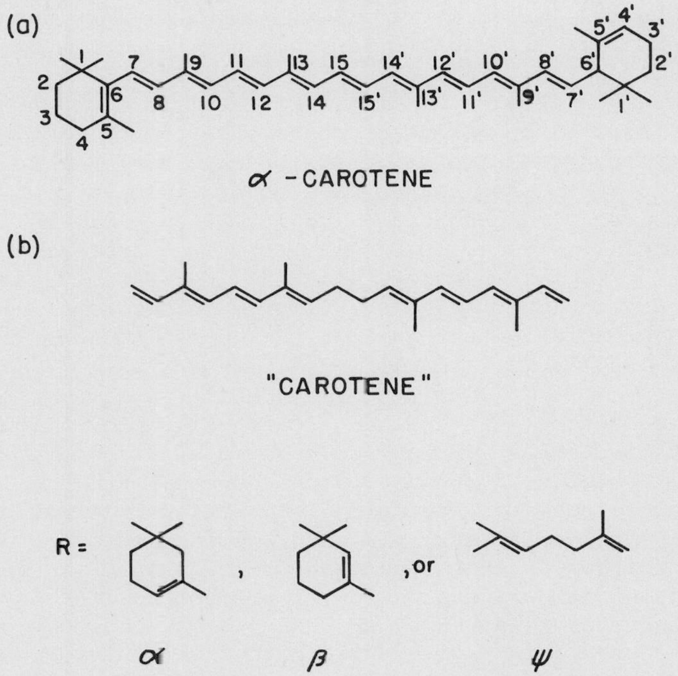

(a)

α -CAROTENE

(b)

"CAROTENE"

R =

α β ψ

Figure 9.3   a) α-Carotene, b) Structure that forms the
basis of the semi-systematic nomenclature of carotenoids.

Xanthophylls may contain hydroxyl, epoxy, carbonyl, or
carboxyl functions. The structures of some carotenoids
found in fungi are given in Figure 9.2.

OCCURRENCE IN FUNGI

Carotenoid levels vary in fungi depending on the species
and growth conditions. Carotenoids generally range from 0 to

30 micrograms per gram (μg/g) dry weight of most fungi, but
may range from 1100 to 1700 in some species. Two of the most
prolific carotenoid producing fungi are *Phycomyces blakesleeanus* and *Blakeslea trispora*. For example, sporangiophores
of a *P. blakesleeanus* mutant contain carotenoids up to 1100
μg/g dry weight (Meissner and Delbruck, 1968). More recently,
strains of *P. blakesleeanus* have been developed that produce
up to 25,342 μg/g β-carotene and 14,600 μg/g dry weight lycopene (Murillo et al., 1978). *B. trispora* can be induced to
produce up to 142 mg carotenoids (92% β-carotene) per 100 ml
growth medium (Ciegler et al., 1964). Other fungi with a
high carotenoid content include the rusts (spores) (Irvine
et al., 1954).

The factors that influence carotenoid production have
been studied in many fungi and have been reviewed by Goodwin
(1972). Carotenoid production can be stimulated by light,
chemicals, regulatory mutations and sexual interaction of
mycelia of opposite sex. In many species, carotenoid synthesis follows a similar pattern; active synthesis to maximum concentration and a period of pigment persistence with
little change, followed by gradual pigment degradation.
Carotenogenesis is favored by a high carbon/nitrogen (C/N)
ratio, 25 C to 30 C, light, and a pH drop from 5.2-7.6 to
2.6-3.0 during growth. The best carbon source for carotenogenesis varies depending on the species, but glucose is often
used in culture media and asparagine is the usual nitrogen
source.

The most widely studied inhibitor of carotenoid synthesis in fungi is diphenylamine, which blocks the conversion
(desaturation) of phytoene to phytofluene. Other inhibitors
include nicotene and 2-(p-chloro-phenylthio)triethylamine
(CPTA). Carotenoid production is stimulated when (+) and
(−) strains of some heterothallic fungi are cultured together.
Sexual reproduction is preceded by intense β-carotene synthesis. β-Carotene is a precursor of trisporic acid (Figure
9.4) which is produced by the (+) strain and appears to
derepress synthesis of the enzyme that catalyzes phytoene
formation. Carotene synthesis may be used as a marker determining the initiation of differentiation in *B. trispora*
(Feofilova and Pivovarova, 1976).

Light also stimulates carotenoid production in fungi that
produce carotenoids in the dark. However, in fungi such as
Neurospora crassa and *Fusarium aquaeductuum*, light, in the

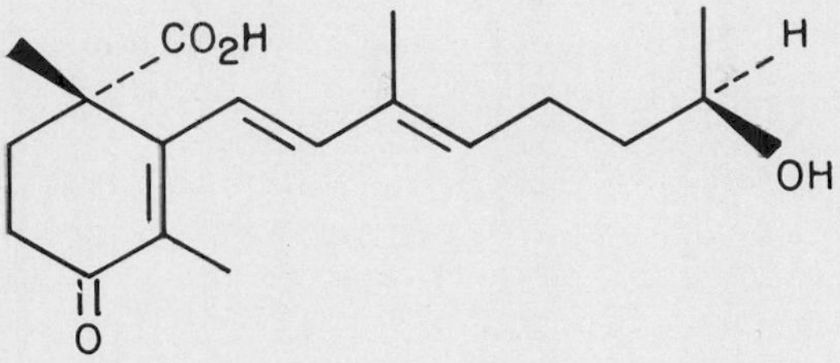

Figure 9.4   Trisporic acid.

presence of oxygen, serves as a photoinducer of carotenogenesis. Studies using inhibitors suggest that these factors
cause induction at the transcription level. Phytoene accumulates in the dark and disappears upon illumination. It appears that both phytoene synthesis and dehydrogenation (conversion to pigmented carotenoids) are photoregulated independently (Lansbergen et al., 1976). The action spectrum
for carotenogenesis has maxima at 450 and 481 nm in the visible range of the electromagnetic spectrum and 280 and 370 nm
in the ultraviolet (Defabo et al., 1976). This action spectrum resembles the absorption spectrum of β-carotene which
has been proposed as both a photoreceptor and regulator of
carotenoid biosynthesis.

    P. blakesleeanus grown in the presence of vitamin A may
produce carotenoids up to 2000 µg/g dry weight (Eslava et
al., 1974), and up to 4000 µg/g when grown in the presence
of β-ionone (Lilly et al., 1960). A large number of different carotenoids occur in fungi and are not restricted to
certain taxa. Carotenoids have been identified in over 200
fungal species and they are reportedly absent from another 135
species. Valadon (1976) has cautioned that some of the early·
carotenoid identifications based on only absorption spectra
and co-chromatography are in error.

    Goodwin (1972) has made the following generalizations
concerning the distribution of carotenoids in fungi: (a)
many fungi produce only carotenes (and not xanthophylls) and
this is particularly true for Phycomycetes, (b) β-carotene is

widely distributed and is generally the principal carotene
present; but it is not present in some Chytridrales and Blas-
tocladiales; γ-carotene is the major carotene in these fungi
and is more widely distributed in fungi than in any other
main group of carotenogenic plants, (c) torulene appears in
red yeasts and certain Ascomycetes, but rarely in Basidiomycetes and apparently never in Phycomycetes, (d) no carotenoids with α-ionone rings have been detected in fungi (α-carotene has been reported but not confirmed), (e) no xanthophylls, characteristic of green tissues, 3-hydroxy derivatives, 5,8-epoxides, or allenes have been clearly shown in
fungi, (f) characteristic fungal xanthophylls are carboxylic
acids involving oxidation of one of the *gem*-methyl groups
(e.g. torularhodin), (h) unique xanthophylls that occur in
Discomycetes, such as phillipsiaxanthin and plectaniaxanthin
and their derivatives, resemble carotenoids of photosynthetic
bacteria, (i) ketocarotenoids are occasionally found in fungi,
(j) acetylenic carotenoids have not been reported in fungi.

    Valadon (1976) has discussed the taxonomic implications
of fungal carotenoids and cited taxa where they have been
used successfully in this regard. For example, species of
the phycomycetous order Mucorales have β-carotene as the predominant carotenoid while species of Chytridiales and Blasto-
cladiales have γ-carotene. An easy way to distinguish species
of the lower Ascomycetes *Protomyces* and *Taphrina* which appear
very similar in culture, is on the basis of the presence or
absence of carotenoids. *Protomyces* has β-carotene as the
predominant carotenoid and *Taphrina* contain no carotenoids.
The presence of carotenoids has been used as the basis of
forming a new family of Discomycetes, the Aleuriaceae. Also,
the distribution of carotenoids can be used to separate species of the yeast genera *Rhodotorula* and *Cryptococcus*. The
former genera contains torulene or torularhodin, while the
latter contains β-carotene as the major carotenoid. Species
of the subgenera *Cantharellus* can be differentiated from
Phaeocantharellus on the basis of β-carotene being the major
pigment of the former and neurosporene or lycopene or both
being predominant in the latter.

    The principal carotenoid of the red-pigmented yeast
Phaffia rhodozyma is astaxanthin (3,3'-dihydroxy-β,β-caro-
tene-4,4'-dione), representing 83.9% of the total carotenoids
(Andrewes et al., 1976). This xanthophyll has the 3R,3'R
configuration which is the opposite of that of the same pigment from non-fungal sources (Andrewes and Starr, 1976).

Numerous carotenoids have been identified in species of
the genera *Amanita*, *Suillus*, and *Boletus*. Eight *Amanita*
species analyzed contain 6 to 11 different carotenoids, mainly
xanthophylls (Czeczuga, 1976). Zeaxanthin is one of the
principal carotenoids of *A. citrina*, *A. muscaria*, *A. panthe-
rina*, and *A. vaginata*. Neurosporaxanthin is the major carotenoid of *A. spissa* and *A. verna*; capsanthin is the major
carotenoid of *A. rubescens*, and unidentified pigments are
predominant in *A. phallorides*.

Boletus species analyzed contain 5 to 11 carotenoids,
ranging from ca. 1.9 to 4.3 µg/g fresh weight of the fungal
tissue and consists of mainly xanthophylls (Czeczuga, 1978).
Each species contains a widely different carotenoid composition (Table 9.4). Twenty-three carotenoids are produced by
5 *Suillus* species that range from 0.3 to 0.4 µg/g fresh
weight of the fungal material (Czeczuga, 1977). The carotenoid composition of fungi representing the major fungal
taxa is given in Table 9.4.

POLYPRENOLS

Polyprenols (poly-isoprene-ol) are high molecular weight
polyisoprenoid substances with 5 to 24 isoprene residues.
They may have all *trans* double bonds, but most have both *cis*
and *trans* (not more than 4) bonds. They may have 1 to 3
saturated isoprene units and an exomethylene dihydro-isoprene
unit. Polyprenols have the following general structure:

$$H \left[CH_2\underset{\underset{CH_3}{|}}{C} = CHCH_2 \right]_N OH$$

They are widely distributed in nature and usually occur
as groups, or families, the members of which differ by the
number of *cis*-isoprene residues, and are given trivial names
that give little clue of their structure. A system of naming
polyprenols has been suggested by Hemming (1974) (Table 9.5).
A particular family of polyprenols is defined by Arabic numerals indicating the range of isoprene residues making up
95% of the mixture. Details of the molecule are shown by
designating the relevant isoprene residues by Roman numerals
starting from the hydroxyl end of the chain and, in the case
of a family of polyprenols, using the smallest polyprenol in

TABLE 9.4 Carotenoids of Selected Fungi[a]

Species	Phytoene	Phytofluene	ξ-Carotene	Neurosporene	β-Zeacarotene	γ-Carotene	β-Carotene	Lycopene	Torulene	Torularhodin	Reference
PHYCOMYCETES											
Rhizophlyctis rosea						++		+			Davies (1961)
Blastocladiella spp.						++					cited in Valadon (1976)
Allomyces macrogynus						++					Emerson and Fox (1940)
Blakeslea trispora			+		+	+	+	+			Anderson et al. (1958)
B. trispora	+			+		+	++	+			Feofilova and Redikina (1975)
Mucor hiemalis				+	+	+	++	+			Herber (1974) cited in Valadon (1976)
M. azygospora	+	+	+	+	+	+	+	+			Van Eijk (1972)
M. inaequisporus	+	+	+	+	+	+	+	+			Van Eijk (1972)
Phycomyces blakes-leeanus			+	+	+	+	++	+			Goodwin (1952)
P. blakesleeanus	++	+	+		+	+	+	+			Aragon et al. (1976)
ASCOMYCETES											
Aspergillus giganteus						++	++	+			Zurzycka (1963)
Penicillium multicolor					+	+	+	+			Mase (1957)
Protomyces inundatus						+	++				Valadon (1964)
Neurospora crassa[b]				+		+	+		+	+	Zalokar (1957)

Organism	Markers	Reference
N. crassa[c]	++ + + +	Hughes and Subden (1973)
N. crassa[b]	+ + + + +	Renaud et al. (1976)
Aleuria aurantia[c,d]	+ + ++ +	Valadon and Mummery (1968)
Pyronema confluense	+ + +	Carlile and Friend (1956)
BASIDIOMYCETES		
Ustilago zeae	++ + +	cited in Goodwin (1972)
Gymnosporangium juniperi-virginianae	++ ++ +	Smits and Paterson (1942)
Puccinia graminis[g]	+ + + +	Hougen et al. (1958)
Cantharellus infundibuliformis[h]	+ + +	Valadon and Mummery (1975)
Pistillaria micans	+ + +	++ cited in Goodwin (1972)
Clitocybe venustissima	++ + + +	cited in Goodwin (1972)
Botetus betulicolus[o-v]	+ +	Czeczuga (1978)
B. edulis[w-z]	+ + +	Czeczuga (1978)
Amanita citrina[p,s,v,z, aa,bb,cc]	+ +	Czeczuga (1976)
A. muscaria[f,q,v,z,dd, ee,ff,gg]	+ +	Czeczuga (1976)
DEUTEROMYCETES		
Epicoccum nigrum[e,f]	+ + +	Foppen and Gribanovski-Sassu (1967)
		Gribanovski-Sassu and Foppen (1967)
Fusarium aqueductuum[f]	++ + +	Gribanovski-Sassu and Foppen (1968)
Verticillium albro-atrum	+ + + + +	Valadon and Mummery (1966)

TABLE 9.4 Continued

Species	Phytoene	Phytofluene	ζ-Carotene	Neurosporene	β-Zeacarotene	γ-Carotene	β-Carotene	Lycopene	Torulene	Torularhodin	Reference
Rhodotorula glutinis[e]	+	+	+	+		+	+		+		Valadon and Mummery (1969)
R. gracilus	+	+				+	+		++		Praus and Dyr (1958) cited in Goodwin (1972)
Sporidiololus johnsonii[e]						+	+	+	+		Fiasson (1967)
Phaffia rhodozyma[j-n]		+		+			+				Andrewes et al. (1976)

[a] + = carotenoid present; ++ = major carotenoid
[b] Neurosporaxanthin; 15,15'-cis-β-carotene (tentative)
[c] 3,4-Dehydrolycopene
[d] Rubixanthin, α-carotene, an aleuiaxanthin ester
[e] Torularhodin
[f] Rhodoxanthin
[g] Cryptoxanthin
[h] An epoxy carotenoid was predominant
[i] Several strains

[j] Echinenone
[k] 3-Hydroxyechinenone
[l] 3-Hydroxy-3',4'-didehydro-β,ψ-carotene-4-one
[m] Phoenicoxanthin
[n] Astaxanthin
[o] Celaxanthin
[p] 1'-Hydroxy-spirilloxanthin
[q] Rhodovibrin
[r] Rhodopin
[s] Rubixanthin
[t] Sarcinoxanthin
[u] Torularhodin

[v] Zeaxanthin
[w] Auroxanthin
[x] 5,6-Dihydro-5,6-dihydroxy-lycopene
[z] 1,2,1',2'-Tetrahydro-1,1'-dihydroxylycopene
[aa] Isocryptoxanthin
[bb] Isozeaxanthin
[cc] Isozeaxanthin
[dd] Saproxanthin
[ee] Flexixanthin
[ff] Capsanthin
[ff] Aleuriaxanthin ester
[gg] 2-Ketorhodovibrin

the range indicated. For example, the trivial name dolichol
refers to a family of widely occurring polyprenols described
by I-dihydro-XII,XIII-di*trans*-poly*cis*-prenols-14→18. This
expression indicates a mixture of polyprenols with 14 to 18
isoprene units with a saturated α-isoprene residue (I). The
12th and 13th isoprene units of the smallest polyprenol (14
isoprene units) of the mixture have the *trans* configuration.
Other components of the mixture also have two *trans*-isoprene
units adjacent to the ω-isoprene residue. Some families of
polyprenols isolated from various sources are given in Table
9.5.

Polyprenols (as phosphates) serve as intermediate carriers in the transfer of monosaccharides from sugar nucleotides to the oligosaccharide chains of glycoproteins and
proteoglycans. It appears that the polyprenol phosphates
aid in the movement of the sugar molecules through the lipid-
rich membrane, outside of which the polymer is formed. Polymer synthesis involving lipid-linked intermediates in bacterial, fungal, and mammalian systems has been reviewed by
Hemming (1974).

Polyprenols have been characterized in a few fungal
species. Dolichols-9 through -17 have been identified by mass
spectrometry in lipid extracts of the pythiaceous fungus
Phytophthora cactorum (Richards and Hemming, 1972). [^{14}C]
from MVA accumulates in dolichols-14 and -15 in this species.
Dolichols-14 through -18 have been detected in baker's yeast
(Dunphy et al., 1967) and possibly dolichol-22 has been tentatively identified in lipid extracts of *Aspergillus fumigatus* (Burgos et al., 1963). More extensive analyses of *A.
fumigatus* polyprenols using MS, NMR, and IR has resulted in the
identification of hexahydroprenols-18 through -24 (Stone et
al., 1967). Each homologue contains a saturated hydroxy-
terminal isoprene residue, a saturated ω-terminal isoprene
residue, and a saturated ψ-isoprene residue (penultimate).
These polyprenols have only two *trans*-isoprene residues of
unknown location.

TABLE 9.5   Nomenclature and Structure of Some Polyprenols[a]

Trivial Name	Structure[b]	Chemical Name
Betulaprenols	ω-T-T-C-[C]$_{1\rightarrow4}$-C-OH	IV,V-ditrans-,polycis-prenols-6→9 (ditrans-,polycis-prenols)
Bacterial prenols	ω-T-T-C-[C]$_{5\rightarrow7}$-C-OH	VIII,IX-ditrans-,polycis-prenols-10→12 (ditrans-,polycis-prenols)
Dolichols	ω-T-T-C-[C]$_{9\rightarrow13}$-S-OH	I-dihydro-,XII,XIII-ditrans-polycis-prenols-14→18
	ω-T-T-C-[C]$_{12\rightarrow16}$-S-OH	I-dihydro-,XV,XVI-ditrans-polycis-prenols-17→21 (dihydro-,ditrans-,polycis-prenols)
Ficaprenols, Castaprenols, etc.	ω-T-T-T-[C]$_{4\rightarrow8}$-C-OH	VI,VII,VIII-tritrans-,polycis-prenols-9→13 (tritrans-,polycis-prenols)
Hexahydropolyprenols	S-S-T-T-[C]$_{14\rightarrow18}$-S-OH	I,XVIII,XIX-hexahydro-,XVI,XVII-ditrans-polycis-prenols-19→23 (hexahydro-ditrans-,polycis-prenols)
Exo-methylene*-hexahydropolyprenols	S*-S-T-T-[C]$_{14\rightarrow18}$-S-OH	XIX exo-methylene-,I,XVIII,XIX-hexahydro-XVI,XVII-ditrans-,polycis-prenols-19→23 (methylene-,hexahydro-,ditrans-,polycis-prenols)

[a]Taken from Hemming (1974).   [b]T = trans isoprene residue, ω = omega isoprene residue, C = cis isoprene residue, S = saturated isoprene residue.

CHAPTER 10   BIOSYNTHESIS OF STEROLS, CAROTENOIDS,

AND POLYPRENOLS

INTRODUCTION TO STEROL BIOSYNTHESIS

Since the demonstration that acetate is the source of
carbon for cholesterol produced by rat liver tissue (Bloch,
1951) and ergosterol in fungi (*Neurospora*) (Ottke et al.,
1950), a large volume of information concerning sterol biosynthesis has been published, and various aspects have been
reviewed on several occasions (Cornforth, 1959; Wolstenholme
and O'Connor, 1959; Popjak and Cornforth, 1960; Heftmann and
Moseltig, 1960; Bloch, 1965; Clayton, 1965; Frantz and
Schroepfer, 1967; Goad, 1967; Goodwin, 1971). Early progress
(1950's) in establishing the pathway of sterol synthesis was
greatly facilitated by the availability of [^{14}C] labeled substrates, and an understanding of the reaction mechanisms was
made possible with the more recent synthesis of stereospecifically labeled mevalonic acid. The early studies were
conducted mainly with mammalian (rat liver) and fungal (yeast
and *Neurospora*) systems. Only more recently has interest in
the synthesis of sterols by higher plants and algae developed. In all organisms, terpenes are produced via the isoprenoid pathway which initially involves the condensation of
two pyrophosphorylated C_5 units, followed by the addition of
C_5 units to produce the various classes of terpenoids (Figure 10.1). A key intermediate in the formation of sterols is
squalene which is oxygenated and then cyclized to lanosterol
(animals and fungi) or cycloartenol (plants). The first
cyclic intermediates in sterol synthesis undergo certain
modifications that give rise to the predominant 4-desmethyl
sterols of a particular type of organism, such as cholesterol
in mammals, sitosterol in plants, and ergosterol in fungi.

261

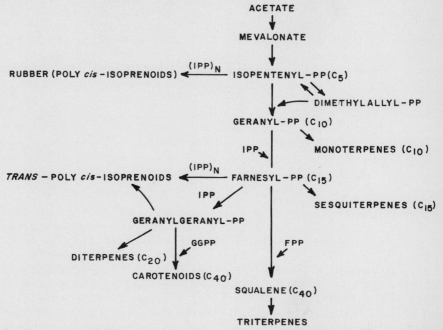

ACETATE

MEVALONATE

RUBBER (POLY *cis* -ISOPRENOIDS) ←—$(IPP)_N$—— ISOPENTENYL−PP(C_5)

DIMETHYLALLYL−PP

GERANYL−PP (C_{10})

IPP   MONOTERPENES (C_{10})

TRANS − POLY *cis*-ISOPRENOIDS ←—$(IPP)_N$—— FARNESYL−PP (C_{15})

IPP   SESQUITERPENES (C_{15})

GERANYLGERANYL−PP

GGPP   FPP

DITERPENES (C_{20})

CAROTENOIDS (C_{40})

SQUALENE (C_{40})

TRITERPENES

Figure 10.1   Outline of isoprenoid pathway.

Fungi, particularly yeast, have played an important role in studies of sterol biosynthesis.

FORMATION OF SQUALENE

The biosynthesis of polyisoprenoids, including squalene, has been recently reviewed by Beytia and Porter (1976). The C_{30} hydrocarbon squalene was first identified by Heilbron et al. (1926). The relationship of this hydrocarbon to sterol synthesis was identified when cholesterol levels increased in rats fed squalene (Channon, 1926). It was subsequently shown that squalene is synthesized from acetate and it is incorporated into cholesterol (Langdon and Bloch, 1953).

Mevalonic acid (MVA) is the first committed intermediate in polyisoprenoid, and hence sterol, biosynthesis (Wagner and Folkers, 1961). Various aspects of MVA formation have been studied in yeast and several other systems. MVA is produced from 3-hydroxy-3-methyl-glutaryl-CoA (HMG) which is synthesized

from acetate by one of three possible routes. In the first
pathway, the initial condensation reaction catalyzed by
β-ketoacyl-ACP snythetase may be associated with the fatty
acid synthetase in the cytosol, and involves the condensation of malonyl-CoA and acetyl-CoA (see Chapter 4) (Figure
10.2a). The product, acetoacetyl-CoA, then condenses with
acetyl-CoA yielding enzyme bound HMG or HMG-CoA. The second
condensation is catalyzed by a soluble enzyme not associated
with the fatty acid snythetase (Figure 10.2c). The synthesis of polyisoprenoids via this pathway has been observed in
the fungi *Blakeslea trispora* (Neujahr and Bjord, 1970) and
baker's yeast (Higgins and Kekwick, 1969).

The second, and most important, pathway of HMG-CoA synthesis involves the sequential condensation of three acetyl-
CoA molecules (Figure 10.2b and 10.2c). The two key enzymes
in this pathway, acetoacetyl-CoA thiolase (β-ketothiolase)
and HMG-CoA synthetase, are present in the mitochondria and
cytosol. They have been studied in detail in mammalian systems, and the synthetase has been studied in yeast (Ferguson
and Rudney, 1959; Stewart and Rudney, 1966). Kinetics of
the reaction catalyzed by β-ketothiolase favors the breakdown of acetoacetyl-CoA, but since the condensation product
is rapidly converted to HMG-CoA the forward direction pre-
dominates. The synthetase accepts acetyl-CoA first, re-
leases HMG-CoA last, and the CoA of HMG-CoA is derived from
acetoacetyl-CoA rather than acetyl-CoA. β-Ketothiolase and
the synthetase in the cytosol are involved in sterol synthesis whereas those in the mitochondria (mammalian systems)
are involved in ketogenesis.

A third pathway of HMG-CoA synthesis has been proposed
to explain the incorporation of radiolabeled leucine and
valine into carotenoids by fungi such as *Phycomyces blakesleeanus*, *Rhodotorula shibatana* and possibly *Blakeslea
trispora* (Goodwin and Lijinsky, 1952; Ishii, 1952). In this
pathway, the deamination product of leucine is decarboxylated
and converted to isovaleryl-CoA which is dehydrogenated to
dimethylacrylyl-CoA. This substance is carboxylated at the
terminal methyl group to form 3-methylglutaconyl-CoA which
is hydrated to HMG-CoA as outlined by Goodwin (1972) (Figure 10.2d). The relative importance of this pathway in providing HMG-CoA for sterol synthesis is not known.

The reaction in which HMG-CoA is converted to MVA is
one of the most important reactions in sterol biosynthesis,

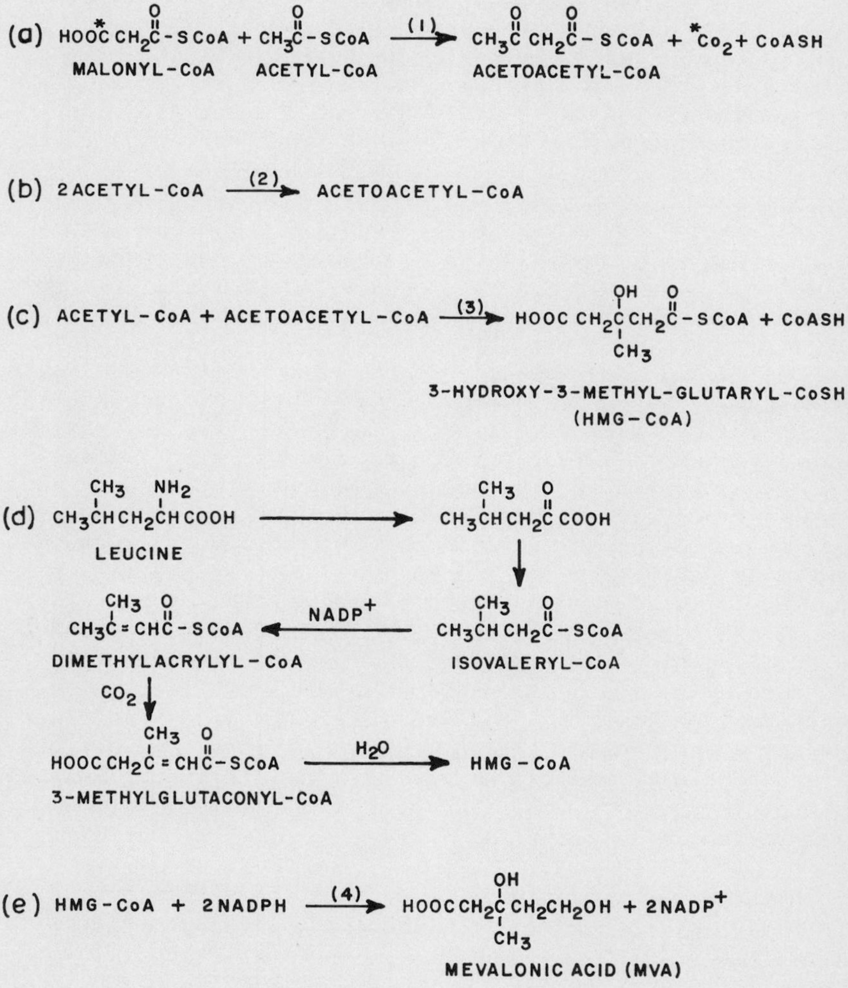

(a) $\overset{*}{HOOC}\,CH_2\overset{O}{\overset{\|}{C}}-SCoA + CH_3\overset{O}{\overset{\|}{C}}-SCoA \xrightarrow{(1)} CH_3\overset{O}{\overset{\|}{C}}\,CH_2\overset{O}{\overset{\|}{C}}-SCoA + \overset{*}{C}O_2 + CoASH$

     MALONYL-CoA   ACETYL-CoA   ACETOACETYL-CoA

(b) $2\,ACETYL-CoA \xrightarrow{(2)} ACETOACETYL-CoA$

(c) $ACETYL-CoA + ACETOACETYL-CoA \xrightarrow{(3)} HOOC\,CH_2\overset{OH}{\underset{CH_3}{C}}CH_2\overset{O}{\overset{\|}{C}}-SCoA + CoASH$

                                  3-HYDROXY-3-METHYL-GLUTARYL-CoSH

                                      (HMG-CoA)

(d) $CH_3\overset{CH_3}{\overset{|}{CH}}CH_2\overset{NH_2}{\overset{|}{CH}}COOH \longrightarrow CH_3\overset{CH_3}{\overset{|}{CH}}CH_2\overset{O}{\overset{\|}{C}}COOH$

     LEUCINE

$CH_3\overset{CH_3}{\overset{|}{C}}=CH\overset{O}{\overset{\|}{C}}-SCoA \xleftarrow{NADP^+} CH_3\overset{CH_3}{\overset{|}{CH}}CH_2\overset{O}{\overset{\|}{C}}-SCoA$

DIMETHYLACRYLYL-CoA   ISOVALERYL-CoA

$CO_2\downarrow$

$HOOC\,CH_2\overset{CH_3}{\overset{|}{C}}=CH\overset{O}{\overset{\|}{C}}-SCoA \xrightarrow{H_2O} HMG-CoA$

3-METHYLGLUTACONYL-CoA

(e) $HMG-CoA + 2\,NADPH \xrightarrow{(4)} HOOC\,CH_2\overset{OH}{\underset{CH_3}{C}}CH_2CH_2OH + 2\,NADP^+$

                                  MEVALONIC ACID (MVA)

(1)   ACYL-ACP:MALONYL-ACP LIGASE (CONDENSING ENZYME)

(2)   ACETYL-CoA:ACETYL-CoA C-ACETYTRANSFERASE (β-KETOTHIOLASE)

(3)   3-HYDROXY-3-METHYLGLUTARYL-CoA ACETOACETYL-CoA-LYASE (HMG-CoA SYNTHETASE)

(4)   MEVALONATE:NADP OXIDOREDUCTASE (HMG-CoA REDUCTASE)

Figure 10.2   Biosynthesis of HMP-CoA and MVA.

particularly in mammalian systems, and has been intensely studied. The enzyme catalyzing this reaction, HMG-CoA reductase, appears to be the principal site of regulation in the synthesis of sterols (see below). HMG-CoA reductase is associated with the mitochondria of yeast (Shimizu et al., 1971) and endoplasmic reticulum of liver cells (Linn, 1967). In yeast, the reductase has a molecular weight of 260,000 to 270,000 and is composed of four identical subunits (Qureshi et al., 1971). The two-step reaction catalyzed by the yeast enzyme is believed to begin with the binding of NADPH and HMG-CoA to the reductase, followed by the formation of mevaldic hemithioacetyl which may be cleaved to mevaldate and CoASH (Qureshi et al., 1972). A second NADPH molecule binds to the enzyme and the products MVA and NADP$^+$ are released after reduction (Figure 10.2e). Lynen (1970) showed that mevaldate can act as an intermediate in MVA synthesis, but it has not been isolated as a free product of HMG-CoA reduction. A hydride ion from the A-side of the pyridine ring of NADPH is transferred in both reduction reactions (Dugan and Porter, 1971). $[3R,(5S)-5-^3H]$ MVA is produced when $[(4R)-4-^3H]$ NADPH is supplied for the second reduction (Blattmann and Retey, 1971). The 3R stereoisomer of MVA is the product of this reduction and natural intermediate for sterol biosynthesis (Figure 10.3a).

Squalene synthesis from MVA occurs in three steps: (1) Conversion of MVA to the C_5 isoprenoid building block (isoprene) precursor, isopentenyl pyrophosphate (IPP), (2) formation of farnesyl pyrophosphate (FPP) through the head-to-tail condensation of three C_5 units, and (3) formation of squalene via the tail-to-tail condensation of two FPP molecules.

The synthesis of IPP from MVA (Figure 10.3b) has been studied in several organisms, including yeast (Rilling and Bloch, 1959; de Waard et al., 1959; Bloch et al., 1959) and involves two sequential phosphorylations followed by a dehydration at C-3 and decarboxylation (see footnote). The enzymes catalyzing this reaction sequence are soluble. MVA kinase catalyzes the phosphorylation of MVA (Figure 10.4a) and has been studied in yeast (Tchen, 1958). This enzyme has a molecular weight of about 100,000, prefers ATP and

---

To follow the discussion of squalene formation from MVA it may be helpful to review the change in carbon numbering when MVA is converted to IPP (Figure 10.3a and 10.3b).

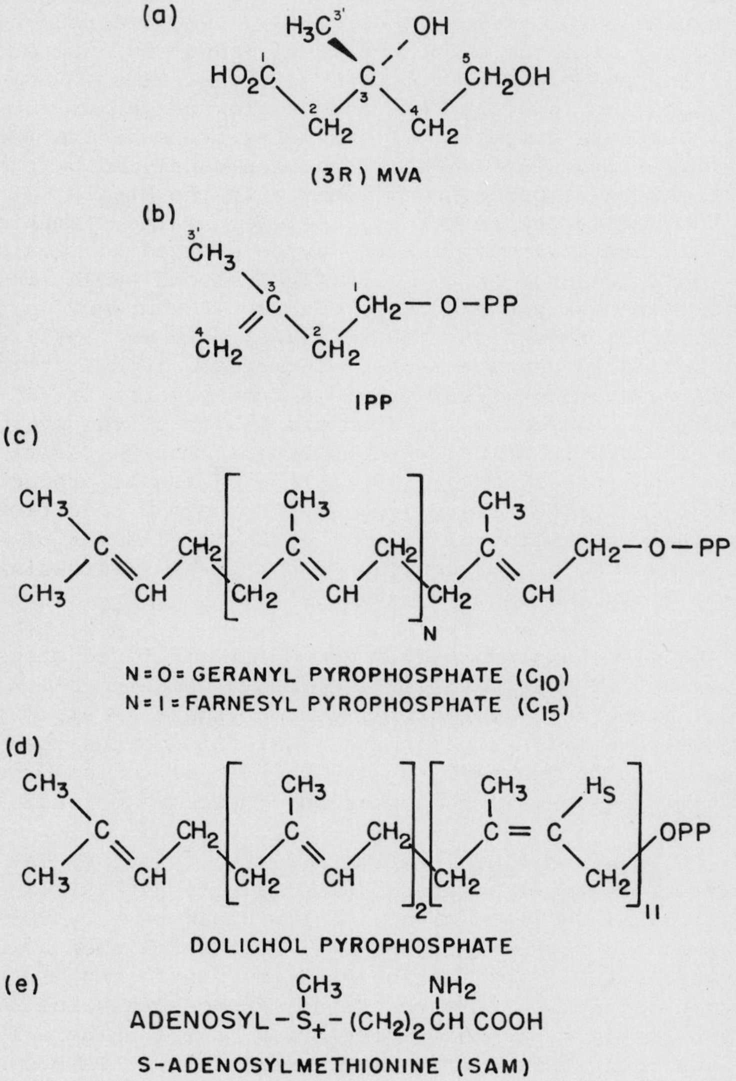

(a)

$$H_3C^{3'} \quad OH$$

$$HO_2C \overset{1}{} \quad \overset{3}{C} \quad \overset{5}{CH_2OH}$$

$$\overset{2}{CH_2} \quad \overset{4}{CH_2}$$

(3R) MVA

(b)

$$\overset{3'}{CH_3}$$

$$\overset{3}{C} \quad \overset{1}{CH_2} - O - PP$$

$$\overset{4}{CH_2} \quad \overset{2}{CH_2}$$

IPP

(c)

$$CH_3 \qquad CH_3 \qquad CH_3$$

$$C \quad CH_2 \quad \left[\begin{array}{c} C \quad CH_2 \end{array} \right] \quad C \quad CH_2 - O - PP$$

$$CH_3 \quad CH \quad [CH_2 \quad CH] \quad CH_2 \quad CH$$

$$N$$

N=0= GERANYL PYROPHOSPHATE (C$_{10}$)

N=1= FARNESYL PYROPHOSPHATE (C$_{15}$)

(d)

$$CH_3 \qquad CH_3 \qquad CH_3 \quad H S$$

$$C \quad CH_2 \quad \left[\begin{array}{c} C \quad CH_2 \end{array} \right] \quad \left[\begin{array}{c} C = C \end{array} \right] OPP$$

$$CH_3 \quad CH \quad [CH_2 \quad CH]_2 \quad [CH_2 \quad CH_2]_{11}$$

DOLICHOL PYROPHOSPHATE

(e)

$$\qquad\qquad CH_3 \qquad NH_2$$

$$ADENOSYL - \overset{}{S_+} - (CH_2)_2 \overset{}{C}H \; COOH$$

S-ADENOSYLMETHIONINE (SAM)

Figure 10.3   Structures of MVA, IPP, GPP, FPP, and dolichol
pyrophosphate.

Mg^{+2} but can be activated by Mn^{+2} or Ca^{+2}, has a pH optimum
between 5.5 and 7.8 depending on the source, and requires
the 3R-isomer of MVA. The second phosphorylation is catalyzed by 5-phosphomevalonate kinase (Figure 10.4b) which
has been studied in yeast (Bloch et al., 1959; Lynen et al.,
1958). The enzyme requires ATP and is activated by Mg^{+2} or
Mn^{+2}. The final reaction in the formation of IPP is catalyzed by 5-pyrophosphomevalonate decarboxylase (Figure 10.4c).
This reaction also requires ATP and the hypothetical intermediate 3-phospho-5-pyrophosphate mevalonate has been proposed (Bloch et al., 1959) but not confirmed unequivocally.
The reaction requires Mn^{+2} and involves a *trans* elimination
of the carboxyl group in the formation of IPP (Popjak and
Cornforth, 1966).

IPP undergoes isomerization to dimethylallyl pyrophosphate (DMAPP) which serves as the C_5 starter molecule for
the synthesis of isoprenoids. The role of DMAPP in isoprenoid synthesis may be considered analogous to that of
acetate in fatty acid synthesis (see Chapter 4). IPP contains the C_5 unit used as the building block in the formation of polyisoprenoids. DMAPP is formed by the enzymatically
catalyzed (IPP isomerase) isomerization of IPP (Figure
10.4d). The equilibrium ratio of this reaction is 1:9
(DMAPP:IPP) (Agranoff et al., 1960). This reaction involves
the addition of a proton from the medium to the *re* face
(Cornforth et al., 1972) of the C-4 (vinylic carbon) of IPP
(Birch et al., 1962) and loss of a proton from the C-2 of
IPP. Using stereospecifically labeled MVA, $[(4S)-4^3H]$ or
$[(4R)-4^3H]$, it has been shown that the proton lost from IPP
is the original 4S hydrogen of MVA (Cornforth, 1973).

The next two reactions in the biosynthesis of squalene
are catalyzed by prenyl transferases (prenyl-pyrophosphate
synthetases) and involve two consecutive head-to-tail condensations. First, IPP condenses with DMAPP giving rise to
geranyl-pyrophosphate (GPP) (Figure 10.5a). This reaction
is followed by the condensation of GPP with another molecule
of IPP giving rise to all-*trans* farnesyl pyrophosphate (FPP)
(Figure 10.5b). This reaction sequence is believed to be
catalyzed by a single C_{15}-specific prenyl transferase which
is soluble, requires Mg^{+2}, and has been studied in yeast
(Lynen et al., 1959; Eberhardt and Rilling, 1975). The
reactions appear to be initiated by a nucleophilic attack
(presumably by the enzyme, X^-) on IPP, followed by the addition of the allylic group to the *si* side of the vinylic

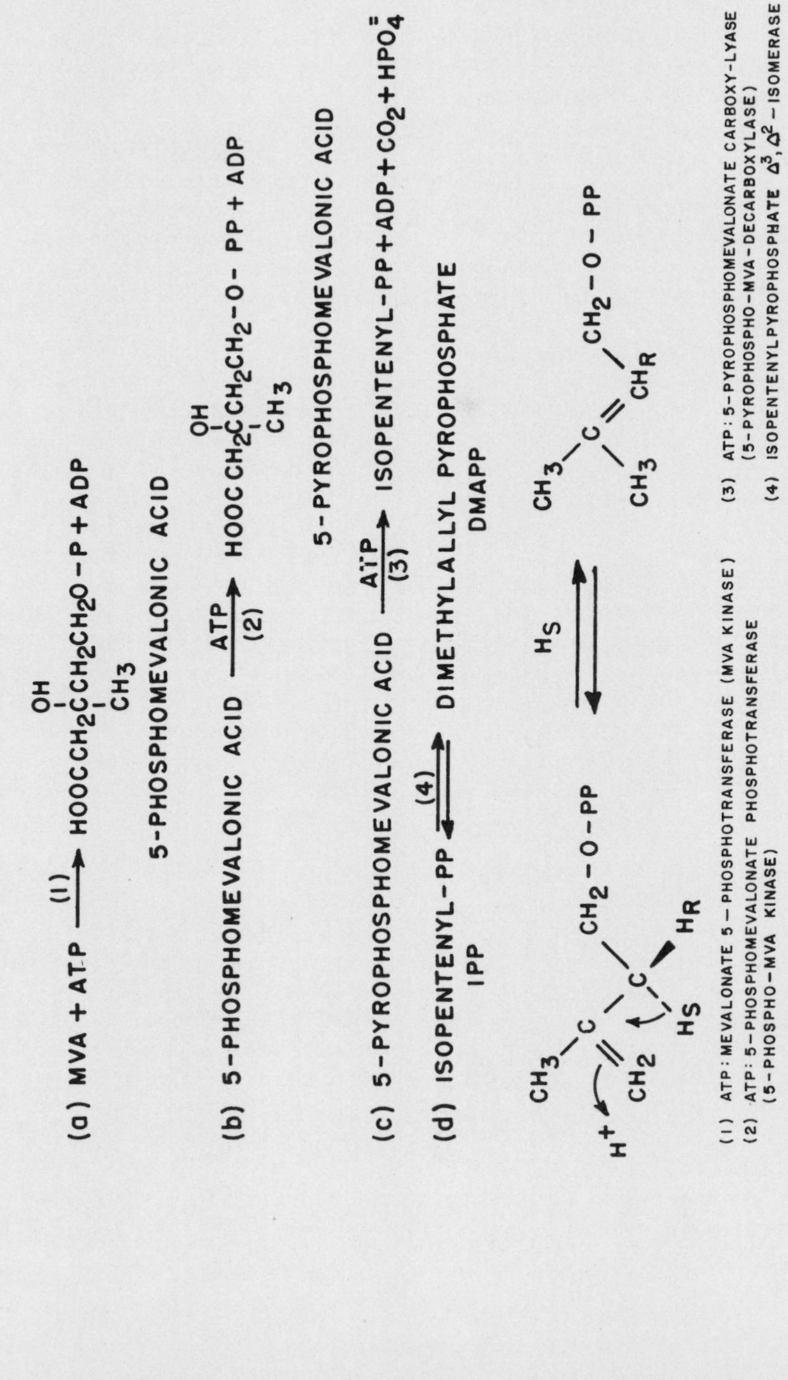
(a) MVA + ATP —(1)→ HOOCCH2C(OH)(CH3)CH2CH2O – P + ADP
5-PHOSPHOMEVALONIC ACID
(b) 5-PHOSPHOMEVALONIC ACID —ATP (2)→ HOOCCH2C(OH)(CH3)CH2CH2–O– PP + ADP
5-PYROPHOSPHOMEVALONIC ACID
(c) 5-PYROPHOSPHOMEVALONIC ACID —ATP (3)→ ISOPENTENYL-PP + ADP + CO2 + HPO4=
(d) ISOPENTENYL-PP ⇌(4) DIMETHYLALLYL PYROPHOSPHATE
IPP
DMAPP
H+
CH3
C
CH2
CH2 – O – PP
HS
HR
CH3
CH3
C
CHR
CH2 – O – PP
(1) ATP: MEVALONATE 5 – PHOSPHOTRANSFERASE (MVA KINASE)
(2) ATP: 5-PHOSPHOMEVALONATE PHOSPHOTRANSFERASE (5-PHOSPHO-MVA KINASE)
(3) ATP: 5-PYROPHOSPHOMEVALONATE CARBOXY-LYASE (5-PYROPHOSPHO-MVA-DECARBOXYLASE)
(4) ISOPENTENYLPYROPHOSPHATE Δ3,Δ2 – ISOMERASE

Figure 10.4 Conversion of MVA to IPP and DMAPP.

carbon. There is an inversion of the configuration of C-1 of IPP with formation of the C-C bond between the two condensing molecules (Cornforth et al., 1966) (Figure 10.5). Using doubly and stereospecifically labeled MVA, it has been shown that the 2-pro-S-hydrogen of IPP is eliminated in the formation of GPP and FPP as described for the IPP→DMAPP isomerization (Cornforth et al., 1966). This is accomplished using $[2-^{14}C,(4R)-4-^{3}H]$ and $[2-^{14}C,(4S)-4-^{3}H]$ MVA and measuring the isotopic ratio in squalene (see footnote). If the 2-pro-R-hydrogen of IPP (4S of MVA, see footnote) is lost during condensation, the $^{14}C/^{3}H$ ratio of the product GPP or FPP, and hence squalene (see below), would be expected to be the same as that of the substrate if $[2-^{14}C,(4R)-4-^{3}H]$ MVA is used. Indeed, the isotopic ratio of squalene is the same as the 4R-MVA substrate in both condensation reactions catalyzed by the C_{15}-specific prenyl transferase (Popjak and Cornforth, 1966) via a *trans* hydrogen elimination (Figure 10.5). Lynen and his associates (1958; 1959) have identified the C_{10} and C_{15} squalene synthesis intermediates in yeast, and have shown that they are formed as described above.

The final step in squalene synthesis is catalyzed by the microsomal enzyme squalene synthetase. This enzyme catalyzes the tail-to-tail condensation of two FPP molecules and apparently the reduction of an intermediate in a reaction requiring NADPH and Mg^{+2} (Figure 10.6a). In yeast, this enzyme occurs in an aggregate from which carries out both reactions, and a protomeric form with a molecular weight of 450,000 to 460,000 that catalyzes only the formation of the intermediate presqualene-pyrophosphate (Qureshi et al., 1972; 1973; Shechter and Bloch, 1971).

Several mechanisms for the condensation of two FPP molecules to squalene have been proposed, but the details have not yet been fully elucidated (Beytia and Porter, 1976). Rilling (1966) has isolated a stable C_{30} intermediate called "presqualene-pyrophosphate" from yeast squalene synthesizing subcellular particles deficient in NADPH. The structure of presqualene-pyrophosphate is 2-(2,6,10-trimethyl-1,6,9-undecatriene)-3-methyl-3-(4,8-dimethyl-3,7-nonadiene)-cyclopropylcarbinyl-pyrophosphate (Figure 10.6a) (Epstein and Rilling, 1970). Presqualene pyrophosphate is converted

---

The C-4 of MVA becomes the C-2 of IPP and the 2R hydrogen of IPP corresponds to the 4S hydrogen of MVA.

(a) ISOPENTENYL-PP + DIMETHYLALLYL-PP $\rightleftharpoons$ GERANYL-PP + PP

(b) GERANYL-PP + ISOPENTENYL-PP $\rightleftharpoons$ FARNESYL-PP + PP

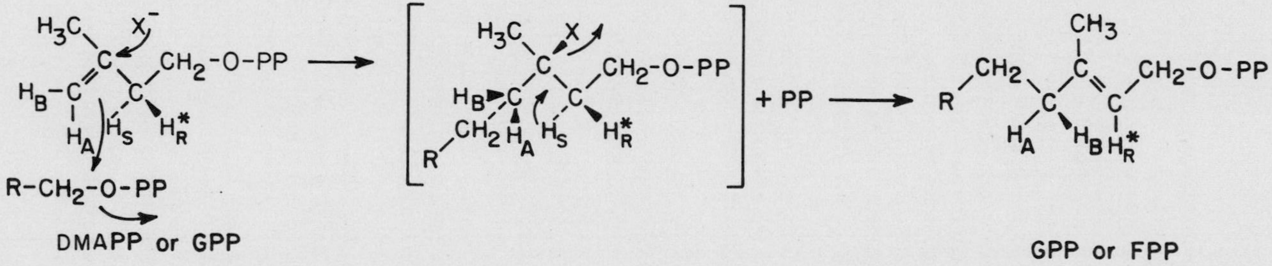

H_S = PRO-S-HYDROGEN OF MVA USED IN THE SYNTHESIS OF IPP

H_R = PRO-R-HYDROGEN OF MVA USED IN THE SYNTHESIS OF IPP

Figure 10.5 Biosynthesis of GPP and FPP.

to squalene by yeast subcellular particles in the presence
of NADPH and $MgCl_2$ (Altman et al., 1971).

Although there is no net loss of hydrogen from the C-1
of FPP during condensation, hydrogen exchange does occur in
the final reaction of squalene synthesis. The hydride ion
from the B-side (4-pro-S hydrogen of the reduced pyridine
ring) of NADPH is inserted at one of the central carbon
atoms of squalene (Cornforth et al., 1966; Popjak et al.,
1962) and occupies the pro-R-position (Cornforth et al.,
1963). Figure 10.6b illustrates the origin of hydrogen
atoms in the center of the squalene molecule. The pro-S-
hydrogen of the C-1 (pro-S-hydrogen of C-5 of MVA) of one of
the two FPP molecules is lost (Cornforth et al., 1966).
There is an inversion of the configuration at the C-1 of
the FPP that does not lose a hydrogen (Donninger and Popjak,
1966).

A cell-free extract of *Rhizopus arrhizus* has been reported as a very efficient system for producing squalene
(Campbell and Weete, 1977).

CONVERSION OF SQUALENE TO LANOSTEROL

In early studies of cholesterol biosynthesis it was
proposed that a hydroxylated form of squalene may be involved in the formation of the first cyclic intermediate in
4-desmethyl sterol synthesis. The oxygenated intermediate has
been subsequently identified as 2,3-oxidosqualene(2,3-epoxy-
squalene) which undergoes cyclization to lanosterol in animals and fungi, and cycloartenol in plants (Corey and Russey,
1966a; 1966b; van Tamelin et al., 1966; Benveniste and Massey-
Weotroop, 1967). Squalene is oxygenated by molecular oxygen;
this is followed by cyclization of the epoxy intermediate to
a postulated intermediate (Mulheirn and Caspi, 1971) as shown
in Figure 10.7. Cyclization is believed to be initiated by
an enzyme that catalyzes opening of the oxirane ring resulting in an electron deficiency at C-2. Conserted cyclization
follows. This is followed by a series of hydrogen and methyl
migrations giving rise to the first stable sterols and precursors to the major 4-desmethyl sterols of plants and animals. The enzyme 2,3-oxidosqualene-sterol cyclase has been
isolated from mammalian (Shechter et al., 1970) and fungal
systems (Shechter et al., 1970; Dean et al., 1967; Mercer
and Johnson, 1969). The functionally identical enzymes have

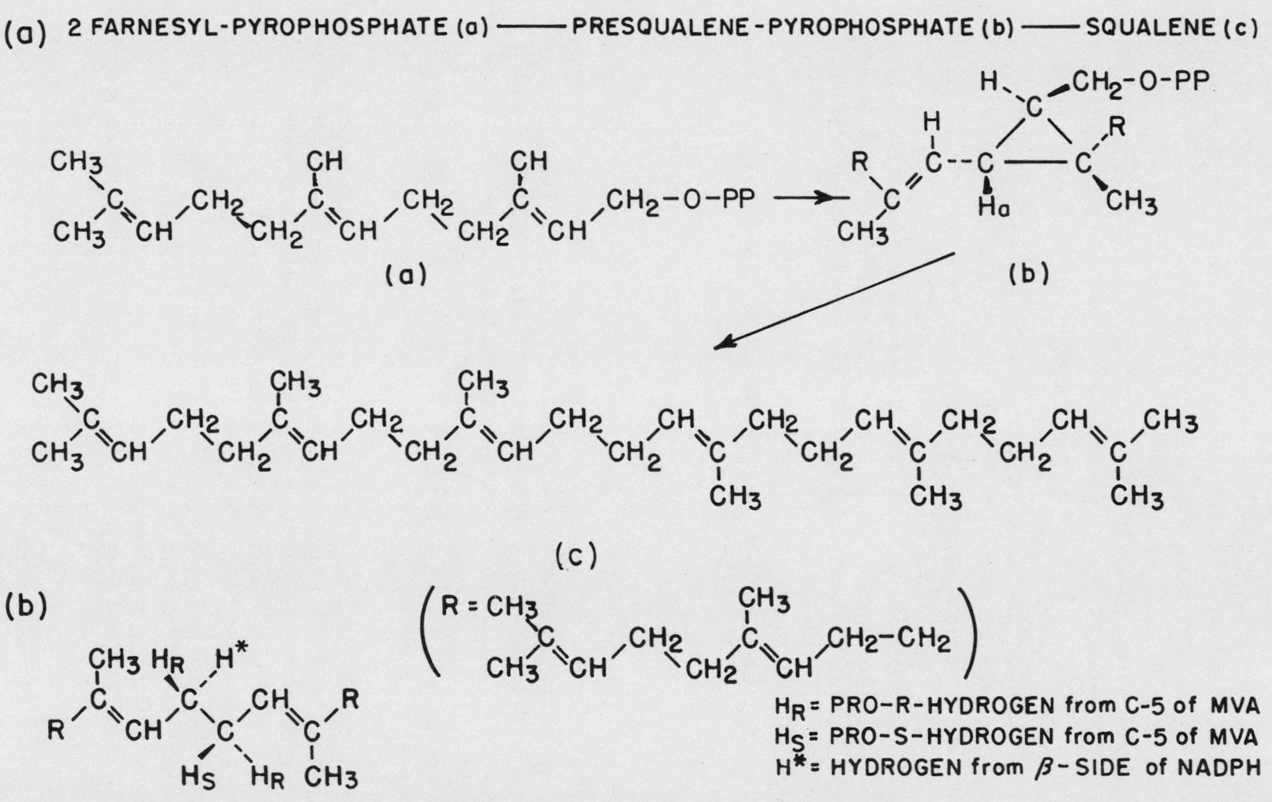
(a) 2 FARNESYL-PYROPHOSPHATE (a) —— PRESQUALENE-PYROPHOSPHATE (b) —— SQUALENE (c)
(a)
(b)
(c)
(b)
R = CH3
HR = PRO-R-HYDROGEN from C-5 of MVA
HS = PRO-S-HYDROGEN from C-5 of MVA
H* = HYDROGEN from β-SIDE of NADPH

Figure 10.6 Biosynthesis of squalene.

markedly different properties (Shechter et al., 1970). For example, the yeast enzyme is soluble and optimally active at low ionic strengths while the enzyme from rat liver is microsomal and active at high ionic strengths.

10,11-Oxidosqualene has been isolated from dark grown shake cultures of *Sclerotinia fructicola*, but has no biological activity (Katayama and Marumo, 1976).

As noted above, lanosterol is the first cyclic intermediate in ergosterol biosynthesis in fungi and cholesterol formation in animals, and its 9β,19-cyclopropane derivative, cycloartenol, is the first cyclic intermediate in phytosterol formation. However, lanosterol has been identified in a few higher plants (Goad, 1967), but cycloartenol has not been found in fungi (yeast) (Ponsinet and Ourisson, 1965). The formation of lanosterol in yeast has been demonstrated (Schwenk, 1955; Kodicek, 1959). Schwenk and Alexander (1958) have shown that it can be incorporated into ergosterol.

The formation of lanosterol from 2,3-oxidosqualene involves the loss of a proton from C-9 and migration of the methyl group at C-14 to the C-13 position giving rise to the C-18 carbon of the final product. The methyl group originally at C-8 fills the vacant C-14 position and becomes the 14α-methyl (Woodward and Block, 1953; Ruzicka, 1953; 1959; Mudgal et al., 1958; Cornforth et al., 1959; 1965) (Figure 10.7). In the formation of cycloartenol, the same methyl migrations occur, but instead of being lost, the C-9 hydrogen migrates to C-8 and a 9β,19-cyclopropane ring is formed with the loss of a proton from C-19. The loss and retention of hydrogen atoms in the formation of these two sterols can be demonstrated using stereospecifically labeled MVA. The carbon atom of squalene that becomes the C-9 of lanosterol originates from C-4 of MVA and is linked to the 4R hydrogen. Lanosterol produced biosynthetically from $[2-^{14}C-(4R)-4-^3H_1]$ MVA has a $^{14}C/^3H$ ratio of 6/5, as would be expected with the loss of a hydrogen from C-9. Cycloartenol produced similarly has a 6/6 isotopic ratio which is consistent with retention of a hydrogen at C-9 (Goodwin, 1971). It has also been shown that cycloartenol is a direct cyclization product rather than an isomerization product of lanosterol (Goad and Goodwin, 1969; Rees et al., 1968).

As noted in Chapter 9, pythiaceous fungi (*Pythium* and *Phytophthora* species) are incapable of producing sterols, in

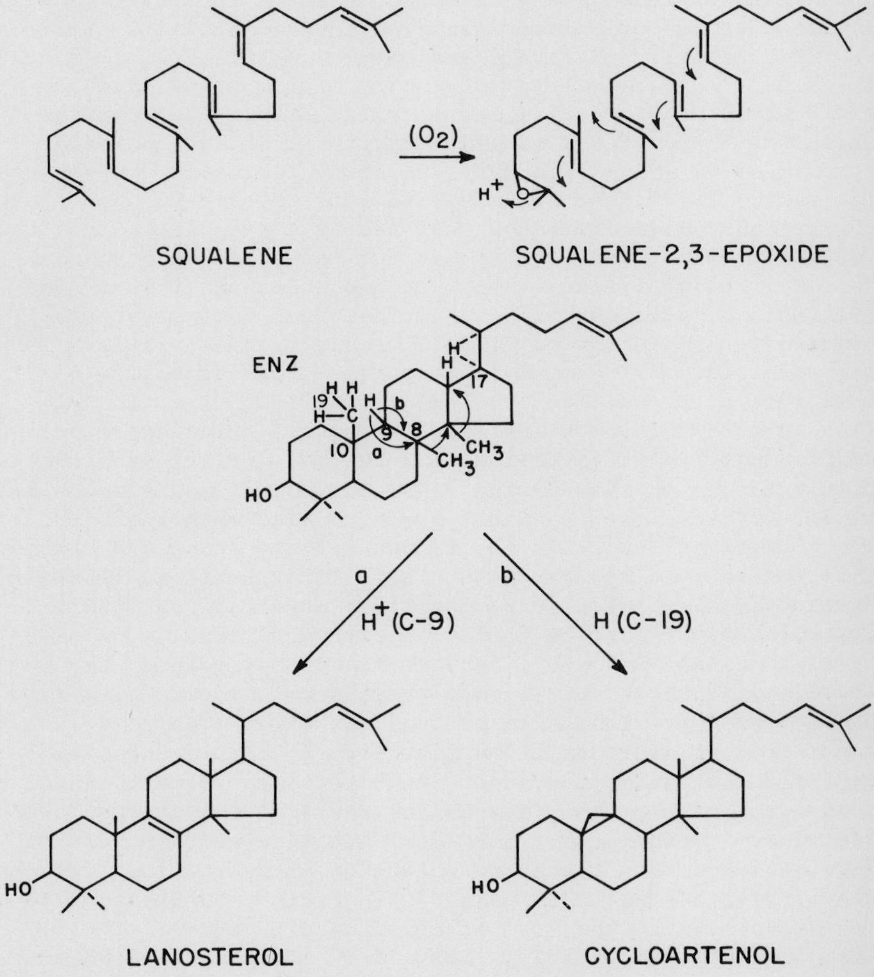

SQUALENE   SQUALENE-2,3-EPOXIDE

ENZ

LANOSTEROL   CYCLOARTENOL

Figure 10.7   Cyclization of 2,3-oxidosqualene to lanosterol and cycloartenol.

spite of the fact that they are required by these fungi for reproduction. Intermediates early in the sterol synthesis pathway such as GPP, FPP, and squalene have been detected in several pythiaceous species (Schlösser et al., 1969; Wood and Gottlieb, 1978a). It appears that the metabolic block preventing sterol synthesis in these fungi lies between squalene and 2,3-oxidosqualene (Gottlieb et al., 1978; Wood and Gottlieb, 1978a; 1978b). However, it appears that enzymes of the later stages of sterol synthesis are also absent, since neither the epoxide nor lanosterol can be converted to ergosterol by these fungi.

TERMINAL REACTIONS OF STEROL BIOSYNTHESIS

The cyclization of 2,3-oxidosqualene completes the biosynthesis of sterols; however, lanosterol or cycloartenol generally do not tend to accumulate in tissues. Instead they are metabolized to the principal sterols of biological systems such as cholesterol, sitosterol, or ergosterol. This final stage in the biosynthesis of sterols involves three major steps: (1) Demethylation. The principal sterols in nature are 4- and 14-desmethyl sterols arising from the first cyclic intermediate. (2) Reduction or alkylation at C-24. The biosynthesis of cholesterol requires reduction of the C-24 double bond of lanosterol, cycloartenol, or a subsequent intermediate. In fungi and plants, sterol biosynthesis intermediates are alkylated at C-24. This reaction does not occur in higher animals. Additional modification of the sterol side-chain occurs in ergosterol formation with introduction of the Δ^{22} double bond. (3) Change in nuclear double bond location (ring B). This involves isomeration of the C-8(9) double bond to C-7 and in most cases introduction of a double bond at C-5. These reactions in the final phase of sterol biosynthesis have been reviewed on several occasions (Bloch, 1965; Clayton, 1965; Frantz and Schroepfer, 1967; Goad, 1970; Schroepfer et al., 1972; Fiecchi et al., 1972).

Demethylation. The conversion of lanosterol to cholesterol, C_{30} to C_{27}, involves the loss of three carbon atoms; Olsen et al. (1957) have first shown that three moles of carbon dioxide are liberated during this conversion. The same loss of carbon occurs during ergosterol and sitosterol formation. The carbon atoms lost are those of the methyl groups at C-4 and C-14 of lanosterol. Removal of the 4-*gem* methyl groups are catalyzed by a membrane bound, multi-enzyme

system containing an oxidase, decarboxylase, and reductase
(Gaylor, 1974). The 4α-methyl is oxidized to a carboxyl
group by a sterol methyl oxidase. A 4α-carboxylic acid has
been isolated and identified from yeast (Miller and Gaylor,
1970a; 1970b; Hornby and Boyd, 1970), and can be converted
to 4-desmethyl sterols (Schroepfer et al., 1972). The reaction process involves hydroxymethyl and formyl intermediates,
and requires 3 moles NAD(P)H and 3 moles oxygen. In rat
liver, the oxidation reactions are carried out by a microsomal, cyanide sensitive, carbon monoxide insensitive mixed
function oxidase containing a cyanide binding haemoprotein
similar to but slightly different from P-450 (see Chapter 8).
Carbon dioxide is subsequently liberated by decarboxylation
which is catalyzed by an NAD-dependent oxidoreductase (decarboxylase). The product of decarboxylation is a 3-keto
steroid. Before the second demethylation at C-4 can occur,
epimerization of the 4β-methyl (4β to 4α) must occur and the
3-keto group must be reduced (Rahman et al., 1970; Sharpless
et al., 1968; 1969; Rahimtula and Gaylor, 1972; Lindberg and
Bloch, 1963; Sumdell and Gaylor, 1968). The 3-keto group is
essential for epimerization. Reduction of the 3-keto requires NADPH. After epimerization and reduction the oxidation, decarboxylation, and reduction reactions are repeated
in the removal of the second methyl group at C-4 in the production of a 4-desmethyl sterol. The reactions of the second demethylation seem to be metabolized by the same enzymes
that attack the 4,4-dimethyl substrate (ex. lanosterol). Moore
and Gaylor (1968) have described the properties of a yeast
microsomal preparation containing sterol demethylase activity
and have compared them to those of mammalian tissues (liver,
skin, and testes). The yeast system closely resembles that of
mammalian tissues in that stoichiometic amounts of [^{14}C]
carbon dioxide are liberated from lanosterol. The yeast
system requires glutathione, NAD, and has optimum activity
at a neutral pH. The systems differ, however, in that the
yeast system is incapable of demethylating Δ^7-sterols and is
stimulated by Mg^{+2}, which can be substituted with S-adenosylmethionine. This latter characteristic is probably related
to C-24 alkylation (see below) which does not occur in mammalian systems. The same demethylation reactions appear to
occur in the conversion of cycloartenol to 4-desmethyl sterols
in plant tissues.

The other methyl group lost in the metabolism of lanosterol and cycloartenol to 4-desmethyl sterols is that occurring at C-14. In animals, the 14α-methyl is removed prior to

those at C-4 while the reverse appears to be true in fungi and photosynthetic plants. It is considered the first methyl group removed in the biosynthesis of cholesterol because, aside from lanosterol, no 14α-methyl sterols have been detected in higher animals, and 14-desmethyl lanosterol can be converted to the C_{27} sterol (Gautschi and Bloch, 1957; 1958). The demethylase seems to be relatively non-specific since 14α-methyl-cholest-7-enol can be converted to cholesterol (Knight et al., 1966) and 4,14α-dimethyl-cholesta-8,24-dienol can be converted to ergosterol by yeast (Barton et al., 1968).

Less is known about 14α-methyl removal; it has been assumed that this methyl group is lost as carbon dioxide by reactions similar to those in the removal of C-4 methyl groups. However, there is evidence that removal of the 14α-methyl occurs at the aldehyde, rather than carboxyl, state of oxidation, resulting in the release of formic acid (Alexander et al., 1972). Also, the α-hydrogen (2S of MVA) is lost at C-15 during 14α-demethylation in rat liver (Canonica et al., 1968; Akhtar et al., 1968; 1969; Fiecchi et al., 1972), yeast (Caspi and Ramm, 1969), and higher plants (Gibbons et al., 1968). A C-15 hydroxylated intermediate is also involved in the demethylation reaction (Schroepfer et al., 1972). Moreover, a Δ^{14} intermediate results from dehydration at C-15. The Δ^{14} double bond is reduced by a hydrogen from the medium to C-15 and a hydride ion from NADPH to C-14 (Akhtar et al., 1972). Further evidence for the involvement of Δ^{14} sterols in the formation of cholesterol has come from studies with the drug AY-9944 which inhibits Δ^{14} reduction. Similar results have been obtained with green algae (Dickson and Patterson, 1972). Yeast converts ergosta-8,14-dienol to ergosterol (Akhtar et al., 1969). This, coupled with the identification of Δ^{14} sterols in fungi (Weete et al., 1973; Barton and Bruun, 1951), lends further support to Δ^{14} sterols being products of 14α-demethylation and intermediates in cholesterol and ergosterol biosynthesis.

C-24 Alkylation and Other Side-Chain Modifications. Unlike the other carbon atoms, those at C-24 of fungal and plant sterols do not originate from acetate (Hanahan and Wakil, 1953). It was first shown that methionine (Danielsson and Bloch, 1957; Alexander and Schwenk, 1957; 1958; Alexander et al., 1957; 1958) and S-adenosylmethionine (SAM) (Parks, 1958) could serve as the methyl donor for C-24 alkylation in ergosterol synthesis in yeast. Studies on the mechanism of

C-24 alkylation in fungal and plant sterol biosynthesis have
been reviewed on several occasions (Lederer, 1964; 1969; Goad
et al., 1974; Goodwin, 1977). Alkylation at C-24 involves
the electrophilic transfer of a C_1 group from the positively
charged donor S-adenosylmethionine (SAM) to the electron rich
C-24(25) double bond site on the sterol side chain (Van Aller
et al., 1969). In this case, the final 4-desmethyl sterol
product contains 28 carbon atoms e.g. ergosterol. C_{29} sterols
have a C_2 alkyl group at C-24 and are produced via a second
transmethylation involving SAM as the methyl donor and a C_{28}
alkyl acceptor. The variety of sterol side chain structures
occurring in nature is illustrated in Figure 9.1 (Chapter 9).
The mechanism of methyl transfer has been determined using SAM
deuterated at the S-methyl (R-S-CD_3). Ergosterol produced
by *Neurospora crassa* contains only two of the original three
deuterium atoms of the CD_3 group (Jaurequiberry et al., 1965).
Similar results have been obtained with *Gliocladium roseum*
(Lenfant et al., 1969), *Polyporus sulphureus* (Villanueva et
al., 1967), *Daedalea quercina* (Lederer, 1969), and *Oospora
virescens* (Vareune et al., 1971). The presence of only two
deuterium atoms suggests that a 24-methylene sterol may be
intermediate in the pathway of ergosterol synthesis. The
presence of 24-methylene sterols in fungi (see Chapter 9)
and the fact that radiolabeled 24-methylene substrates (e.g.
24-methylene-lanost-8-enol) are converted to ergosterol by
yeast and *P. sulphureus* supports this pathway (Akhtar et al.,
1966; Barton et al., 1966) (Figure 10.8).

Reduction of the C-24(28) double bond is catalyzed by a
sterol methylene reductase specific for NADPH (Neal and
Parks, 1977), and leads to the formation of a 24β-methyl
sterol. Lanosterol or cycloartenol biologically produced
from $[2\text{-}^{14}C, (4R)\text{-}4\text{-}^3H]$ MVA have the same $^{14}C/^3H$ ratio as the
substrate (MVA), and a tritium atom in the β-position of C-24.
When these sterols are converted to their respective C-24
alkylated forms, the $^{14}C/^3H$ ratio remains unchanged suggesting retention of the C-24 tritium atom. This has been shown
in the fungi *Aspergillus fumigatus* (Stone and Hemming, 1965),
S. cerevisiae (Akhtar et al., 1967), and *P. sulphureus* (Lenfant et al., 1967) as well as higher plants (Raab et al.,
1968) and algae (Goad and Goodwin, 1965; 1969). The absence
of the tritium at C-24 has been confirmed using an alga that
produces fucosterol (24-ethylidene) which can not have a second substituent at C-24, but does have the tritium at C-25
when produced from stereospecifically labeled MVA as above
(Goodwin, 1971). This suggests that the initial stages of

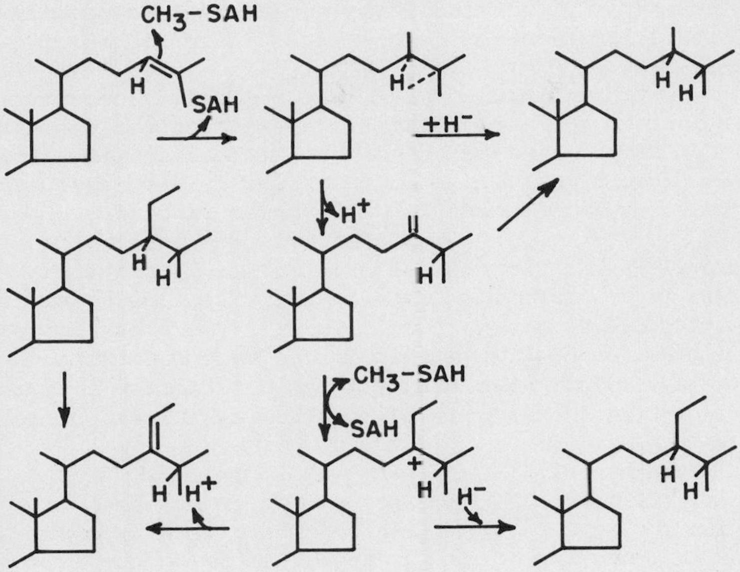

Figure 10.8   C-24 alkylation (SAH = S-adenosylhomocysteine; CH$_3$-SAH = S-adenosylmethionine).

alkylation involves the migration of a hydrogen from C-24 to C-25 (Figure 10.8).

An alternate pathway of C-24 alkylation involving the transfer of an intact methyl group from SAM occurs in maize and some algae (Goad et al., 1974) (Figure 10.8).

The production of C$_{29}$ sterols appears to be restricted to certain groups of fungi i.e. aquatic phycomycetes and rust fungi. Although the origin of the C-29 carbon of sterols from these fungi is SAM, via a second transmethylation, the mechanism of methyl transfer in fungi has not been established. However, the occurrence of 24-ethylidene sterols in certain rust species and aquatic phycomycetes suggest that a Δ24(28) sterol may be the substrate for the second transmethylation reaction in these fungi (Figure 10.8). In this case, only four of the original deuterium atoms of the CD$_3$ of SAM would be expected to the present in the C$_2$ alkyl group of C$_{29}$ sterols of these organisms. See Goad et al. (1974) and Goodwin (1977) for alternate pathways of sterol side chain biosynthesis as they occur in higher plants and algae.

S-adenosylmethionine:Δ^{24}-sterol methyl transferase has been isolated from yeast and studied in considerable detail, particularly by L. W. Parks and his associates. This enzyme occurs in promitochondria and mitochondria (Thompson et al., 1974) and has been solubilized from microsomes (Moore and Gaylor, 1969). The enzyme is also present in promitochondria of anaerobically grown yeast cells and is not repressed by glucose. It is produced on cytoplasmic ribosomes, coded for by nuclear DNA, and incorporated into the inner mitochondrial membrane matrix (Thompson et al., 1974). Zymosterol is the preferred substrate for the methyl transferase (Moore and Gaylor, 1969; Thompson et al., 1974; Fryberg et al., 1975), which is non-competitively inhibited by ergosterol (Thompson and Parks, 1974). The enzyme is also inhibited by monovalent cations (Bailey et al., 1974) and homocysteine (Hatanaka et al., 1974).

In cholesterol synthesis, reduction rather than alkylation occurs at C-24. The Δ^{24}-reductase is microsomal and requires NADPH, but does not appear to be substrate specific with respect to nuclear double bond position (Avigan et al., 1963). The hydrogen introduced at C-24 during Δ^{24}-reduction becomes the 24-pro-S-hydrogen of cholesterol (Caspi et al., 1969) and originates from the medium. A hydride ion is transferred from NADPH to C-25 to complete the reduction.

Biosynthesis of ergosterol and certain phytosterols involves further modification of the sterol side chain, i.e. introduction of the *trans*-Δ^{22} double bond. Formation of this double bond is independent of C-24 alkylation (Akhtar et al., 1968). The 23-pro-S-hydrogen is eliminated in Δ^{22} formation during ergosterol biosynthesis by *Aspergillus fumigatus* and *Blakeslea trispora* (Bimpson et al., 1969). The stereochemistry of hydrogen removal in Δ^{22} formation during poriferisterol and ergosterol biosynthesis by the alga *Ochromonas malhamensis* and C-22(23) desaturation of cholesterol by the protozoan *Tetrahymena pyriformis* is opposite to that in fungi, the 23-pro-R-hydrogen is eliminated (review by Goodwin, 1971).

Nuclear Double Bond Shift, Formation, and Reduction. Several modifications of the sterol ring structure occurs with respect to the shift, formation, and reduction of double bonds during conversion of the first cyclic intermediate to the major sterol product of the organism. The nuclear double bond of lanosterol occurs at the C-8(9) position while principal sterols in nature have ethylenic bonds at C-5, C-7, or both. In cholesterol synthesis, ring modification follows an established sequence: Isomerization of the $\Delta^{8(9)}$ double bond to the Δ^7 position, introduction of a double bond at C-5 giving rise to a $\Delta^{5,7}$ sterol, and reduction of the double bond at C-7 (Δ^8 to Δ^7 to $\Delta^{5,7}$ to Δ^5). Generally, the same sequence of reactions occurs in the synthesis of other sterols, but in some cases the stereochemistry of the reactions differ between major groups of organisms.

The $\Delta^8 \rightarrow \Delta^7$ isomerization in cholesterol synthesis has been studied using MVA stereospecifically labeled at C-2 which becomes the C-7 of lanosterol, or cycloartenol, and the major sterol product. This reaction involves loss of the 7β-hydrogen (2S of MVA) while the 7α-hydrogen is retained (Canonica et al., 1968; 1969; Caspi et al., 1968; Gibbins et al., 1968). The Δ^7 double bond of poriferasterol is formed in the same manner by the alga *Ochromonas malhamensis* (Smith et al., 1968) and higher plant sterols in species tested (Goodwin, 1971). However, the 7α-hydrogen is lost in the $\Delta^8 \rightarrow \Delta^7$ isomerization in the synthesis of ergosterol by yeast (Caspi and Ramm, 1969).

Introduction of the Δ^5 double bond has been studied in mammalian and yeast systems, yet the exact mechanism is not known. Oxygen is required for this reaction in the formation of cholesterol (Frantz et al., 1959) and ergosterol (Akhtar and Parvez, 1968). Introduction of the Δ^5 double bond during both cholesterol and ergosterol synthesis involves removal of the α-hydrogens or an overall *cis*-elimination of the hydrogens at C-5 and C-6 (Bimpson et al., 1969; Akhtar and Parvez, 1968; Akhtar and Marsh, 1967; Goad et al., 1969; Paliokas and Schroepfer, 1967; 1968). Two mechanisms have been proposed for formation of the Δ^5 double bond (Akhtar and Parvez, 1968; Dewhurst and Akhtar, 1967): (1) Hydroxylation-dehydration in which the first step occurs under aerobic conditions and second step under anaerobic conditions, and (2) dehydration. The use of various oxygenated sterols (hydroxyl, keto, epoxy) with rat liver cell-free extracts for cholesterol synthesis have met with little success in determining a natural oxygenated intermediate as would be required in the oxygen-dependent step of Δ^5 double bond formation. In yeast, 3α-^{3}H-ergosta-7,22-dienol and 3α-^{3}H-ergosta-7,22-dien-3β,5α-diol are converted to ergosterol only under aerobic conditions (Akhtar and Parvez, 1968). However, a membrane-bound 5α-hydroxysterol dehydrase has been isolated from yeast that catalyzes the conversion of ergosta-7,22-dien-3β,5α-diol to ergosterol (Topham and Gaylor, 1967; 1970; 1972). Additionally, the 3α-^{3}H is removed during conversion of only the diol to ergosterol. A cyclopropane ring between C-3 and C-5 could explain the loss of the 3α-^{3}H during Δ^5 double bond formation (Topham and Gaylor, 1972). Also ergosterol peroxide (5α,8α-epidioxyergosta-6,22-dienol) is converted to ergosterol by this enzyme preparation, which is in accordance with the results of Hamilton and Castrejon (1966). Although several C-5 oxygenated sterols can serve as substrates for the yeast dehydrase, the natural intermediate in Δ^5 double bond formation by animal or fungal systems remains unknown.

The final reaction of cholesterol formation involves the reduction of a Δ^{5,7} diene to the corresponding Δ^5 sterol. This reaction involves a *trans* addition of the hydride ion from the B side of NADPH to the 7α-position and a proton to the 8β-position. The hydrogen originally at the 7α-position of lanosterol takes the 7β-position in cholesterol (Wilton et al., 1968). Although 7-dihydro sterols occur, most fungal sterols possess the Δ^7 double bond.

A summary of the comparative aspects of sterol biosynthesis in plants, animals, and fungi is given in Table 10.1.

PATHWAY OF ERGOSTEROL BIOSYNTHESIS

With the few exceptions noted in Chapter 9, sterols are universally distributed throughout the plant and animal kingdoms. The reactions and intermediates in the isoprenoid pathway leading to 2,3-oxidosqualene occur in a restricted sequence in all sterol-producing organisms. The reactions in the production of principal sterols from 2,3-oxidosqualene are also similar in different organisms, but the reaction sequence appears to be less restricted resulting in alternate pathways of sterol biosynthesis. There is now considerable evidence for alternate pathways of ergosterol biosynthesis in fungi, particularly yeasts. Possible pathways of ergosterol formation have been discussed by Weete (1973; 1976), Fryberg et al. (1973; 1975a), and Barton et al. (1973). The ability of fungi to produce ergosterol by more than one pathway may be explained, at least in part, on the basis of the low substrate specificity of enzymes in the later stages of its biosynthesis. This has certainly added to the frustration in determining the 'preferred' sterol synthesis pathway

TABLE 10.1   Summary of Comparative Aspects of Sterol Biosynthesis in Plants, Animals and Fungi

Sterol Intermediate or Reaction	Animals[a]	Fungi	Plants
Predominant sterols First cyclic intermediate	$(\Delta^5)C_{27}$ Lanosterol	$(\Delta^{5,7})C_{28}$ Lanosterol	$(\Delta^5)C_{29}$ Cycloartenol
C-24 Alkyl sterols	No	Yes	Yes
Configuration of C-24 alkyl groups	--	24β(R) (Ergosterol)	24α(R)-Higher plants certain algae 24β(S)-Some algae
Presence of Δ^{22} double bond	No	Yes	Yes
Stereochemistry of Δ^{22} desaturation	--	23-pro-S-hydrogen elimination	23-pro-R-hydrogen elimination (algae)
Stereochemistry of $\Delta^8 \rightarrow \Delta^7$	Loss of 7β-hydrogen	Loss of 7α-hydrogen (yeast)	Loss of 7β-hydrogen (algae, higher plants, protozoa)
Stereochemistry of $\Delta^7 \rightarrow \Delta^{5,7}$	cis-Hydrogen elimination (loss of 6α-hydrogen)	cis-Hydrogen elimination (loss of 6α-hydrogen A. funigatus)	--
Stereochemistry of $\Delta^{5,7} \rightarrow \Delta^5$	Addition of hydride ion to 7α and proton to 8β	--	--

[a] Mammals

in a particular organism. The principal criteria used to
establish a substance as an intermediate in a biochemical
pathway include: (1) Isolation and identification of the
proposed intermediate from the biological system, (2) conversion of the proposed intermediate to the next intermediate
and end-product of the pathway, and (3) isolation and char-
acterization of the enzyme catalyzing conversion of the proposed intermediate to the next in the pathway. Additional
criteria required to establish the 'preferred' intermediates
in a biochemical pathway are outlined by Schroepfer et al.
(1972).

In spite of the fact that most potential intermediates,
reactions, and reaction mechanisms of ergosterol biosynthesis
are known, the precise pathway, if indeed a single pathway
exists, is not well established in fungi. Most of our knowledge of ergosterol biosynthesis comes from studies with
yeast, including mutants with deficiencies in enzymes that
catalyze specific reactions in the later stages of the pathway. Specific information has come from the indentification
of naturally occurring sterols, and the use of radiolabeled
sterol synthesis intermediates and inhibitors. It has been
found that several fungicides inhibit specific reactions in
the conversion of lanosterol to ergosterol, and results from
studies using these substances have been particularly informative about the pathway of ergosterol biosynthesis (Table
10.2). The following discussion focuses on the 'preferred'
intermediates in ergosterol formation and the sequence of
major reactions, i.e. demethylation at C-4 and C-14, C-24
alkylation, $\Delta^{24(28)}$ and $\Delta^{14(15)}$ reductions, and nuclear
double bond alteration.

As in cholesterol biosynthesis, demethylation at C-4
and C-14 precedes nuclear double bond alteration in the formation of ergosterol. *In vitro*, yeast demethylases seem to
prefer Δ^8 over Δ^7 substrates. The 14α-methyl group is removed first in the series of demethylation reactions occurring in cholesterol formation. This is based on the inabil-
ity to detect 14α-methyl-sterols other than lanosterol in
mammalian tissues, and the fact that demethylation of exogenous C-4 methyl sterols does not occur unless the C-14-
methyl group is absent (Mitropoulos et al., 1976). There is
evidence that demethylation at C-14 also precedes that at C-4
during ergosterol formation in yeast. This is supported by
the identification of sterols in yeast extracts that have the
4,4-*gem*-dimethyl or 4α-methyl structure, but no C-14 methyl
group (Barton et al., 1972; Trocha et al., 1974; Fryberg et

al., 1973; 1975a; Pierce et al., 1978). In addition, lanosterol is not readily metabolized by *S. cerevisiae* grown in
the presence of the fungicide Denmert, which blocks C-14 demethylation (Kato and Kawase, 1976). It appears that this
demethylation sequence may also occur in *Neurospora crassa*
since no C-14 methyl sterols other than lanosterol have been
detected in this fungus (Renaud et al., 1978). On the other
hand, other fungi may not follow the C-14→C-4 demethylation
sequence observed in *S. cerevisiae*. Treatment of *Monilinia
fructigena*, *Candida albicans*, *Ustilago maydis*, and *Aspergillus fumigatus*, which represent the major fungal taxa, with
the C-14 demethylation inhibitors Denmert (Kato et al.,
1975), Miconazole (Van den Bossche et al., 1977; 1978), Triaimol
(Ragsdale, 1975; Sherald and Sisler, 1975), and Triforine
(Sherald and Sisler, 1975), respectively, results in the
accumulation of partially and fully C-4 demethylated sterols
containing the C-14 methyl group. Whether sterol formation
in these fungi follow the C-4→C-14 demethylation sequence
under natural conditions, or these results of inhibitor
studies only reflect a low substrate specificity of the
demethylase(s) is not known.

The next consideration in establishing the pathway of
ergosterol biosynthesis is the point at which C-24 alkylation occurs. In fungi, the product of C-24 alkylation is
a 24-methylene sterol which is reduced at some point in the
formation of ergosterol. It appears that C-24 alkylation
can occur as the initial step in the conversion of lanosterol
to ergosterol since 24,25-dihydro-24-methylenelanosterol has
been identified as a product of several fungi, including
U. maydis (Ragsdale, 1975), *Phycomyces blakesleeanus*, *A.
fumigatus* (Goulston et al., 1967), *N. crassa* (Renaud et al.,
1978), and some yeast strains (Barton et al., 1970). Moreover, this sterol is converted *in vivo* to ergosterol by yeast
(Akhtar et al., 1966; 1969; Barton et al., 1966). However,
lanosterol is not a substrate for yeast sterol 24-methyl-
transferase *in vitro* (Fryberg et al., 1975). Based on
trapping experiments with yeast, it appears that 24,25-
dihydro-24-methylenelanosterol is almost certainly not a
principal intermediate in ergosterol biosynthesis (Barton
et al., 1970). *In vitro* studies have shown that both sterol
demethylase and C-24 methyltransferase from yeast can accept
a variety of sterol substrates, and C-24 alkylation can occur
prior to complete demethylation (Moore and Gaylor, 1968;
1970). For example, the relative *in vitro* efficiencies of
4,4-dimethylzymosterol, 4α-methylzymosterol, and zymosterol

TABLE 10.2   Some Inhibitors of Sterol Biosynthesis

Inhibitor	Site(s) of Action	Fungi
Azasterol (15-aza-24-methylene-D-homocholestadiene)	$\Delta^{24(28)}$ reductase $\Delta^{14(15)}$ reductase C-24 methyltransferase	Yeast[b] Yeast Yeast
Sinefungin [6,9-diamino-1-(6-amino-9H-purin-9-yl)-1,5,6,7,8,9-hexadeoxy-β-D-ribo-decofuranuronic acid]	C-24 methyltransferase	Yeast[c]
Homocysteine	C-24 methyltransferase	Yeast[d]
AY-9944 [trans-1,4-bis-(2-chlorobenzylamino-methyl)cyclohexane dihydrochloride]	Δ^5 formation $\Delta^{8}{\to}\Delta^7$ isomerase Δ^{14} reductase Δ^7 reductase	Chlorella ellipsoidea[e] Chlorella ellipsoidea Chlorella ellipsoidea Rat liver[f]
Denmert (S-1358) (S-n-butyl s'-p-tert-butylbenzyl-N-3-pyridyldithiocarbonimidate)	C-14 demethylase	Saccharomyces cerevisiae[g]

Compound	Enzyme/Process	Organism
Triforine (CELA W524) [N,N'-bis-(1-formamide-2,2,2-trichloroethyl)-piperazine]	C-14 demehtylase	*Aspergillus fumigatus*[h]
Miconazole Nitrate [1-(2(2,4-dichlorophenyl)-2-[(2,4-dichlorophenyl)methoxy)-IH-imidazole nitrate]	C-14 demethylase	*Candida albicans*[i]
Triarimol (EL-273)[a] [α-(2,4-dichlorophenyl)-α-phenyl-5-pyrimidine-methanol]	C-14 demethylase $\Delta^{22(23)}$ dehydrogenase $\Delta^{24(28)}$ reductase	*Ustilago maydis*[j] (sporidia)
Imazalil [1,2-(2,4-dichlorophenyl)-2-(2-propenyloxy)ethyl)-1H-imidazole]	C-14 demethylase	*Penicillium expansum*[k]
23-Azacholesterol	$\Delta^{24(28)}$ reductase (1 µm) C-24 methyltransferase (10 µm)	*S. cerevisiae*[l]
25-Azacholesterol	C-24 methyltransferase (1 µm)	*S. cerevisiae*[m]

TABLE 10.2 Continued

Inhibitor	Site(s) of Action	Fungi
Triadimenol [1-(4-chlorophenoxy)-3,3-dimethyl-1-(1,2,4-triazol-1-yl) butan-2-ol]	C-14 demethylase	*Ustilago avenae*[n]
Triadimefon [1-(4-chlorophenoxy)-3,3-dimethyl-1-(1H-1,2,4-triazol-1-yl)-2-butanone)]	Later stages of ergosterol synthesis	*U. avenae*[l]

[a]C-14 demethylation is also inhibited by fenarimol [α-(2-chlorophenyl)-α-(4-chlorophenyl-5-pyrimidine methanol], nuarimol [α-(2-chlorophenyl)-α-(4-fluorophenyl)-5-pyrimidine methanol], and triadimefon [1-(4-chlorophenoxy)-3,3-dimethyl-1-(1,2,4-triazol-1-yl)-2-butanone] in *Botrytis cinerea, Penicillium expansum,* and *Ustilago maydis* (Leroux and Gredt, 1978). [b]Hays et al., 1977a; Hays et al., 1977b; [c]Turner et al., 1977; [d]Hatanaka et al., 1974; Thompson and Parks, 1974; [e]Dickson and Patterson, 1972; [f]Horlick, 1966; [g]Kato and Kawase, 1976; [h]Sherald and Sisler, 1975; [i]Van den Bossche et al., 1978a; Van den Bossche et al., 1978b; [j]Ragsdale, 1975; [k]Leroux and Gredt, 1978; [l]Pierce et al., 1978; [m]Pierce et al., 1979; [n]Buchenauer, 1978; [o]Buchenauer, 1976.

as substrates for the yeast C-24 methyltransferase is 2:5:100
(Fryberg et al., 1975a). This seems contradictory to the
fact that nuclear demethylation is enhanced by S-adenosylmethionine, suggesting that C-24 alkylated substrates are
favored by the demehtylase(s). Further evidence that zymosterol is the preferred substrate for C-24 alkylation is
that this sterol accumulates when yeast is grown in the
presence of C-24 methyltransferase inhibitors such as 25-
azacholesterol (Pierce et al., 1979).

Although 24-methyl sterols are preferentially produced
in fungi, a C-24 alkyl group may not be required by some
of these organisms for further transformation of the side
chain or ring structure. This is supported by the presence
of cholesterol and other C_{27} sterols in certain aquatic phyco-
mycetes and C_{27} sterols with $\Delta^{5,7}$, $\Delta^{8,22,24}$, $\Delta^{7,22,24}$, and
$\Delta^{7,24}$ double bond arrangements in a yeast treated with a C-24
methyltransferase inhibitor (Avruch et al., 1976; Pierce et
al., 1979) and other yeast strains (Barton et al., 1974;
Bailey et al., 1976).

Fungi in which C-24 alkylation occurs as the first step
in the conversion of lanosterol to ergosterol, such as *N.
crassa* (Renaud et al., 1978) and *Candida* species (Fryberg
et al., 1975b), 24-methyl-lanosterol has been detected and
can be converted to ergosterol by yeast (Akhtar et al., 1968).
This suggests that $\Delta^{24(28)}$ reduction can occur as the second
step in ergosterol formation from lanosterol; however, the
expected 24-methyl, 4α- and 4,4-dimethyl intermediates have
not been detected in these fungi. With zymosterol as the
substrate, fecosterol [ergosta-8,24(28)-dienol] is the
product of C-24 alkylation and side chain reduction of this
sterol is ergost-8-dienol. This sterol has been detected in
several fungi. $\Delta^{24(28)}$ reduction does not appear to be required for nuclear double bond alteration or Δ^{22} formation,
since all the expected 4,14-desmethyl sterol intermediates
have been isolated from fungi as 24-methylene derivatives.
The presence of ergosta-5,7,22,24(28)-tetraenol (Barton et
al., 1972; Katsuki and Bloch, 1967; Pierce et al., 1978)
suggests that $\Delta^{24(28)}$ may be the final step in ergosterol
formation in some fungi. $\Delta^{24(28)}$ reduction is not reversible (Fryberg et al., 1973). Also, the Δ^8 bond is not required for Δ^{22} formation since ergost-7-enol can be converted
to ergosterol by *S. cerevisiae* (Akhtar et al., 1969).

The Δ^{22} double bond does not occur in C-4 and C-14
methyl sterols. Tracing probable intermediates back from
ergosterol, the least advanced intermediate with a Δ^{22} double
bond is ergosta-8,22-dienol (Trocha et al., 1974; Parks et
al., 1972), suggesting that Δ^{22} formation is not necessarily
associated with C-24 alkylation and occurs after demethylation. Since 24,25-dihydro-24-methyl-lanosterol can be converted to ergosterol, it appears that the $\Delta^{24(28)}$ double bond
is also not required for Δ^{22} formation (Akhtar et al., 1969).

Nuclear double bond alteration in ergosterol biosynthe-
sis is generally believed to follow the $\Delta^{8 \to} \Delta^{7 \to} \Delta^{5,7}$ sequence
as in cholesterol formation. Although a Δ^{7} sterol appears
to be the preferred substrate for the C-5 desaturase, the
occurrence of sterols with double bonds in the C-5 and
C-8(9) positions of the same molecule from *N. crassa* (Renaud
et al., 1978) and from a yeast mutant deficient in $\Delta^{8 \to} \Delta^{7}$
isomerase (Pierce et al., 1979) suggests that a Δ^{7} double
bond is not required for Δ^{5} formation.

A fourth ring modification given less consideration with
respect to when it occurs relative to other double bond alterations is $\Delta^{14(15)}$ reduction. The formation of $\Delta^{14(15)}$ is
associated with demethylation at C-14. Since the expected
4,4-dimethyl- and 4α-methyl-Δ^{14} intermediates have not been
detected under natural conditions, it appears that the
$\Delta^{14(15)}$ double bond is reduced soon after its formation.
However, 4,4-dimethyl- and 4α-methyl-$\Delta^{8,14,24(28)}$ sterols
accumulate in yeast grown in high relative concentrations
of 23-azacholesterol, an inhibitor of C-24 methyltransferase
and $\Delta^{24(28)}$ reductase (Pierce et al., 1978). In the presence
of $\Delta^{14(15)}$ reductase inhibitor 15-aza-24-methylene-D-homo-
cholesta-8,14-dienol, yeast accumulates ergosta-8,14-dienol
(Hays et al., 1977). If the loss of the C-14 methyl is the
first demethylation reaction in the metabolism of lanosterol
to ergosterol as is generally believed to occur in yeast,
4,4-dimethyl-cholesta-8,14-dienol, or the 24-methylene
derivative, would be the expected product and would be expected to accumulate if the pathway is blocked at the $\Delta^{14(15)}$
step. However, because of the low substrate specificity of
the C-4 demethylase, demethylation at C-4 probably continues
in the presence of the inhibitor, resulting in the accumulation of ergosta-$\Delta^{8,14}$-dienol rather than the 4,4-demethyl
and 4α-methyl intermediates. An alternative explanation
would be that demethylation at C-4 precedes that at C-14 and
the final demethylation would result in the formation of

ergosta-$\Delta^{8,14}$-dienol. The fact that $\Delta^{8,14}$ sterols accumulate in the presence of $\Delta^{14(15)}$ reductase inhibitors in fungi and algae (Patterson et al., 1974) suggests that $\Delta^{8} \rightarrow \Delta^{7}$ isomerization is favored by $\Delta^{14(15)}$ reduction. However, the $\Delta^{14(15)}$ double bond apparently does not block the isomerization, since ergosta-8,14-dienol can be converted to ergosterol by yeast (Akhtar et al., 1969) and $\Delta^{5,7,14}$ (Weete et al., 1973) and $\Delta^{5,7,14,22}$ C_{28} (Barton and Bruun, 1951) sterols have been tentatively identified in two filamentous fungi.

As noted above, the alteration of ring double bond positions during ergosterol biosynthesis follows the $\Delta^{8} \rightarrow \Delta^{7} \rightarrow \Delta^{5,7}$ sequence. Generally, it appears that initiation of ring double bond transformation follows demethylation, C-22 desaturation, and $\Delta^{14(15)}$ reduction. However, based on the identification of 4α-methyl-ergosta-7,24(28)-dienol in *A. fumigatus* (Goulston et al., 1967), it appears that complete demethylation at C-4 is not essential under all circumstances for the $\Delta^{8} \rightarrow \Delta^{7}$ isomerization.

It is clear that ergosterol biosynthesis by fungi cannot be discussed in terms of a single pathway. Instead it appears that most fungi possess the potential to produce ergosterol by several alternate routes. The problem remains, therefore, to elucidate the 'preferred' pathway by a particular fungal species or strain under certain conditions. In this connection, results of experiments with sterol synthesis inhibitors should be interpreted with caution since accumulating sterols may represent intermediates involved in an alternate rather than primary pathway of ergosterol biosynthesis. The most informative data should come from trapping experiments and determination of enzyme substrate specificities. An outline of ergosterol biosynthesis as it may occur in yeast is given in Figure 10.9. Sufficient studies on the biosynthesis of ergosterol by non-yeast fungi have not been conducted to allow further comments on if or how sterol formation may differ from that in yeast.

CAROTENOID BIOSYNTHESIS

The biosynthesis of carotenoids has been thoroughly studied in higher plants, particularly tomato, and in fungi, particularly *Blakeslea trispora*, *N. crassa*, and *Phycomyces blakesleeanus*. Mutants of these and other fungi have played a particularly important role in studies of carotenogenesis.

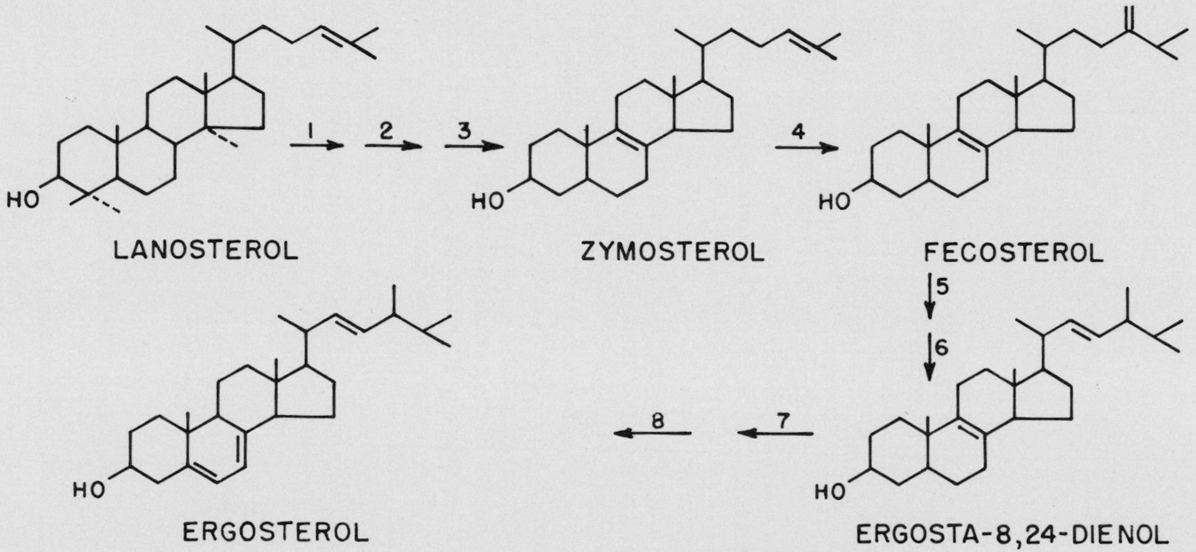

Figure 10.9   Outline of ergosterol biosynthesis in fungi.

These studies have been reviewed by Beytia and Porter (1976) and Goodwin (1971, 1972). Carotenoid biosynthesis is described only briefly in this chapter with the inclusion of some results published since these reviews.

The early stages of carotenoid biosynthesis follow the sequence of reactions for farnesyl pyrophosphate formation as shown in Figures 10.2, 10.4, and 10.5. Rather than condensing with another molecule of farnesyl pyrophosphate as in squalene formation, the C_{15} intermediate condenses with another molecule of isopentenyl pyrophosphate to produce geranylgeranyl pyrophosphate (C_{20}) (GGPP). A C_{20}-specific prenyl transferase has been reported for yeast (Grob et al., 1961) as well as several other systems. Mn^{+2} is a better activator of the enzyme than Mg^{+2}. As in the elongation reactions leading to FPP, the original 4S hydrogen of MVA is lost in the third condensation reaction (Figure 10.5). The next intermediate in carotenoid biosynthesis is formed by the tail-to-tail condensation of two GGPP molecules. One hydrogen atom, 5S of the original MVA, from each of the GPP molecules is lost during the condensation to form phytoene. Phytoene is not the C_{40} homologue (lycopersene) of squalene because it has a double bond in the center of the molecule. In higher plants, cis-phytoene is the isomer formed by the phytoene synthetase and involves retention of the two pro-R-hydorgens of the C-1 of the two condensing C_{20} molecules. However, trans-phytoene has been identified in certain micro-organisms and involves retention of one pro-S and one pro-R hydrogen of the condensing molecules (Gregonis and Rilling, 1974). NADPH is not required for trans-phytoene formation; ATP stimulates the reaction, but not through phosphorylation of the enzyme. Prephytoene pyrophosphate (analogous to presqualene pyrophosphate) appears to be an intermediate in phytoene biosynthesis, and has been identified in N. crassa mutants (Kushwaha et al., 1978). Yeast squalene synthetase catalyzes the synthesis of prephytoene pyrophosphate and lycopersene when supplied GPP (Qureshi et al., 1973).

The next reactions in carotenogenesis involve four dehydrogenations, and lead to the conversion of phytoene to lycopene (Figure 10.10). The reactions can occur in the dark and require FAD and possibly NADP. In each dehydrogenation step, 5R and 2S hydrogens originally present in MVA are removed (Goodwin, 1971). Polyene products of the four respective dehydrogenation reactions are phytofluene, ξ-carotene, neurosporane, and lycopene in order of their formation.

Although little is known about the enzymology of these reactions, it is generally believed that four copies of the dehydrogenase are assembled as an enzyme complex and act sequentially in the conversion of phytoene to lycopene (Aragon et al., 1976). This conclusion is based on the analyses of carotenes and nuclear proportions in *Phycomyces* heterokaryons. Similar results have been obtained for *Ustilago violacea* using ultraviolet light induced pigmented mutants (Garber et al., 1975).

The principal carotenes in nature are cyclic, containing either a single terminal ring or one at each end of the molecule. Rings may be of the α- or β-ionone types with the latter being more common (see Chapter 9). Since β-carotene contains two terminal rings and is the principal carotene in many fungi, it might be considered the end-product of carotene biosynthesis in these organisms, and other cyclic carotenes may be viewed as intermediates in its synthesis. Based on the structures of these intermediates, there are two possible routes of β-carotene synthesis (Figure 10.11). The predominant pathway in fungi is not well established, but may depend on the species. The earliest step in the desaturation sequence (Figure 10.11) that cyclization occurs seems to be at neurosporene which is the immediate precursor to β-zeacarotene. β-Zeacarotene would then undergo desaturation to γ-carotene which would in turn cyclize to β-carotene. However, in most species tested, lycopene accumulates when grown in the presence of certain carotene synthesis inhibitors, and is quantitatively converted to β-carotene when the inhibitor is removed. Species in which β-zeacarotene appears to be a precursor of β-carotene include *Rhodotorula rubra*, *Phycomyces blakesleeanus*, and *Rhizophlyctis rosea* (Goodwin, 1972). Species following the lycopene pathway include *Blakeslea trispora*, *P. blakesleeanus*, and certain bacteria. More recent results suggest that both pathways have equal quantitative importance in *P. blakesleeanus* (Bramley et al., 1977). About equal reductions in the incorporation of $[^{14}C]$ neurosporene into β-carotene results from the addition of either unlabeled lycopene or β-zeacarotene to cell-free extracts of a *P. blakesleeanus* mutant.

The mechanism of cyclization to form the α- and β-ionone rings of cyclic carotenes has been suggested on the basis of studies with *Phycomyces* using stereospecifically labeled MVA and is illustrated in Figure 10.12. Details of the stereochemistry of carotenoid cyclization have been described by Goodwin (1971).

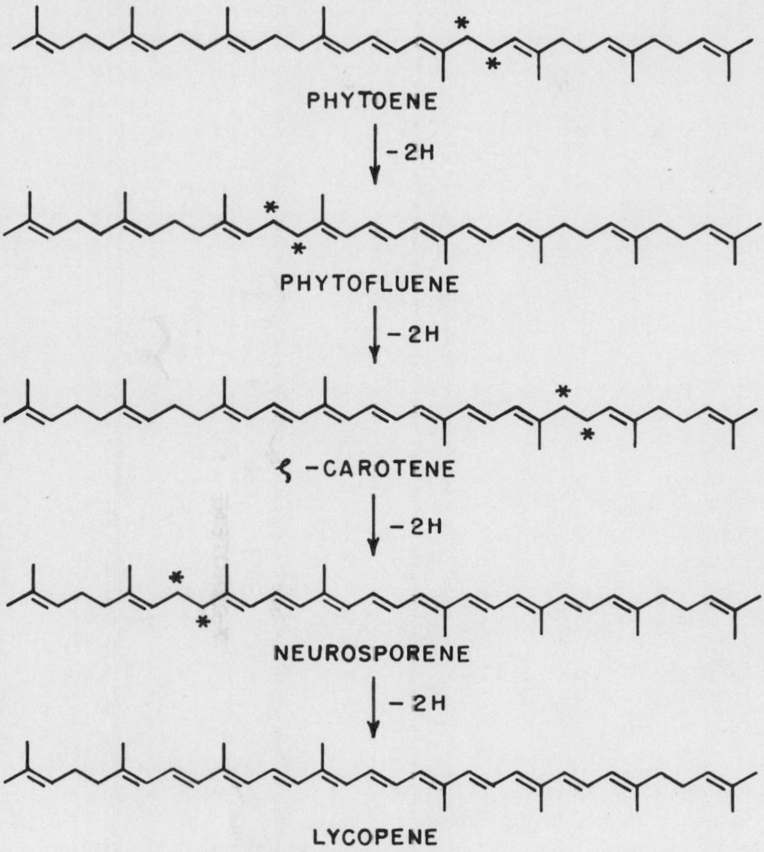

Figure 10.10   Conversion of phytoene to lycopene.   Asterisks indicate carbons from which hydrogens are lost.

There are a large number of oxygenated carotenoids (xanthophylls) produced by fungi, although they are not widely distributed among a large number of species (see Chapter 9). Relatively little is known about the biosynthesis of these substances. Torulene and torularhodin formation have been studied in *Rhodotorula* species. It is believed that torulene is produced by dehydrogenation of γ-carotene

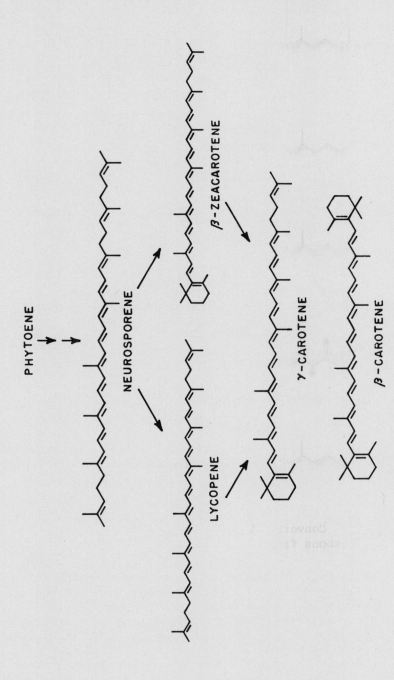

Figure 10.11 Biosynthesis of β-carotene.

at the 3',4'-position and torularhodin is produced via the stepwise oxidation of the C-17' methyl of torulene to a carboxyl group (Goodwin, 1972). The initial oxidation is catalyzed by a mixed function oxidase and one atom of O_2 is incorporated into one molecule of torulene to form torularhodin.

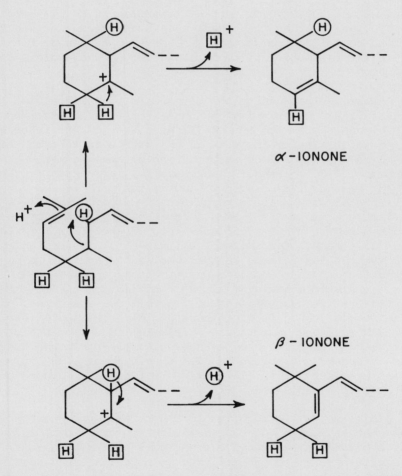

Figure 10.12   Mechanism of α- and β-ionone ring formation in carotenoid biosynthesis. H* indicates [^{3}H] from [(4R)-4-^{3}H$_1$]MVA, H indicates [^{3}H] from [2-^{3}H]MVA. Redrawn from Goodwin (1971).

BIOSYNTHESIS OF POLYPRENOLS

Like sterols and carotenoids, polyprenols have their origin from the isoprenoid pathway (Figure 10.1). Chain elongation by five-carbon increments ceases at the C_{15} level for squalene (sterols) synthesis and at the C_{20} level for carotenoid formation. Each head-to-tail condensation involves the loss of the 2S hydrogen of IPP (4S hydrogen of MVA becomes the 2S hydrogen of IPP) giving rise to a *trans* double bond in the product. If this double bond is saturated, the presence of the 2R hydrogen characterizes it as biogenetically *trans*, rather than chemically *trans*. Chain elongation during sterol and carotenoid synthesis continues by the tail-to-tail condensation of two C_{15} and two C_{20} molecules, respectively.

Less is known about the biogenesis of polyprenols, but the available information has been summarized by Hemming (1974) and Beytia and Porter (1976). Most of the interest in these substances has been in their role as intermediates involved in glycosyl transfer during glycan biosynthesis (Hemming, 1974). Rather than cease polymerization at the C_{15} or C_{20} level as described above, polyprenols are formed by repeated condensation reactions involving IPP until 6 to 24 isoprene units are linked. Products of the polyprenol synthetase are usually di*trans*-, poly*cis*-polyprenols rather than all-*trans*. Generally, the first 2 to 3 condensations involve the loss of the 2S hydrogen of IPP leading to the formation of *trans* double bonds as before. However, at this point the stereochemistry of the condensation reaction changes so that the 2R hydrogens of the IPP molecules are lost giving rise to *cis* double bonds (Figure 10.13). The stereochemistry of each isoprene unit of the polyprenol is determined at polymerization rather than by subsequent *trans-cis* isomerization. The saturated α-isoprene residue is biogenetically *cis*. Stereochemistry of the ω-isoprene residue is ambiguous and not usually considered.

A polyprenol pyrophosphate is probably the product of the synthetase and the monophosphate the functional form. Interconversion between the functional and free forms in bacteria appears to be catalyzed by a phosphatase and ATP-dependent prenol phosphokinase, respectively. Many questions concerning the biosynthesis of polyprenols remain unanswered. The synthesis of a dolichol is outlined in Figure 10.13.

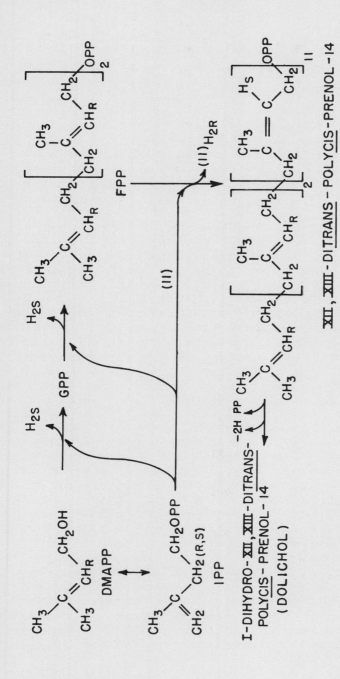

Figure 10.13   Biosynthesis of dolichol.   DMAPP = Dimethylallyl pyrophosphate, IPP = Iso-pentenyl pyrophosphate, GPP = Geranyl pyrophosphate, FPP = Farnesyl pyrophosphate, H_R = 2R Hydrogen of IPP, 2S Hydrogen of IPP.

CHAPTER 11   LIPID METABOLISM DURING

FUNGAL DEVELOPMENT[a]

INTRODUCTION

Fungal growth and development can be divided into three distinct phases: spore germination, vegetative growth, and reproduction. The purpose of this chapter is to organize the available information on the involvement of lipids in the germination and reproductive processes. Lipid metabolism in relation to environmental and nutritional factors during vegetative growth has been discussed in Chapter 2.

SPORE GERMINATION

Introduction. Fungal spore germination has been the subject of numerous reviews, some of which have dealt specifically with physiological aspects of the process (Gottlieb, 1950; Cochrane, 1958; 1960; Shaw, 1964; Sussman and Land, 1965; Merrill, 1970; Allen, 1965; Smith and Berry, 1978). Spore germination has also been emphasized at two symposia (Madelin, 1966; Weber and Hess, 1976). With the exception of one article (Reisner, 1976), these reviews have given relatively little attention to lipids.

Lipophilic Stimulators of Spore Germination. Factors affecting fungal spore germination have been widely studied in many fungi. Germination of spores from many species occurs simply by exposure to water, whereas spores of other

[a]Author: Darrell J. Weber, Department of Botany, Brigham Young University, Provo, Utah 84602.

species require special treatment or chemicals to break
dormancy such as heat and certain nutrients, or organic
chemicals. It is well-known that certain low molecular
weight lipophilic substances stimulate spore germination in
some fungal species. For example, low concentrations of
n-nonanal and n-nonanone stimulate *Puccinia graminia* uredospore germination (French and Gillimore, 1972; Kihara, 1962;
Skucas, 1968), and 2,3-dimethyl-1-pentene, a natural product
of *Agaricus campestris*, stimulates basidiospore germination
(McTeague et al., 1959). *A. bisporus* basidiospore germination is stimulated by vapors of short chain fatty acids,
particularly isovaleric and isocaproic acids (Losel, 1967;
O'Sullivan and Losel, 1971). Straight chain fatty acid from
5 to 11 carbons in chain length have shown no stimulatory
activity. 2-Propenal and other unsaturated aldehydes and
ketones stimulate differentiation (infection structures) in
Puccinia graminis (Macko et al., 1978). The activation mechanism by which these volatile chemicals stimulate fungal
spore germination and differentiation is not known.

Electron Microscopic Observations of Lipid Bodies in
Fungal Spores. The process of spore germination can be
divided into three phases: (1) swelling, (2) increased
metabolic activity, and (3) germ tube formation. The swelling of spores in the early stage of germination has been
generally attributed to the passive uptake of water. However, there is apparently some metabolic activity associated
with this process (Gottlieb and Tripathi, 1968). The overall
germination process is accompanied by a high rate of metabolic activity which has been extensively studied, but relatively little attention has been given to lipids. Fungal
spores contain variable amounts of lipid depending on the
species, and perhaps on the conditions under which they are
formed. There are numerous reports of spore lipid analyses
(see Chapter 2). Lipid bodies or globules have been observed
in spores of most fungal species examined by electron microscopy. The lipid bodies are often described as being associated with membranous portions of the cytoplasm, and in some
cases they appear to be associated with mitochondria and
glyoxysomes. Also, crystalline structures believed to be
protein are often found in association with lipid bodies
which may be bound by membrane-appearing elements (Gardner
and Hess, 1977). Glycogen is also often observed as a component of fungal spores; it presumably serves as a storage
material for germination. During germination, lipid bodies
tend to be reduced in size or disappear; but become promi-
nant in sporagenous hyphae and developing spores (Manocha

and Shaw, 1967). The results of some ultrastructure studies where lipids of dormant and germinating fungal spores have been observed are given in Table 11.1.

Lipid Metabolism During Germination. The accumulation of lipid as globules in fungal spores has been described as reserve material, particularly for energy, and this is supported by observations that the lipid bodies disappear during germination. However, carbohydrates such as trehalose and glycogen are also present in dormant fungal spores and are utilized during germination. The relative importance of these two storage materials in providing energy and substrates for anabolic processes during spore germination is not well-established for many fungal species. However, it appears to vary among different fungal groups.

There are several fungal species for which the involvement of lipids during spore germination has been studied more thoroughly than others, and these will be stressed here. The results of these studies and others are summarized in Table 11.2. It appears well-established that lipids are metabolized during fungal spore germination. In *Blastocladiella emersonii* zoospores, there are about a dozen lipid granules attached to a single large mitochondrion. Prior to germination the zoospore may swim for several hours and apparently utilizes lipid during this period (Suberkropp and Cantino, 1973). Triacylglycerides, then phospholipids, are used during swimming, but glycolipids are decreased during encystment (Mills et al., 1974). In germlings, phospholipids, monoacylglycerides, and sterol esters exhibit a marked increase.

Lipid metabolism has been studied during the germination of *Neurospora* ascospores and conidia. Ascospores of *N. tetrasperma* contain 27% lipid, 70% of which are acyl lipids, and 33% carbohydrates, 40% of which is trehalose. Based on a respiratory quotient (RQ) of 0.6, it appears that lipids are preferentially utilized for endogenous metabolism by the dormant spores. This has also been observed with *Fusarium solani* (Cochrane et al., 1963b). However, upon activation and subsequent germination of *N. tetrasperma* ascospores the RQ progressively increases to 1.2 and it appears that trehalose is the primary respiratory substrate. As the spore trehalose content declines, the metabolism shifts from predominately glycolytic to oxidative, and at this point germination becomes visible. An exogenous carbon sourse is required at this time and, if not available, endogenous lipid is utilized. Although 95% of the lipid of *N. crassa* conidia is phospholipid (Bianchi and Turian, 1967), germinating conidia have an RQ of 0.74 suggesting that lipids are being utilized as the main source of energy (Owens, 1955a; 1955b).

Spore germination in rust fungi has been studied extensively. Rust spores have high lipid contents (see Chapter 2), and apparently lipid is an important energy reserve in spores of these fungi since an RQ of 0.7 has been recorded for germinating uredospores of *Puccinia graminis* f. sp. *tritici* and *Uromyces phaseoli* (Caltrider et al., 1963; Gottlieb and Caltrider, 1963). As has been observed with several fungi (Table 11.2), fatty acid levels decrease during uredospore germination, and short chain fatty acids stimulate respiration of these spores (Reisener et al., 1961). A lipase has been isolated from uredospores of *P. graminis* (Knoche and Horner, 1970).

Neutral glycerides are generally considered storage lipid whereas phospholipids are the principal lipid components of membranes (see Chapter 2). However, phospholipids are extensively degraded in the very early stages of uredospore (Langenbach and Knoche, 1971) and conidiospore (Nishi, 1961) germination suggesting that they are also utilized for energy and perhaps substrates for anabolic pathways. In *Aspergillus*, the catabolism of phospholipids is concomitant with an increase in sugar phosphates and nucleotides, with some of the phosphorus being incorporated into RNA. Also, oxygen consumption has been correlated with phospholipid catabolism in germinating spores (Maheshwari and Sussman, 1970). Phospholipid utilization is followed by a period of synthesis. A continuous increase in lipid phosphorus has been observed during the first 6 hours of *Melampsora lini* uredospore germination (Jackson and Frear, 1967), but an initial catabolic phase may have been overlooked since the first germinating spore sample was taken at 2 hours after the initiation of germination.

Although it has been demonstrated that lipid can serve as an endogenous source of energy for spore germination in some species, this may not be the only, or in some cases, the most important role of lipid in this process. It is well-documented that isocitrate lyase and malate synthetase are present in spores of some species suggesting that lipids may be converted to carbohydrates via the glyoxylate pathway

TABLE 11.1 Fungi for Which Lipid Bodies Have Been Observed in Spores by Electron Microscopy

Fungus	Reference	Comments
PHYCOMYCETES		
Allomyces sp.	Skucas (1968)	sporangia
Allomyces macrogynus	Hill (1968)	motile and encysted spores more abundant in encysted spores--lipid bodies scattered throughout periphery of nuclear cap
Blastocladiella emersonii	Cantino & Truesdell (1970)	lipid sac
Phytophthora palmivara	Bimpong & Hickman (1975)	zoospore cytoplasm contained lipid bodies--crystalline vesicles containing lipid and other substances; no change in lipid bodies or vesicle during 6 hr motile period--vescile and lipid bodies disrupted 30 min after germination begins
Endophlyctis sp.	Powell (1976)	association of ER, lipid bodies and mitochondria and glyoxysomes
Mucor	Takeo (1974)	lipid droplets have multi-layer shells, more highly developed than in mycelial cells

Rhizopus stolonifer, R. arrhizus	Buckley et al. (1968)	lipids and proteins present and associated with membranes
Mucor rouxii	Barknicki-Garcia et al. (1968a; 1968b); Hawker et al. (1970)	disappearance of lipid bodies during germination
ASCOMYCETES		
Fusarium culmorum	Marchant (1966)	macroconidia with lipid bodies
Claviceps purpurea	Komersova et al. (1969)	abundant lipid bodies
Erysiphe graminis hordei	McKeen (1970a; 1970b); McKeen et al. (1966)	osmiophilic bodies in conidiophores, conidia, and mycelium but not haustorial mother cells--they are intracytoplasmic and intravacuolar
Penicillium chrysogenum	McCoy et al. (1971)	lipid bodies containing protein
Sphaerotheca macularis	Mitchell & McKeen (1970)	lipid bodies disappear upon germination of conidia
Stemphylium sarcinaeforme	Murray & Maxwell (1976)	lipid bodies in close association with vacuoles during germination
Botrytis fabae	Richmond & Pring (1971a; 1971b)	lipid bodies occur either close to the plasmalemma or in association with vacuoles
BASIDIOMYCETES		
Lenzites saepiaria	Hyde & Walkinshaw (1966)	osmiophilic bodies fewer and larger in number in germinating spores--bodies associated with vacuole and vesicular membrane

TABLE 11.1 Continued

Fungus	Reference	Comments
Schizophyllum commune	Aitken & Niederpruem (1970)	lipid bodies abundant in germlings, 3-fold increase in lipid in 12 hr germlings
Psilocybe sp.	Stocks & Hess (1970)	lipid bodies in dormant spores, germinating spores have large amounts of lipid
Coprinus lagopus	Heintz & Niederpruem (1971)	basidiospores have lipid droplets, lipid droplets and vacuoles increase in size as oidia begin to germinate
Cronartium fusiforme	Walkinshaw et al. (1967)	numerous membrane-bound lipid bodies in aeciospores, smaller and fewer in number in germinating spores
Puccinia graminis var. *tritici*	Sussman et al. (1969)	lipid bodies present, appearance changes in heat-shocked uredospores
Melampsora lini	Manocha & Shaw (1967)	numerous lipid globules which tend to disappear as germination progresses; few lipid globules in vegetative mycelium or haustoria; globules increase in sporogenous hyphae and developing uredospores

Tilletia caries	Gardner & Hess (1977)	some lipid bodies surrounded by possibly half-unit membranes; some lipid bodies contain lamellar structures
Agaricus bisporus	Greuter & Rast (1975)	lipid droplets present
MYXOMYCETE		
Arcyria cinerea	Mims (1971)	disappearance of polysaccharide and numerous lipid bodies during germination
Dictystelium discordeum	Cotter et al. (1969)	crystalline structures and dark bodies (lipids) disappear shortly before or during emergence of myxamoebae

TABLE 11.2   Studies on Lipid Metabolism During Fungal Spore Germination

Fungus	Reference	Comments
MYXOMYCETES		
Dictyostelium discoideum	Davidoff & Korn (1963b); Davidoff (1964)	short chain fatty acids appear to serve as a source of carbon and energy whereas long chain fatty acids are incorporated into acyl lipids
PHYCOMYCETES		
Blastocladiella emersonii	Truesdell & Cantino (1970); Cantino (1969)	it is assumed that energy source for zoospore swimming is lipid; glycolipid is also involved in chitin synthesis
Rhizopus arrhizus	Gunasekaran et al. (1972)	lipid decreases during first 2 hours of germination, but increases later until germination is complete; neutral lipid decreases and polar lipid increases
Phycomyces blakesleeanus	Furch et al. (1976)	heat activation does not stimulate lipid metabolism so lipid decomposition does not account for increase in free glycerol during heated-activated germination

ASCOMYCETES AND DEUTEROMYCETES

Organism	Reference	Comments
Neurospora tetrasperma; *N. crassa*	Sussman & Lingappa (1959); Sussman (1954); Budd et al. (1966); Johnston & Paltauf (1970); Geck & Greenwalt (1976); Owens (1955a; 1955b)	see text
Penicillium atrovenetrum	Gottleib & Ramachandran (1960); Van Etten & Gottlieb (1965)	five percent of available glucose is incorporated into lipid during spore germination; lipid content doubles during first 24 hours of germination; this is due mainly to saponifiable lipid
P. roquefortii	Lawrence & Bailey (1970)	octanoic acid oxidized in dormant spores
Aspergillus nidulans	Shepherd (1957)	lipid content of spores does not change during germination
A. niger	Yanagita & Kogane (1963a; 1963b)	little lipid is synthesized from carbon dioxide
Fusarium solani	Cochrane (1958); Cochrane et al. (1963a; 1963b)	induced high lipid content of macroconidia has no effect on respiration during germination; only a slight amount of acetate or glucose is incorporated into lipid during germination; lipid utilization during endogenous respiration of dormant spores accounts for 37%

TABLE 11.2   Continued

Fungus	Reference	Comments
		of weight loss, but only 5% of carbon requirement satisfied by lipid utilization during germination
Microsporum gypseum	Barash et al. (1967)	endogenous fatty acids and soluble carbohydrates utilized during spore germination
BASIDIOMYCETES *Puccinia graminis* var. *tritici*; *Uromyces appendiculatus*	Shu et al. (1954); Caltrider et al. (1963); Gottlieb & Caltrider (1963); Reisener et al. (1961); Staples & Wynn (1965); Daley et al. (1967); Chigrin & Bessmeltzeva (1974)	endogenous fats and proteins furnish substrate for germination of uredospores; fatty acids decrease during germination; respiratory quotient of germinating uredospores is 0.7; short chain fatty acids stimulate respiration; bulk of lipid reserves consumed within first 2 hours of germination
Uromyces phaseoli	Langenbach & Knoch (1970; 1971a; 1971b)	phospholipid synthesis active during uredospore germination; phospholipid levels drop 60% in first 20 minutes of germination, but increase to 80% of pregermination levels 2 to 3 hours after

		germination initiation; PC and PE are metabolized at different rates during germination
Melampsora lini	Jackson & Frear (1968)	lipid phosphorus levels increase continuously during first 6 hours of germination
Ustilago maydis	Bushnell (1972)	fatty acids (free) decrease to almost zero during first 24 hours of germination of teliospores; di- and triacylglycerides and phospholipids also decrease; significant amounts of [^{14}C]acetate is incorporated into diacylglycerides, phospholipids and sterols during germination
Tilletia caries	Weber & Trione (1980)	free fatty acids decrease rapidly during first 12 hours of germination of teliospores

(see Chapter 4). The glyoxylate pathway has been detected in germinating spores of *Penicillium oxalicum* (Ohmori and Gottlieb, 1965) and *P. notatum* (Gottlieb and Ramachandran, 1960), and an increase in carbohydrates has been observed during conidiospore germination of *P. atrovenetum* (Van Etten and Gottlieb, 1965), but not *P. oxalicum* (Gottlieb and Caltrider, 1963). Similar results have been obtained with uredospores of *P. graminis* and *U. phaseoli* (Caltrider et al., 1963). The carbohydrates formed via β-oxidation of fatty acids, the glyoxylate pathway, and reverse glycolysis (see Chapter 4) are probably incorporated into the germ tube cell wall.

In addition to degradation, lipids are also synthesized during fungal spore germination. Significant amounts of $[^{14}C]$acetate are incorporated into diacylglycerides, phospholipids, and sterols during *Ustilago maydis* teliospore germination (Bushnell, 1972). As has been noted above, rapid phospholipid synthesis occurs in germinating rust spores after an initial catabolic phase. On the other hand, it appears that lipid synthesis is not quantitatively significant in germinating conidiospores of *Fusarium solani* (Cochrane et al., 1963b) and *Aspergillus nidulans* (Shepherd, 1957).

REPRODUCTIVE GROWTH

Introduction. Fungal sporulation is a complex process that has been difficult to investigate from biochemical and ultrastructural points of view. One of the factors that handicaps investigations of sporulation is that the process is not always initiated in a predictable way, as compared to the germination process which can be initiated in a synchronized or nearly synchronized manner. Furthermore, the initiation of sporulation often involves the interaction of nutrition, light, temperature, chemical inhibitors and stimulators, and other environmental factors. On the other hand, the sensitivity of a fungus to one or more of these factors has been used to advantage in a limited number of investigations. The development of generalizations about the correlation of certain metabolic events with reproduction is further complicated by the wide variety of types of sporulation in fungi. Consequently, relatively little is known about the factors which initiate and regulate fungal reproduction and the biochemical events that are specific to the sporulation process.

The subject of fungal sporulation has been reviewed by Hawker (1950), Cochrane (1958), Burnett (1968), Ainsworth and Sussman (1966), Moore-Landdecker (1972), and Smith and Berry (1978).

Lipid Metabolism During Sporulation. As noted above, there are a limited number of studies in which aspects of lipid metabolism have been precisely correlated with sporulation. The water mold *Achlya* sp. has proven a useful tool in studying metabolism associated with reproduction, since its life cycle is completed predictably in about 28 hours. The total lipid content of this fungus drops from 10% to 6% eight hours after spore germination (Law and Burton, 1976). This decrease appears to be due to a reduction in the acylglycerides. There is an increase in lipid content during sporangium formation, which is due to increased activity of the fatty acid synthetase. Fatty acid oxidation also increases during sporangium formation. Further evidence for lipid degradation during sporulation has been obtained with *Aspergillus niger* (Lloyd et al., 1972). Increased lipid esterase (lipase) activity has been detected in this fungus only in the conidiophore tip prior to formation of the vesicle and phialides, and in these structures after their formation, relative to vegetative hyphae.

The membrane phospholipid composition changes quantitatively and qualitatively during differentiation in the cellular slime mold *Dictyostelium discoideum*, e.g. when amoebae begin to adhere to each other and during spore formation (Ellingson, 1974). PC, PI, and PA increase while PE and its 1-alkenyl derivative decrease. Two unidentified plasmalogens decrease during aggregation but reappear during culmination and fruiting body formation.

There has been considerable attention given to the role of sterols in fungal reproduction. This subject has been comprehensively reviewed by Hendrix (1970) and, except for a brief review and mention of a few pertinent recent studies, it will not be covered in detail here. Much of the interest in the role of sterols in fungal reproduction has developed from the discovery of the steroid antheridiol as a hormone involved in sexual reproduction of *Achlya bisexualis* (Barksdale, 1969); and the sterol requirement of the non-sterol

TABLE 11.3   Lipid Metabolism During Fungal Sporulation

Fungus	Reference	Comments
MYXOMYCETES		
Dictyostelium discoideum	Krivanek & Krivanek (1958)	lipid metabolism that occurs in the pre-stalk cells appears to be concerned with energy production and intermediates for the synthesis of polysaccharide used in stalk formation
PHYCOMYCETES		
Blastocladiella emersonii	Cantino (1966); Cantino & Turian (1959); Suberkropp & Cantino (1973)	increase in lipid content during resistant sporangium formation; motile spores utilize lipid, mainly triacylglyceride then phospholipids; glycolipids are reduced during encystment, and phospholipids, monoacylglycerides, and sterol esters increase in germlings
Rhizopus arrhizus	Gunasekaran et al. (1972); Weete et al. (1973); Lawler (1972)	sterols increase during vegetative growth but decrease during sporulation
Achlya ap.	Law & Burton (1976)	utilizes lipid during growth; fatty acid synthetase and β-oxidation stimulated during sporangium

Organism	Reference	
Neurospora crassa	Ballou & Branchi (1978)	formation which is also accompanied by increased lipid content; sterols increase during vegetative growth and decrease upon sporulation
ASCOMYCETES AND DEUTEROMYCETES *Sclerotium sclerotinium*	Weete et al. (1970); Weete et al. ( )	
Saccharomyces cerevisiae	Henry & Halvorson (1973); Illingworth et al. (1973)	two phases of lipid synthesis during ascospore formation; first phase involves increased neutral and polar lipid production and is not sporulation-specific; the second phase involves only neutral lipids and is sporulation-specific
Endothia parasitica	McDowell & DeHertogh (1968)	suggest a close relationship among sporulation, increased fatty acid synthesis and *de novo* pigment synthesis
Aspergillus niger	Lloyd et al. (1972)	lipid esterase (lipase) activity increases in conidiophore tip prior to vesicle and phialide formation; only low levels detected in vegetative hyphae

producing pythiacious fungi (*Pythium* and *Phytophthora*
species) for oospore and zoosporangium formation (Haskins,
1963; Hendrix and Apple, 1964). A requirement of sterols
for reproduction has been shown in other fungi such as
Cochliobolus carbonum (Nelson et al., 1967) and *Stemphyllium
solani* (Sproston, 1971; Sproston and Setlow, 1968), but not
others such as *Leptosphaerulina briosiana* (Mayer and Leath,
1976).

The lipid content of mycelium increases rapidly during
the vegetative growth of most species (see Chapter 2), which
is due mainly to triacylglycerides and phospholipids. However, although they are relatively minor components of the
mycelium (see Chapter 9) sterols also increase during vegetative growth as shown in *Neurospora crassa* (Ballou and
Bianchi, 1978) and *Rhizopus arrhizus* (Weete et al., 1973).
In each species, a decrease in sterol content appears correlated with the sporulation process. This has also been shown
for *Gnomonia leptostyla* (Fayret et al., 1979). It has been
suggested that perhaps there is a "sterol threshold" required
for sporulation (Ballou and Bianchi, 1978). The mechanism
of sterol action in relation to fungal reproduction is not
known.

The metabolism of lipids has also been studied during
yeast sporulation. It is well-established that lipid accumulates during ascospore formation by *Saccharomyces cerevisiae*
(Pontefract and Miller, 1962; Esposito et al., 1969), and
there are two phases of lipid synthesis during this process
(Illingworth et al., 1973). Using two yeast strains treated
identically, one haploid (non-sporulation) and one diploid
(sporulating), it has been shown that the first period of
lipid synthesis involves an increase in both neutral and
polar lipids. This occurs in both strains suggesting non-
sporulation-specific synthesis (Henry and Halvorson, 1973).
However, the second phase of lipid synthesis occurs only in
the diploid strain and hence is sporulation-specific. The
increase in lipid is due only to neutral lipid.

Some studies on lipid metabolism during fungal sporulation are given in Table 11.3.

REFERENCES

Adams, B. G. and L. W. Parks. 1967. Biochem. Biophys. Res. Commun. 28:490.

Adams, B. G. and L. W. Parks. 1968. J. Lipid Res. 9:8.

Agranoff, B. W., H. Eggerer, U. Henning, and F. Lynen. 1960. J. Biol. Chem. 235:326.

Aho, L. and R. Kurkela. 1978. Nahrung 22:603.

Aitken, M. M. and J. Sanford. 1969. Nature 223:317.

Akamatsu, Y. and J. H. Law. 1970. J. Biol. Chem. 245:701.

Akhtar, M., A. D. Rahimtula, I. A. Watkinson, D. C. Wilton, and K. A. Munday. 1969. Eur. J. Biochem. 9:107.

Akhtar, M., D. C. Wilton, I. A. Watkinson, and A. D. Rahimtula. 1972. Proc. Royal Soc. Lond. Ser. B 180:167.

Akhtar, M., I. A. Watkinson, A. D. Rahimtula, D. C. Wilton, and K. A. Munday. 1968. Chem. Commun. 1406.

Akhtar, M. and M. A. Parvez. 1968. Biochem. J. 108:527.

Akhtar, M., M. A. Parvez, and P. F. Hunt. 1966. Biochem. J. 100:38C.

Akhtar, M., M. A. Parvez, and P. F. Hunt. 1968. Biochem. J. 106:623.

Akhtar, M., M. A. Parvez, and P. F. Hunt. 1969. Biochem. J. 113:727.

Akhtar, M., P. F. Hunt, and M. A. Parvez. 1967. Biochem. J. 103:616.

Akhtar, M. and S. Marsh. 1967. Biochem. J. 102:462.

Akhtar, M., W. A. Brooks, and I. A. Watkinson. 1969. Biochem. J. 115:135.

Albro, P. W. 1971. J. Bacteriol. 108:213.

Albro, P. W. 1976. *In* Chemistry and biochemistry of natural waxes (P. E. Kolattukudy, ed.). Elsevier, Amsterdam, p. 419.

Albro, P. W. and J. C. Dittmer. 1969a. Biochemistry 8;394.

Albro, P. W. and J. C. Dittmer. 1969b. Biochemistry 8:1913.

Albro, P. W. and J. C. Dittmer. 1969c. Biochemistry 8:3317.

Albro, P. W. and J. C. Dittmer. 1970. Lipids 5:320.

Alexander, G. J., A. M. Gold, and E. Schwenk. 1957. J. Am. Chem. Soc. 79:2967.

Alexander, G. J. and J. Schwenk. 1958. J. Biol. Chem. 232:611.

Alexander, K., M. Akhtar, R. B. Board, J. F. McGhie, and D. H. R. Barton. 1972. Chem. Commun. 383.

Alford, J. A., D. A. Pierce, and F. G. Suggs. 1964. J.
    Lipid Res. 5:390.
Allen, L. A., N. H. Barnard, M. Fleming, and B. Hollis.
    1964. J. Appl. Bacteriol. 27:27.
Altman, L. J., R. C. Kowerski, and H. C. Rilling. 1971. J.
    Am. Chem. Soc. 93:1782.
Anderson, J. A., F. Kang Sun, J. K. McDonald, and V. H.
    Cheldelin. 1964. Arch. Biochem. Biophys. 107:37.
Anderson, R. F., M. Arnold, G. E. N. Nelson, and A. Ciegler.
    1958. J. Agric. Food Chem. 6:543.
Anding, C., R. D. Brandt, G. Ourisson, R. J. Pryce, and M.
    Rohmer. 1972. Proc. Roy. Soc. (Lond.) B. 180:115.
Ando, K., A. Kato, G. Tamura, and K. Arima. 1969. J.
    Antibiot. 22:23.
Ando, K., S. Suzuki, K. Suzuki, K. Kodama, A. Kato, G.
    Tamura, and K. Arima. 1969. J. Antibiot. 22:18.
Andrewes, A. G., H. J. Phaff, and M. P. Starr. 1976. Phyto-
    chemistry 15:1003.
Andrewes, A. G. and M. P. Starr. 1976. Phytochemistry 15:
    1009.
Angeletti, A. and G. Tappi. 1947. Gazz. Chim. Ital. 77:112.
Angus, W. W. and R. L. Lester. 1972. Archs. Biochem. Bio-
    phys. 151:483.
Applegarth, D. H. and G. Bozoian. 1968. Can. J. Microbiol.
    14:489.
Appleton, G. S., R. J. Kieber, and W. J. Payne. 1955. Appl.
    Microbiol. 3:249.
Aragon, M. G., F. J. Murillo, M. D. de la Guardia, and E.
    Cerda-Olmedo. 1976. Eur. J. Biochem. 63:71.
Arditti, J., R. Ernst, M. H. Fisch, and B. H. Flick.
    1972. J. Chem. Soc. (Lond.), Chem. Comm. 1217.
Arvidson, G. A. E. 1968. Eur. J. Biochem. 5:415.
Avigan, J., D. J. Goodman, and D. Steinberg. 1963. J. Biol.
    Chem. 238:1283.
Avruch, L. A., S. Fischer, H. Pierce, and A. C. Oehlschlager.
    1976. Can. J. Biochem. 54:657.
Babij, T., F. J. Moss, and B. J. Ralph. 1969. Biotech.
    Bioeng. 11:593.
Bailey, R. B., E. D. Thompson, and L. W. Parks. 1974. Bio-
    chim. Biophys. Acta 334:127-136.
Bailey, R. B., L. Miller, and L. W. Parks. 1976. J. Bacte-
    riol. 126:1012.
Baker, K. and G. A. Strobel. 1965. Proc. Montana Acad. Sci.
    25:83.
Baker, N. and F. Lynen. 1971. Eur. J. Biochem. 19:200.
Ballance, P. E. and W. M. Crombie. 1961. Biochem. J. 80:170

Bamberger, M. and A. Landsiedl. 1905. Monatsh. Chem. 26: 1109.

Bangham, A. D. and R. M. C. Dawson. 1959. Biochem. J. 72: 486.

Baptist, J. N., R. K. Gholson, and M. J. Coon. 1963. Biochem. Biophys. Acta 69:40.

Barash, I., M. L. Conway, and D. H. Howard. 1967. J. Bacteriol. 93:656.

Baraud, J., A. Maurice, and C. Napias. 1970. Bull. Soc. Chem. Biol. Paris 52:421.

Baraud, J., C. Cassagne, L. Genevois, and M. Joneau. 1967. C. R. Acad. Sci. Paris 265:83.

Bard, M., R. A. Woods, and J. M. Haslam. 1974. Biochem. Biophys. Res. Commun. 56:324.

Barksdale, A. W. 1969. Science 166:831.

Barran, L. R. and I. de la Roche. 1979. Trans. Br. Mycol. Soc. 73:166.

Barron, E. J. and D. J. Hanahan. 1961. J. Biol. Chem. 231: 493.

Barron, E. J. and L. A. Mooney. 1970. Biochemistry 9:2143

Bartnicki-Garcia, S. 1968. Annu. Rev. Microbiol. 22:87.

Bartnicki-Garcia, S. and E. Reyes. 1964. Arch. Biochem. Biophys. 108:125.

Bartnicki-Garcia, S. and E. Reyes. 1968. Biochem. Biophys. Acta 165:32.

Barton, D. H. R., D. M. Harrison, and D. A. Widdowson. 1968. Chem. Commun. 17.

Barton, D. H. R., D. M. Harrison, and G. P. Moss. 1966. Chem. Commun. 595.

Barton, D. H. R., D. M. Harrison, G. P. Moss, and D. A. Widdowson. 1970. Chem. Soc. (Lond.) C 6:775.

Barton, D. H. R. and J. D. Cox. 1948. J. Chem. Soc. 1354.

Barton, D. H. R., J. E. T. Corrie, and D. A. Widdowson. 1974. J. Chem. Soc. Perkin Trans. I, 1326.

Barton, D. H. R., J. E. T. Corrie, P. J. Marshall, and D. A. Widdowson. 1973. Bio-organic Chem. 2:363.

Barton, D. H. R. and T. Bruun. 1951. J. Chem. Soc. 2728.

Barton, D. H. R., Y. M. Kempe, and D. A. Widdowson. 1972. J. Chem. Soc. Perkin Trans. I, 513.

Bass, A. and J. Hospudka. 1952. Chem. Listy 46:243.

Basu, S., B. Kaufman, and S. Roseman. 1968. J. Biol. Chem. 243:5802.

Bean, G. A., G. W. Patterson, and J. J. Motta. 1972. Comp. Biochem. Physiol. B43:935.

Beare, J. L. and M. Kates. 1967. Can. J. Biochem. 45:101.

Bednarz-Prashad, A. J. and C. E. Mize. 1978. Biochemistry 17:4178.

Benjamin, J. A. and B. W. Agranoff. 1969. J. Neurochem.
16:513.

Bennett, A. S. and F. W. Quackenbush. 1969. Arch. Biochem.
Biophys. 130:567.

Bentley, R., W. V. Lavate, and C. C. Sweeley. 1964. Comp.
Biochem. Physiol. 11:263.

Benveniste, P., M. J. E. Hewlins, and B. Fritig. 1969. Eur.
J. Biochem. 9:526.

Benveniste, P. and R. A. Massey-Westropp. 1967. Tetrahedron
Letters 37:3553.

Bergelson, L. D., V. A. Vaver, N. V. Prokozova, A. N. Ushakov,
and G. A. Popkova. 1966. Biochim. Biophys. Acta 116:511.

Berger, Y., N. Bottelbergs-Dahner, and J. Jabot. 1978.
Phytochemistry (in press).

Bergman, W. 1954. Annu. Rev. Plant Physiol. 4:383.

Bernard, K. and H. Albrecht. 1948. Helv. Chim. Acta 31:977.

Bernhard, L., L. Abisch, and H. Wagner. 1958. Helv. Chim.
Acta 41:850.

Bernhauer, K. and G. Patzelt. 1935. Biochem. Z. 280:388.

Bertoli, E., G. Barbaresi, and A. Castelli. 1971. Bioener-
getics 2:135.

Beytia, E. D. and J. W. Porter. 1976. Annu. Rev. Biochem.
45:113.

Bharucha, K. E. and F. D. Gunstone. 1956. J. Sci. Food
Agric. 7:606.

Bianchi, D. E. 1967. Antonie van Leeuwenhoek, J. Microbiol.
Serol. 33:324.

Bianchi, D. E. and G. Turian. 1967. Nature 214:1344.

de Bievre, C. and C. Jourd'huy. 1974. C. R. Acad. Sci.
Paris 278 Ser. D:53.

Bills, C. E., O. N. Massengale, and P. S. Puckett. 1930.
J. Biol. Chem. 87:259.

Bimpong, C. E., C. J. Hickman. 1975. Can. J. Bot. 53:1310.

Bimpson, T., L. J. Goad, and T. W. Goodwin. 1969. Chem.
Commun. 297.

Birch, A. J., M. Kocor, N. Sheppard, and J. Winter. 1962.
J. Chem. Soc. 1502.

Blank, F., F. E. Shorland, and G. Just. 1962. J. Invest.
Dermatol. 39:91.

Blattmann, P. and J. Retey. 1971. Hoppe-Seyler's Z.
Physiol. Chem. 352:369.

Bloch, K. 1951. Rec. Prog. Hormone Res. 6:111.

Bloch, K. 1965. Science 150:19.

Bloch, K. 1969. Accounts in Chem. Res. 2:129.

Bloch, K. and D. Vance. 1977. Annu. Rev. Biochem. 46:263.

Bloch, K., S. Chaykin, A. H. Phillips, and A. DeWaard. 1959. J. Biol. Chem. 234:2595.

Bloomfield, D. K. and K. Bloch. 1960. J. Biol. Chem. 235: 537.

Bloxham, D. P., D. C. Wilton, and M. Akhtar. 1971. Biochem. J. 125:625.

Bohonos, N. and W. H. Peterson. 1943. J. Biol. Chem. 149: 295.

Borgstrom, B. and R. L. Ory. 1970. Biochim. Biophys. Acta 212:521.

Borrow, A., E. G. Jefferys, R. H. J. Kessell, E. C. Lloyd, P. B. Lloyd, and I. S. Mixson. 1961. Can. J. Microbiol. 7:227.

Botham, P. A. and C. Ratledge. 1978. Biochem. Soc. Trans. 6:383.

Boulton, A. A. 1965. Exp. Cell. Res. 37:343.

Bours, J. and D. A. A. Mossel. 1969. Antonie van Leeuwen-hoek, J. Microbiol. Serol. Suppl. 35:129.

Bowman, R. D. and R. O. Mumma. 1967. Biochim. Biophys. Acta 144:501.

Bracco, V. and H. R. Muller. 1969. Rev. Fr. Corps Gras 16:573.

Bradbeer, C. and P. K. Stumpf. 1960. J. Lipid Res. 1:214.

Brady, R. N., S. J. DiMari, and E. E. Snell. 1969. J. Biol. Chem. 244:491.

Brady, R. O. 1958. Proc. Nat. Acad. Sci. (USA) 44:993.

Brady, R. O. and G. J. Koval. 1958. J. Biol. Chem. 223:26.

Brady, R. O., J. V. Formica, and G. J. Koval. 1958. J. Biol. Chem. 233:1072.

Bramley, P. M., A. Than, and B. H. Davies. 1977. Phyto-chemistry 16:235.

Brandt, R. D., G. Ourisson, and R. Pryce. 1969. Biochem. Biophys. Res. Commun. 37:399.

Braun, P. E. and E. E. Snell. 1967. Proc. Natl. Acad. Sci. (USA) 58:298.

Braun, P. E. and E. E. Snell. 1968. J. Biol. Chem. 243: 3775.

Braun, P. E., R. N. Brady, and E. E. Snell. 1968. Federa-tion Proc. 27:458.

Breivak, O. N., J. L. Owades, and R. F. Light. 1954. J. Org. Chem. 19:1734.

Brennan, P. J., M. P. Flynn, and P. F. S. Griffin. 1970. FEBS Letters 8:322.

Brennan, P. J., P. F. S. Griffin, D. M. Losel, and D. Tyrrell. 1974. *In* Progress in the chemistry and biochem-istry of fats and other lipids (R. T. Holman, ed.). Pergamon Press, Ltd. Vol. 14, Part 2, p. 51.

Bretscher, M. S.   1973.   Science 181:622.

Brett, D., D. Houling, L. J. Morris, and A. T. James.   1971.
    Arch. Biochem. Biophys. 143:535.

Brockerhoff, H.   1965.   J. Lipid Res. 6:10.

Brockerhoff, H.   1967.   J. Lipid Res. 8:167.

Brown, A. C., B. A. Knights, and E. Conway.   1969.   Phyto-
    chemistry 8:543.

Brown, C. M. and A. H. Rose.   1969a.   J. Bacteriol. 97:261.

Brown, C. M. and A. H. Rose.   1969b.   J. Bacteriol. 99:371.

Brown, C. M. and B. Johnson.   1970.   J. Gen. Microbiol. 64:
    279.

Brown, L. C. and J. J. Jacobs.   1975.   Aust. J. Chem. 28:2317.

Buchenauer, H.   1976.   Pflanzenschutz-Nachrichten 3:281.

Buchenauer, H.   1978.   Pestic. Sci. 9:507.

Budd, K., A. S. Sussman, and F. I. Eilers.   1966.   J. Bacte-
    riol. 91:551.

Budzikiewicz, H. C., H. Djerassi, and D. H. Williams.   1964.
    Structural elucidation of natural products by mass spec-
    tometry.   Holden-Doz, London, Vol. 2, p. 145.

Bull, A. T.   1970.   J. Gen. Microbiol. 63:75.

Bu'lock, J. D.   1966.   *In* Biosynthesis of antibiotics (J. F.
    Sneel, ed.).   Academic Press, New York, p. 141.

Bu'lock, J. D. and G. N. Smith.   1967.   J. Chem. Soc. 332.

Burgos, J., F. W. Hemming, and J. F. Pennock.   1963.   Proc.
    Roy. Soc. Ser. B 158:291.

Burton, R. M.   1974.   *In* Fundamentals of lipid chemistry.
    B.I.-Science Publications Div., Webster Groves, MO.
    p. 373.

Burton, R. M. and F. C. Guerra.   1974.   Fundamentals of lipid
    chemistry.   B.I.-Science Publications Div., Webster Groves,
    MO.   p. 697.

Burton, R. M., M. A. Sodd, and R. O. Brady.   1958.   J. Biol.
    Chem. 233:1053.

Bushnell, J. L.   1972.   Lipids of fungal spores: Identifica-
    tion and metabolism.   Thesis, Brigham Young University.

Cahn, R. S., C. K. Ingold, and V. Prelog.   1966.   Angew
    Chem. 5:385.

Calam, D. H.   1969.   Nature 221:856.

Caltrider, P. G., S. Ramachandran, and D. Gottlieb.   1963.
    Phytopathology 53:86.

Cambie, R. C. and P. W. LeQuesne.   1966.   J. Chem. Soc. (C)
    72.

Campbell, O. A. and J. D. Weete.   1978.   Phytochemistry.
    17:431.

Cannon, H. J. and A. C. Chibnall.   1929.   Biochemistry J.
    23:168.

Canonica, L., A. Fiecchi, M. G. Kienle, A. Scala, G. Galli,
    E. G. Paoletti, and R. Paoletti. 1968. J. Am. Chem.
    Soc. 90:3597.
Canonica, L., A. Fiecchi, M. G. Kienle, A. Scala, G. Galli,
    E. G. Paoletti, and R. Paoletti. 1968. Steroids 11:749.
Canonica, L., A. Fiecchi, M. G. Kienle, A. Scala, G. Galli,
    E. G. Paoletti, and R. Paoletti. 1969. Steroids 12:445.
Cantino, E. C. 1969. Phytopathology 59:1071.
Cantrell, H. F. and W. M. Dowler. 1971. Mycologia 63:31.
Carlile, M. J. and J. Friend. 1956. Nature 178:369.
Carlson, C. A. and K.-H. Kim. 1974. Arch. Biochem. Biophys.
    164:490.
Carmack, C. L., J. D. Weete, and W. D. Kelley. 1976.
    Physiol. Plant Path. 8:43.
Carter, H. E., D. R. Strobach, and N. J. Hawthorne. 1969.
    Biochemistry 8:383.
Carter, H. E. and J. L. Koob. 1969. J. Lipid Res. 10:363.
Carter, H. E., K. Ohno, S. Nojima, C. L. Tipton, and N. Z.
    Stanacev. 1961a. J. Lipid Res. 2:215.
Carter, H. E., R. A. Hendry, and N. Z. Stanacev. 1961b. J.
    Lipid Res. 2:223.
Carter, H. E., P. Johnson, and E. J. Weber. 1965. Annu.
    Rev. Biochem. 34:109.
Caspi, E., J. B. Greig, P. J. Ramm, and K. R. Varma. 1968.
    Tetrahedron Letters 35:3829.
Caspi, E., K. R. Varma, and J. B. Greig. 1969. Chem. Commun.
    45.
Caspi, E. and P. J. Ramm. 1969. Tetrahedron Letters 3:181.
Cassagne, C. and C. R. Larrouqene. 1972. Acad. Sci. Paris,
    Ser. D 275:1711.
Casselton, P. J. 1976. *In* The filamentous fungi (J. E.
    Smith and D. R. Berry, eds.). John Wiley & Sons, New
    York. p. 121.
Castelli, A., G. P. Littaru, and G. Barbaresi. 1969a. Arch.
    Mikrobiol. 66:34.
Castelli, A., G. Barbaresi, E. I. Bertoli, and P. Orlando.
    1969b. Ital. J. Biochem. 18:78.
Castelli, A., G. Barbaresi, and E. I. Bertoli. 1969c. Ital.
    J. Biochem. 18:91.
Catalano, S., A. Marsili, I. Morelli, and M. Pacchiani. 1976.
    Phytochemistry 15:1178.
Cattaway, F. W., M. R. Holmes, and A. J. E. Barlow. 1968.
    J. Gen. Microbiol. 51:367.
Cavillito, C. J. 1944. Science 100:333.
Cejkova, A. and V. Jirku. 1978. Folia Microbiologica 23:372.
Cerniglia, C. E. and J. J. Perry. 1974. J. Bacteriol. 118:
    844.

Challinor, S. W. and N. W. R. Daniels. 1955. Nature 176:
  1267.
Chamberlin, J. W., M. O. Chaney, S. Chen, P. V. Demarco,
  N. D. Jones, and J. L. Occolowitz. 1974. J. Antibiot.
  27:992.
Chang, S. B. and R. S. Matson. 1972. Biochem. Biophys.
  Res. Commun. 46:1529.
Channon, H. J. 1926. Biochem. J. 20:400.
Chassang, A., M. Roger, F. Vezenhet, and P. Galzy. 1972.
  Folia Microbiol. (Prague) 17:241.
Chavant, L., M. Sancholle, and C. Montant. 1978. Biochem.
  Systematics Ecol. 6:261.
Chavant, L., P. Mazliak, and M. Sancholle. 1978. Physiol.
  Veg. 16:607.
Chavant, L., P. Mazliak, and M. Sancholle. 1979. Ann. Pharm.
  Francaises 37:55.
Chen, Y. S. and L. H. Haskins. 1963. Can. J. Chem. 41:1647.
Chesters, C. G. and J. F. Peberdy. 1965. J. Gen. Microbiol.
  41:127.
Chibnall, A. C., S. H. Piper, and E. F. Williams. 1953.
  Biochem. J. 55:711.
Chigrin, V. V. and L. M. Bessmeltzeva. 1974. Fiziol. Rast.
  21:289.
Chisholm, M. J. and C. Y. Hopkins. 1957. Can. J. Chem. 35:
  358.
Chu, I. M., M. A. Wheeler, and C. E. Holmlund. 1972. Bio-
  chim. Biophys. Acta 270:18.
Chung, C. W. and W. J. Nickerson. 1954. J. Biol. Chem.
  208:395.
Ciegler, A., Z. Pazola, and H. H. Hall. 1964. Appl. Micro-
  biol. 12:150.
Clark, T. and D. A. M. Walkins. 1978. Phytochemistry
  (in press).
Clarke, S. M. and M. McKenzie. 1967. Nature 213:504.
Clayton, R. B. 1965. Quart. Rev. 19:168.
Clenshaw, E. and I. Smedley-Maclean. 1929. Biochem. J.
  23:107.
Cobon, G. S. and J. M. Haslam. 1973. Biochem. Biophys. Res.
  Commun. 52:320.
Cobon, G. S., P. D. Crowfoot, and A. W. Linnane. 1974.
  Biochem. J. 144:265.
Cochrane, J. C., V. W. Cochrane, F. G. Simon, and J. Spaeth.
  1963. Phytopathology 53:115.
Cochrane, V. W. 1958. Physiology of fungi. John Wiley,
  New York. p. 45.

Cochrane, V. W., J. C. Cochrane, C. B. Collins, and F. G. Serafin. 1963. Am. J. Botany 50:906.

Colli, W., P. C. Hinkle, and M. G. Pullman. 1969. J. Biol. Chem. 244:7432.

Colotelo, N., J. L. Sumner, and W. S. Voegelin. 1971. Can. J. Microbiol. 17:1189.

Combs, T. J., J. J. Guarneri, and M. A. Pisano. 1968. Mycologia 60:1232.

Cook, R. C. and D. T. Mitchell. 1970. Trans. Br. Mycol. Soc. 54:93.

Cooney, D. G. and R. Emerson. 1964. Thermophillic fungi. W. H. Freeman & Co., San Francisco.

Corey, E. J. and W. E. Russey. 1966a. J. Am. Chem. Soc. 88:4751.

Corey, E. J. and W. E. Russey. 1966b. J. Am. Chem. Soc. 88:4750.

Cornforth, J. W. 1959. J. Lipid Res. 1:3.

Cornforth, J. W. 1973. Chem. Soc. Rev. 2:1.

Cornforth, J. W., K. Clifford, R. Mallaby, and G. T. Phillips. 1972. Proc. Roy. Soc. (Lond.) Ser. B 182:277.

Cornforth, J. W., R. H. Cornforth, A. Peter, M. G. Horning, and G. Popjak. 1959. Tetrahedron Letters 5:311.

Cornforth, J. W., R. H. Cornforth, C. Donninger, and G. Popjak. 1966. Proc. Roy. Soc. (Lond.) Ser. B 163:492.

Cornforth, J. W., R. H. Cornforth, C. Donninger, G. Popjak, G. Ryback, and G. J. Schroepfer. 1963. Biochem. Biophys. Res. Commun. 11:129.

Cornforth, J. W., R. H. Cornforth, C. Donninger, G. Popjak, G. Ryback, and G. J. Schroepfer. 1966. Proc. Roy. Soc. (Lond.) Ser. B 163:436.

Cornforth, J. W., R. H. Cornforth, C. Donninger, G. Popjak, Y. Shimizu, S. Ichii, E. Forchielli, and E. Caspi. 1965. J. Am. Chem. Soc. 87:3224.

Crocken, B. J. and J. F. Nye. 1964. J. Biol. Chem. 239:1727.

Cullimore, D. R. and M. Woodbine. 1961. Nature 190:1022.

Czeczuga, B. 1976. Acta Mycol. 12:256.

Czeczuga, B. 1977. Acta Mycol. 13:158.

Czeczuga, B. 1978. Qual. Plant.--Pl. Fds. Hum. Nutr. XXVIII, 1:37.

Daly, J. M., H. W. Knoche, and M. V. Weise. 1967. Plant Physiol. 42:1633.

Danielsson, H. and K. Bloch. 1957. J. Am. Soc. Chem. 79:500.

Davenport, J. B. and A. R. Johnson. 1971. *In* Biochemistry and methodology of lipids (A. R. Johnson and J. R. Davenport, eds.). Wiley Interscience, New York.

Davidoff, F.   1964.   Biochem. Biophys. Acta 90:414.

Davidoff, F. and E. D. Korn.   1963.   J. Biol. Chem. 238:3210.

Davies, B. H.   1961.   Phytochemistry 1:25.

Davis, J. B.   1967.   Petroleum microbiology.   Elsevier Publishing Co., Amsterdam.

Davis, J. B. and D. M. Updegraff.   1954.   Bacteriol. Rev. 18:215.

Dawson, P. S. S. and B. M. Craig.   1966.   Can. J. Microbiol. 12:775.

Dawson, R. M. C. and H. Hauser.   1967.   Biochim. Biophys. Acta 137:518.

Dean, P. D. G., P. R. Ortiz de Montellano, K. Bloch, and E. J. Corey.   1967.   J. Biol. Chem. 242:3014.

DeBell, R. M. and R. Cecil Jack.   1975.   J. Bacteriol. 124:220.

de Bievre, C.   1974.   Ann. Microbiol. 125A:309.

de Bievre, C. and C. Jound'hry.   1974.   C. R. Acad. Sc. Paris. 278 Ser. D:53.

DeDeken, R. H.   1966.   J. Gen. Microbiol. 44:149.

DeFabo, E. C., R. W. Harding, and W. Schropshire.   1976. Plant Physiol. 57:440.

Deierkauf, F. A. and H. L. Booij.   1968.   Biochim. Biophys. Acta 150:214.

Demel, R. A. and B. DeKruyff.   1976.   Biochim. Biophys. Acta 457:109.

DeWaard, A., A. H. Phillips, and K. Bloch.   1959.   J. Amer. Chem. Soc. 81:2913.

Dewhurst, S. M. and M. Akhtar.   1967.   Biochem. J. 105:1187.

Dickson, L. G. and G. W. Patterson.   1972.   Lipids 7:635.

DiMari, S. J., R. N. Brady, and E. E. Snell.   1971.   Arch. Biochem. Biophys. 143:553.

Divakaran, P. and M. J. Mod.   1968.   Experimentia 24:1102.

Domer, J. E. and J. G. Hamilton.   1971.   Biochim. Biophys. Acta 231:465.

Domer, J. E., J. G. Hamilton, and J. C. Harkin.   1967.   J. Bacteriol. 94:466.

Donninger, C. and G. Popjak.   1966.   Proc. Roy. Soc. (Lond.) Ser. B 163:465.

Dugan, R. E. and J. W. Porter.   1971.   J. Biol. Chem. 246:5361.

Dulaney, E. L., E. O. Staply, and K. Simf.   1954.   Appl. Microbiol. 2:371.

Dunphy, P. J., J. D. Kerr, J. F. Pennock, K. J. Whittle, and J. Feeney.   1967.   Biochim. Biophys. Acta 136:136.

Eastcott, E. V.   1928.   J. Phys. Chem. 32:1094.

Eberhardt, N. L. and H. C. Rilling. 1975. J. Biol. Chem. 250:863.

Eddy, A. A. 1958. Proc. Roy. Soc. (Lond.) Ser. B 149:425.

Eglington, G. and R. J. Hamilton. 1967. Science 156:1322.

Elinor, N. P. and N. A. Zaikina. 1963. Vopr. Med. Khim. 9: 177.

Ellenbogen, B. B., S. Aaronson, S. Goldstein, and M. Belsky. 1969. Comp. Biochem. Physiol. 29:805.

Ellingson, J. S. 1974. Biochim. Biophys. Acta 337:60.

Elliot, C. G. and B. A. Knights. 1974. Biochim. Biophys. Acta 360:78.

Elliot, C. G., B. A. Knights, and J. A. Freeland. 1974. Biochim. Biophys. Acta 360:339.

Ellis, W. J. 1945. Aust. Council Sci. Ind. Res. 18:314.

Ellman, G. L. and H. K. Mitchell. 1954. J. Amer. Oil Chem. Soc. 76:4028.

Elovson, J. 1975. J. Bacteriol. 124:524.

Emerson, R. and D. L. Fox. 1940. Proc. Roy. Soc. (Lond.) Ser. B 128:275.

Enbressangler, B., H. Sari, and P. Desnuelle. 1966. Biochim. Biophys. Acta 125:597.

Endo, M., M. Kajiware, and K. Nakanishi. 1970. Chem. Commun. 309.

Enebo, L. and H. Iwamoto. 1966. Acta Chem. Scand. 20:439.

Enebo, L., L. J. Anderson, and H. Lundin. 1946. Arch. Biochem. Biophys. 11:383.

Enebo, L., M. Elander, F. Berg, H. Lundin, R. Nillson, and K. Myroack. 1944. IVA 6:1.

Epstein, W. W., E. Aoyagi, and P. W. Jennings. 1966. Comp. Biochem. Physiol. 18:225.

Epstein, W. W. and H. C. Rilling. 1970. J. Biol. Chem. 245:4597.

Esders, T. W. and R. J. Light. 1972a. J. Lipid Res. 13:663.

Esders, T. W. and R. J. Light. 1972b. J. Biol. Chem. 247: 7494.

Esders, T. W. and R. J. Light. 1972c. J. Biol. Chem. 247: 1375.

Eslava, A. P., M. I. Alvarez, and E. Cerda-Olmedo. 1974. Eur. J. Biochem. 48:617.

Esposito, M. S., R. E. Esposito, M. Arnaud, and H. O. Halvorson. 1969. J. Bacteriol. 100:180.

Evans, H. J., R. B. Clark, and S. A. Russell. 1964. Biochim. Biophys. Acta 92:582.

Fabre-Joneau, M., J. Baraud, and C. Cassagne. 1969. C. R. Acad. Sci. Paris t:268.

Fairbairne, D. 1948. J. Biol. Chem. 173:705.

Fathipour, A., K. K. Schlender, and H. M. Sell. 1967. Biochem. Biophys. Acta 144:476.

Fayret, J., L. LaCoste, J. Alais, A. LaBlache-Combier, A. Maquestiau, Y. Van Harverbeke, R. Flammang, and H. Mispreuve. Phytochemistry 18:431.

Fehler, W. G. and R. J. Light. 1970. Biochemistry 9:418.

Feofilova, E. P. and T. M. Pivovarova. 1976. Microbiology 45:854.

Feofilova, E. P. and T. V. Red'kina. 1975. Microbiology 44:153.

Fiasson, J. L. 1967. C. R. Acad. Sci. 264:2744.

Fiecchi, A., M. G. Kienle, A. Scala, G. Galli, E. G. Paoletti, F. Cattabeni, and R. Paoletti. 1972. Proc. Roy. Soc. (Lond.) Ser. B 180:147.

Fieser, L. F. and M. Fieser. 1959. Steroids. Reinhold, New York.

Fishcer, E. 1890. Berichte der Deutschen Chemischen Gesellschaft 23:2114.

Fisher, D. J., P. J. Holloway, and D. V. Richmond. 1972. J. Gen. Microbiol. 72:71.

Fluhasity, A. L. and J. S. O'Brien. 1969. Phytochemistry 8:2627.

Folkers, K., C. H. Skunk, B. O. Linn, F. M. Robinson, P. E. Wittreich, J. W. Huff, J. L. Gilfillan, and H. R. Skeggs. 1959. *In* Biosynthesis of terpenes and sterols. Little, Brown, and Co., Boston.

Foppen, F. H. and O. Gribanovski-Sassu. 1967. Biochem. J. 106:97.

Foster, J. W. 1949. Chemical activities of fungi. Academic Press, New York. p. 125.

Foster, J. W. 1962. *In* Oxygenases (O. Hayaiski, ed.). Academic Press, New York.

Frantz, I. O. and G. J. Schroepfer. 1967. Annu. Rev. Biochem. 36:691.

Frantz, Jr., I. O., A. G. Davison, E. Dulit, and M. L. Mobberly. 1959. J. Biol. Chem. 234:2290.

Froeschl, N. and J. Zellner. 1928. Monatsh. Chem. 50:201.

Fryberg, M., A. C. Oehlschlager, and A. M. Unrau. 1973. J. Amer. Chem. Soc. 95:5747.

Fryberg, M., L. Avruch, A. C. Oehlschlager, and A. M. Unrau. 1975. Can. J. Biochem. 53:881.

Fulco, A. J. 1967. J. Biol. Chem. 242:3608.

Fulco, A. J. and J. F. Mead. 1961. J. Biol. Chem. 236:2416.

Furch, B., J. Poltz, and H. Rudolph. 1976. Biochem. Physiol. Pflanzen 169:S249.

Gaby, W. L., C. Hadley, and Z. C. Kaminski. 1957. J. Biol. Chem. 227:853.

Gad, A. M. and M. M. Hassan. 1964. J. Chem. U.A.R. 7:31.

Gad, M. A. and S. El-Nockrashy. 1960. J. Chem. U.A.R. 3:57.

Gailey, F. B. and R. L. Lester. 1968. Fed. Proc. Fed. Amer. Soc. Exp. Biol. 27:458.

Garber, E. D., M. L. Baird, and D. J. Chapman. 1975. Bot. Gaz. 136:341.

Gaucedo, C., J. M. Gaucedo, and A. Sols. 1968. 5:166.

Gautschi, F. and K. Bloch. 1957. J. Am. Chem. Soc. 79:1343.

Gautschi, F. and K. Bloch. 1958. J. Biol. Chem. 223:1343.

Gaylor, J. L. 1974. *In* Biochemistry of lipids (H. L. Kornberg and D. C. Phillips, eds.). University Press, Baltimore. pp. 1-37.

Gellerman, J. L. and H. Schlenk. 1979. Biochim. Biophys. Acta 573:23.

Gelpi, E., H. Schneider, J. Mann, and J. Oró. 1970. Phytochemistry 9:603.

Gelpi, E., J. Oró, and E. A. Bennett. 1968. Science 161: 700.

Gerard, E. 1892. Compt. Rend. 114:1541.

Gerard, E. 1895a. Compt. Rend. 121:723.

Gerard, E. 1895b. J. Pharm. Chem. 1:601.

Gerard, E. 1898. Compt. Rend. 126:909.

Gerasimova, N. M., Le Zui Lin, and M. N. Bekhtereva. 1975. Microbiology 44:408.

Getz, G. S. 1970. Adv. Lipid Res. 8:175.

Getz, G. S., S. Jakovcic, J. Heywood, J. Frank, and M. Rabinowitz. 1970. Biochim. Biophys. Acta 218:441.

Gholson, R. K., N. J. Baptist, and M. J. Coon. 1963. Biochemistry 2:1155.

Gibbons, G. F., L. J. Goad, and T. W. Goodwin. 1968. Chem. Commun. 1212.

Gibbons, G. F., L. J. Goad, and T. W. Goodwin. 1968. Chem. Commun. 1458.

Gibson, D. M., E. B. Titchener, and S. J. Wakil. 1958. Biochim. Biophys. Acta 30:376.

Gill, C. O. and C. Ratledge. 1973. J. Gen. Microbiol. 78: 337.

Gill, C. O., M. J. Hall, and C. Ratledge. 1977. Appl. & Environ. Microbiol. 33:231.

Goad, L. J. 1967. *In* Terpenoids in plants (J. B. Pridham, ed.). Academic Press, New York.

Goad, L. J. 1970. Biochem. Soc. Symp. No. 29, Natural substances formed biologically from mevalonic acid, 45. Academic Press, New York.

Goad, L. J., A. S. A. Hammon, A. Denis, and T. W. Goodwin. 1966. Nature 210:1322.

Goad, L. J., G. F. Gibbons, L. Lolger, H. H. Rees, and T. W. Goodwin. 1969. Biochem. J. 96:79.

Goad, L. J. and T. W. Goodwin. 1965. Biochem. J. 96:76P.

Goad, L. J. and T. W. Goodwin. 1969. Eur. J. Biochem. 7: 502.

Goldberg, I. H. 1961. J. Lipid Res. 2:103.

Goldie, A. H. and R. E. Subden. 1973. Biochem. Genet. 10: 275.

Goldman, P. and P. R. Vagelos. 1961. J. Biol. Chem. 236: 2620.

Goni, F. M., J. B. Dominguez, and F. Uruburu. 1978. Chem. Phys. Lipids 22:79.

Goodwin, T. W. 1952. Biochem. J. 50:550.

Goodwin, T. W. 1971. Biochem. J. 123:293.

Goodwin, T. W. 1972. Progress in Industrial Microbiology 11:31.

Goodwin, T. W. 1973. *In* Lipids and biomembranes of eukaryotic microorganisms (J. A. Erwin, ed.). Academic Press, Inc., New York. p. 1.

Goodwin, T. W. 1977. Biochem. Soc. Trans. 570th Meeting 5:1252.

Goodwin, T. W. and W. Lijinsky. 1952. Biochem. J. 50:268.

Gorin, P. A. J., J. F. T. Spencer, and A. P. Tulloch. 1961. Can. J. Chem. 39:846.

Gottlieb, D. and P. G. Caltrider. 1963. Nature 197:916.

Gottlieb, D., R. J. Knaus, and S. G. Wood. 1978. Phytopathology 68:1168.

Gottlieb, D. and S. Ramachandran. 1960. Mycologia 52:599.

Goulston, G. and E. I. Mercer. 1969. Phytochemistry 8: 1945.

Goulston, G., E. I. Mercer, and L. J. Goad (cited by Goodwin, T. W.). 1973. *In* Lipids and biomembranes of eukaryotic microorganisms (J. A. Erwin, ed.). Academic Press, New York, 1972, p. 1.

Goulston, G., L. J. Goad, and T. W. Goodwin. 1967. Biochem. J. 102:15C.

Graff, G. L. A., B. Vanderkelen, C. Guening, and J. Humpers. 1968. Soc. Belge Biol. 1635.

Green, M. L., T. Kaneshiro, and J. H. Law. 1965. Biochim. Biophys. Acta 98:582.

Gregonis, D. E. and H. C. Rilling. 1974. Biochemistry 13: 1538.

Gregory, M. E. and M. Woodbine. 1953. J. Exp. Botany 4:314.

Gribanovski-Sassu, O. and F. H. Foppen. 1967. Phytochemistry 6:907.

Gribanovski-Sassu, O. and F. H. Foppen. 1968. Arch. Mikrobiol. 62:251.

Gribanovski-Sassu, O. and J. Beljak. 1971. Ann. Ist. Super. Sanita 7:95.

Griffin, P. F. S., P. J. Brennan, and D. M. Losel. 1970. Biochem. J. 119:11P.

Grob, E. C., K. Kirschner, and F. Lynen. 1961. Chimia 15: 308.

Guarneri, J. J. 1966. Ph.D. Dissertation. St. Johns University, New York, NY.

Gulpta, S. R., L. Viswanathan, and T. A. Venkitasubramanian. 1970. Ind. J. Biochem. 7:108.

Gulz, P. 1968. Phytochemistry 7:1009.

Gunasekaran, M. and D. J. Weber. 1972. Phytochemistry 11:1.

Gunasekaran, M. and D. J. Weber. 1972. Phytochemistry 11: 3367.

Gunasekaran, M., D. J. Weber, and J. L. Bushnell. 1972. Arch. Mikrobiol. 82:184.

Gunasekaran, M., D. J. Weber, and S. L. Hess. 1972. Lipids 6:430.

Gunasekaran, M., D. J. Weber, and W. M. Hess. 1972. Trans. Br. Mycol. Soc. 59:241.

Gunasekaran, M., J. L. Bushnell, and D. J. Weber. 1972. Res. Commun. Chem. Pathol. Pharmacol. 3:621.

Gunasekaran, M., W. M. Hess, and D. J. Weber. 1972. Can. J. Microbiol. 18:1575.

Gurr, M. I. and A. T. James. 1975. Lipid biochemistry, 2nd ed. Cornell University Press, Ithaca, New York.

Gyllenberg, H. and A. Raitio. 1952. Physiol. Plant. 5:367.

Haehn, H. 1921. Zeitschrift fur technische Biologie 9:217.

Hajra, A. K. and B. W. Agranoff. 1968a. J. Biol. Chem. 243: 1617.

Hajra, A. K. and B. W. Agranoff. 1968b. J. Biol. Chem. 243: 3542.

Hall, M. J. and C. Ratledge. 1977. Appl. & Environ. Microbiol. 33:577.

Hall, M. O. and J. F. Nye. 1961. J. Lipid Res. 2:321.

Halsall, T. G. and I. R. Hills. 1971. Chem. Commun. 448.

Hamilton, J. G. and R. N. Castrejon. 1966. Federation Proc. 25:221.

Hamilton-Miller, J. M. T. 1972. J. Gen. Microbiol. 73:201.

Han, J., G. D. McCarthy, M. Calvin, and M. H. Benn. 1968. J. Chem. Soc. C. 2785.

Han, J., H. W. S. Chan, and M. Calvin. 1969. J. Am. Chem. Soc. 91:5156.

Hanahan, D. J. and M. E. Jayko. 1952. J. Am. Chem. Soc. 74:5070.

Hanahan, D. J. and S. J. Wakil. 1953. J. Am. Chem. Soc. 75:273.

Harrison, J. F. and W. E. Trevelyan. 1963. Nature 200:1189.

Hartman, L., I. M. Morice, and F. B. Shorland. 1962. Biochem. J. 82:76.

Hartman, L., J. C. Hawke, F. B. Shorland, and M. F. de Menna. 1959. Arch. Biochem. Biophys. 81:346.

Hartman, L., J. C. Hawke, I. M. Morice, and F. B. Shorland. 1960. Biochem. J. 75:274.

Hartmann, E. and J. Zellner. 1928. Monatsh. Chem. 50:193.

Hartmann, G. R. and D. S. Frear. 1963. Biochem. Biophys. Res. Commun. 10:366.

Harwood, J. L. 1977. Biochem. Soc. Trans. 5:1259.

Haskell, B. E. and E. E. Snell. 1965. Arch. Biochem. Biophys. 112:494.

Haskins, R. H. 1963. Can. J. Microbiol. 9:451.

Haskins, R. H., A. P. Tulloch, and R. G. Micetich. 1964. Can. J. Microbiol. 10:187.

Hatanaka, H., N. Ariga, J. Nagai, and H. Katsuki. 1974. Biochem. Biophys. Res. Commun. 60:787.

Hawthorne, J. N. 1960. J. Lipid Res. 1:225.

Hayes, P. R., L. W. Parks, H. D. Pierce, and A. C. Oehlschlager. 1977a. Lipids 12:666.

Hays, P. R., W. D. Neal, and L. W. Parks. 1977b. Antimicrobiol. Agents Chemother. 12:185.

Hebert, R. J., L. T. Hart, G. T. Dimopoullos, and A. D. Larson. 1973. Abstracts of the Society of Microbiologists Annual Meeting. p. 19.

Heftmann, E. 1974. Lipids 9:626.

Heftmann, E. and E. Moseltig. 1960. Biochemistry of Steroids. Reinhold Publishing Corp., New York.

Heiduschka, A. and H. Lindner. 1929. Z. Physiol. Chem. 181:15.

Heilbron, I. M., E. D. Kamm, and W. M. Owens. 1926. J. Chem. Soc. 1630.

Heinz, E., A. P. Tulloch, and J. F. T. Spenser. 1969. J. Biol. Chem. 244:882.

Heinz, E., A. P. Tulloch, and J. F. T. Spencer. 1970. Biochim. Biophys. Acta 202:49.

Hemming, F. W. 1974. *In* Biochemistry of lipids. Vol. 4 (T. W. Goodwin, ed.). University Park Press, Baltimore, Maryland. pp. 39–97.

Hendrix, J. W. 1970. Annu. Rev. Phytopath. 8:111.

Hendrix, J. W. and G. Rouser. 1976. Mycologia 68:354.

Hendrix, J. W. and J. L. Apple. 1964. Phytopathology 54: 987.

Henry, S. A. and A. D. Keith. 1971. J. Bacteriol. 106:174.

Henry, S. A. and H. O. Halvorson. 1973. J. Bacteriol. 114: 1158.

Herbin, G. A. and P. A. Robins. 1969. Phytochemistry 8: 1985.

Higgins, M. J. P. and R. G. O. Kekwick. 1969. Biochem. J. 113:36p.

Hirshmann, H. 1960. J. Biol. Chem. 235:2762.

Hitchcock, C. and A. T. James. 1966. Biochim. Biophys. Acta 116:413.

Hitchcock, C. and B. W. Nichols. 1971. Plant lipid biochemistry. Academic Press, New York.

Hoffmann, B. and H. J. Rehm. 1976. Eur. J. Appl. Microbiol. 3:31.

Hoffmann, B. and H. J. Rehm. 1978. Eur. J. Appl. Microbiol. 5:189.

Hoffman-Ostenhoff, O., M. Geyer-Fenzl, and E. Wagner. 1961. *In* The enzymes of lipid metabolism (P. Desnuelle, ed.). Pergamon Press, Oxford. p. 39.

Holloway, P. 1971. Biochemistry 10:1556.

Holloway, P. W. and S. J. Wakil. 1970. J. Biol. Chem. 245: 1862.

Holmberg, J. 1948. Svensk Kim. Tidskr. 60:14.

Holtermuller, K. H., E. Ringelmann, and F. Lynen. 1970. Hoppe-Seylers Z. Physiol. Chem. 351:1411.

Holtz, R. B. and L. C. Schisler. 1971. Lipids 6:176.

Holtz, R. B. and L. C. Schisler. 1972. Lipids 7:251.

Holtz, R. B., P. S. Stewart, S. Patton, and L. C. Schisler. 1972. Plant Physiol. 50:541.

Horlick, L. 1966. J. Lipid Res. 7:116.

Hornby, G. M. and G. S. Boyd. 1970. Biochem. Biophys. Res. Commun. 40:1452.

Hossack, J. A., A. H. Rose, and P. S. S. Dawson. 1979. J. Gen. Microbiol. 113:199.

Hougen, F. W., B. M. Craig, and G. A. Ledingham. 1958. Can. J. Microbiol. 4:521.

Howell, D. McB. and C. L. Fergus. 1964. Can. J. Microbiol. 10:616.

Howling, D., L. J. Morris, M. I. Gurn, and A. T. James. 1972. Biochim. Biophys. Acta 260:10.

Hubscher, G. H. 1970. *In* Lipid metabolism (S. J. Wakil, ed.). Academic Press, Inc., New York.

Hunter, K. and A. H. Rose. 1971. *In* The yeasts (A. H. Rose and J. S. Harrison, eds.). Academic Press, New York.

Hunter, K. and A. H. Rose. 1972. Biochim. Biophys. Acta
    260:639.

Hutchins, R. F. N. and M. M. Martin. 1967. Lipids 3:250.

Hutchison, H. T. and J. E. Cronan, Jr. 1968. Biochim.
    Biophys. Acta 164:606.

Ikan, I., E. D. Bergmann, U. Yinon, and H. Shular. 1969.
    Nature 223:317.

Illina, C. D., E. V. Kiva, and S. Sveridovskeya. 1970.
    Microbiol. 39:453.

Illingworth, R., A. Rose, and A. Beckett. 1973. J. Bacte-
    riol. 113:373.

Infante, R., C. Soler-Argilaga, J. Etienne, and J. Polo-
    novski. 1967. C. R. Soc. Biol. 162:50.

Infante, R., C. Soler-Argilaga, and J. Polonovski. 1968.
    C. R. Soc. Biol. 163:54.

Irvine, G. N., M. Golubchuk, and J. A. Anderson. 1954.
    Can. J. Agric. Sci. 34:234.

Ishidate, K., K. Kawazuchi, K. Tagawa, and B. Hagihara.
    1969. J. Biochem. 65:375.

Ishii, R. 1952. J. Ferment. Technol. 30:350.

Itoh, T. and H. Kaneko. 1974. Yukagaku 23:10.

Jack, C. M. 1965. J. Amer. Oil Chem. Soc. 42:1051.

Jack, C. R. and J. A. Laredo. 1968. Lipids 3:459.

Jack, R. C. M. 1966. J. Bacteriol. 91:2101.

Jackson, L. L. and D. S. Frear. 1967. Can. J. Biochem. 45:
    1309.

Jackson, L. L. and D. S. Frear. 1968. Phytochemistry 7:
    651.

Jackson, L. L. and G. L. Blomquist. 1976. *In* Chemistry and
    biochemistry of natural waxes (P. E. Kolattukudy, ed.).
    Elsevier, Amsterdam. p. 201.

Jackson, L. L., L. Dobbs, A. Hildebrand, and R. A. Yokiel.
    1973. Phytochemistry 12:2233.

Jacobson, B. S., C. G. Kannangara, and P. K. Stumpf. 1973.
    Biochim. Biophys. Res. Commun. 52:1190.

Jamsons, V. K. and W. J. Nickerson. 1970. J. Bacteriol.
    104:922.

Jaureuiberry, G., J. H. Law, J. A. McCloskey and E. Lederer.
    1965. Biochemistry 4:347.

Jaworski, J. and P. K. Stumpf. 1974. Arch. Biochem. Bio-
    phys. 162:158.

Jensen, R. J., J. Sampugna, and J. G. Quinn. 1966. Lipids
    1:294.

Jeong, T. M., T. Itoh, T. Tamura, and T. Matsumota. 1974.
    Lipids 9:921.

Johnson, B. and C. M. Brown. 1972. Ant. van Leeuwenhoek
    38:137.
Johnson, B., C. M. Brown, and D. E. Minnikin. 1973. J. Gen.
    Microbiol. 75:x.
Johnson, B., S. J. Nelson, and C. M. Brown. 1972. Ant. van
    Leeuwenhoek J. Microbiol. Serol. 38:129.
Johnston, J. M. and F. Paltauf. 1970. Biochim. Biophys.
    Acta 218:431.
Jollow, D., G. M. Kellerman, and A. W. Linname. 1968. J.
    Cell Biol. 37:221.
Jones, J. G. 1969. J. Gen. Microbiol. 59:145.
Jones, M. E., S. Black, R. M. Lynen, and F. Lipmann. 1953.
    Biochim. Biophys. Acta 12:141.
Jones, P. D., P. W. Holloway, R. O. Peluffo, and S. J. Wakil.
    1969. J. Biol. Chem. 244:744.
Jurtshuk, P. and G. E. Cardini. 1972. CRC Critical Rev.
    Microbiol. 1:239.
Kaneda, T. 1968. Biochemistry 7:1194.
Kaneda, T. 1969. Phytochemistry 8:2039.
Kaneko, H., M. Hosohara, and M. Tanaka. 1976. Lipids 11:
    837.
Kaneshiro, T. and J. H. Law. 1964. J. Biol. Chem. 239:1705.
Kanetsuna, F., L. M. Carbonell, R. E. Moreno, and J. Rod-
    riquez. 1969. J. Bacteriol. 97:1036.
Kannangara, C. G., B. S. Jacobson, and P. K. Stumpf. 1973.
    Plant Physiol. 52:156.
Kannangara, C. G. and P. K. Stumpf. 1972. Arch. Biochem.
    Biophys. 148:414.
Karlsson, K. A. 1966. Acta Chem. Scand. 20:2884.
Karlsson, K. A. 1970a. Lipids 5:878.
Karlsson, K. A. 1970b. Chem. Phys. Lipids 5:6.
Karlsson, K. A. and G. A. L. Holm. 1965. Acta Chem. Scand.
    19:2423.
Katayama, M. and S. Marumo. 1976. Tetrahedron Lett. 16:1293.
Kates, M. 1972. Techniques of lipidology. North-Holland/
    American Elsevier, London, New York. pp. 341.
Kates, M. and B. E. Volcani. 1966. Biochim. Biophys. Acta
    116:264.
Kates, M., J. R. Madeley, and J. L. Beare. 1965. Biochim.
    Biophys. Acta 106:630.
Kates, M. and M. Paradis. 1973. Can. J. Biochem. 51:184.
Kates, M. and M. K. Wassef. 1970. Annu. Rev. Biochem. 39:
    323.
Kates, M. and R. M. Baxter. 1962. Can. J. Physiol. 40:1213.
Kato, T., S. Tanaka, M. Ueda, and Y. Kawase. 1975. Agric.
    Biol. Chem. 39:169.

Kato, T. and Y. Kawase. 1976. Agric. Biol. Chem. 40:2379.

Katsuki, H. and K. Bloch. 1967. J. Biol. Chem. 242:222.

Kaufman, B., S. Basu, and S. Roseman. 1971. J. Biol. Chem. 246:4266.

Kawamoto, S., C. Nozaki, A. Tanaka, and S. Fukui. 1978. Eur. J. Biochem. 83:609.

Kawamoto, S., M. Ueda, C. Nozaki, M. Yamamura, A. Tanaka, and S. Fukui. 1978. FEBS Letters 96:37.

Keenan, R. W. and B. Haegelin. 1969. Biochem. Biophys. Res. Commun. 37:888.

Keenan, R. W. and L. E. Hokin. 1964. J. Biol. Chem. 239:2123.

Kennedy, E. P. and S. B. Weiss. 1956. J. Biol. Chem. 222:193.

Kessell, R. H. J. 1968. J. Appl. Bacteriol. 31:220.

Kessler, G. and W. J. Nickerson. 1959. J. Biol. Chem. 234:2281.

Kim, S. J., K. J. Kwon-Chung. 1974. Antimicrob. Agents Chemother. 6:102.

Kirk, P. W. and P. Catalfomo. 1970. Phytochemistry 9:595.

Kisic, A. and M. Prostenik. 1960. Croat. Chem. Acta 32:229.

Klein, H. P. 1957. J. Bacteriol. 73:530.

Klein, H. P., C. Volkmann, and J. Weibel. 1967. J. Bacteriol. 94:475.

Klein, H. P. and F. Lippman. 1953. J. Biol. Chem. 203:95.

Kleinzeller, A. 1944. Biochem. J. 38:480.

Klug, M. J. and A. J. Markovetz. 1967. J. Bacteriol. 93:1847.

Knight, J. C., P. D. Klein, and P. A. Szczepanik. 1972. Proc. Roy. Soc. (Lond.) 180:179.

Knights, B. A. 1970. Phytochemistry 9:701.

Knights, B. A. and W. Laurie. 1967. Phytochemistry 6:407.

Knobling, A., D. Schiffmann, H. D. Sickinger, and E. Schwerger. 1975. Eur. J. Biochem. 56:359.

Knoche, H. W. 1968. Lipids 3:163.

Knoche, H. W. 1971. Lipids 6:581.

Knoche, H. W. and T. L. Horner. 1970. Plant Physiol. 46:401.

Kodicek, E. 1959. CIBA Foundation Symposium on the Biosynthesis of Terpenes and Sterols, p. 173. Churchill, London.

Kolattukudy, P. E. 1966. Biochemistry 5:2265.

Kolattukudy, P. E. 1967. Biochemistry 6:963.

Kolattukudy, P. E. 1968. Plant Physiol. 43:1466.

Kolattukudy, P. E. 1968. Science 159:498.

Kolattukudy, P. E. 1969. Plant Physiol. 44:315.

Kolattukudy, P. E.   1970a.   Lipids 5:259.

Kolattukudy, P. E.   1970b.   Annu. Rev. Plant Physiol. 21:163.

Kolattukudy, P. E.   1976.   *In* Chemistry and biochemistry of natural waxes (P. E. Kolattukudy, ed.).   Elsevier, Amsterdam.   p. 459.

Kolattukudy, P. E., R. Croteau, and J. S. Buckner.   1976. *In* Chemistry and biochemistry of natural waxes (P. E. Kolattukudy, ed.).   Elsevier, Amsterdam.   p. 289.

Koman, V., V. Betina, and Z. Barath.   1969.   Arch. Microbiol. 65:172.

Kovac, L., J. Subik, G. Russ, and K. Kollar.   1967.   Biochim. Biophys. Acta 144:94.

Kramer, R., F. Kopp, W. Niedermeyer, and G. F. Fuhrmann. 1978.   Biochim. Biophys. Acta 507:369.

Krivanek, J. O. and R. C. Krivanek.   1958.   J. Exp. Zool. 137:89.

Kuhn, L., H. Castorph, and E. Schweizer.   1972.   Eur. J. Biochem. 24:492.

Kuhn, N. J. and F. Lynen.   1965.   Biochem. J. 94:240.

Kullenberg, B., G. Bergstrom, and S. Stallberg-Stenhagen. 1970.   Acta Chem. Scand. 24:1481.

Kulmacz, R. J. and G. J. Schroepfer.   1978.   Biochem. Biophys. Res. Commun. 82:371.

Kumagai, T. and Y. Oda.   1969.   Plant & Cell Physiol. 10:387.

Kumar, S., G. T. Phillips, and J. W. Porter.   1972.   Intern. J. Biochem. 3:15.

Kuntzman, C. P., R. F. Vesonder, and M. J. Smiley.   1974. Mycologia 66:580.

Kushuaha, S. C., M. Kates, J. K. G. Kramer, and R. E. Subden. 1976.   Lipids 11:778.

Kushwaha, S. C., M. Kates, R. L. Renauld, and R. E. Subden. 1978.   Lipids 13:352.

Kusunose, M., J. Matsumato, K. Ichihara, E. Kusunose, and J. Nozaka.   1967a.   J. Biochem. 61:66.

Kusunose, M, K. Ichihara, E. Kusunose, J. Nozaka, and J. Matsumato.   1967b.   Agr. Biol. Chem. 31:990.

Kusunose, M., K. Ichihara, E. Kusunose, and J. Nozaka.   1968. Physiol. Ecol. 15:45.

Labach, J. P. and D. C. White.   1969.   J. Lipid Res. 10:528.

Laboureur, P. and M. Labrousse.   1964.   C. R. Acad. Sci. (Paris) 259:4394.

Laboureur, P. and M. Labrousse.   1966.   Bull. Soc. Chim. Biol. (Paris) 48:747.

Laboureur, P. and M. Labrousse.   1968.   Bull. Soc. Chim. Biol. (Paris) 50:2179.

Laine, R. A., P. F. S. Griffin, C. C. Sweeley, and P. J. Brennen. 1972. Phytochemistry 11:2267.

Lands, W. E. M., R. A. Pieringer, Sister P. M. Salkely, and Z. Zschocki. 1966. Lipids 1:444.

Lands, W. E. M. and Sister P. M. Slakely. 1966. Lipids 1:295.

Langdon, R. G. and K. Bloch. 1953. J. Biol. Chem. 200:129.

Langenbach, R. J. and H. W. Knoche. 1970. Proc. Nebraska Acad. Sci. Affil. Soc. 80:10.

Langenbach, R. J. and H. W. Knoche. 1971a. Plant Physiol. 48:735.

Langenbach, R. J. and H. W. Knoche. 1971b. Plant Physiol. 48:728.

Lansbergen, J. C., R. L. Renauld, and R. E. Subden. 1976. Can. J. Bot. 54:2445.

Laseter, J. L. and D. J. Weber. 1966. Phytopathology 58:886.

Laseter, J. L. and J. D. Weete. 1971. Science 172:864.

Laseter, J. L., J. Weete, and D. J. Weber. 1968. Phytochemistry 7:1177.

Laseter, J. L. and R. Valle. 1971. Environ. Sci. Technol. 5:631.

Laseter, J. L., W. M. Hess, J. D. Weete, D. L. Stochs, and D. J. Weber. 1968. Can. J. Microbiol. 14:1149.

Law, S. W. J. and D. N. Burton. 1973. Can. J. Biochem. 51:241.

Law, S. W. T. and D. N. Burton. 1976. Can. J. Microbiol. 12:1716.

Lawler, G. 1972. Thesis, Louisiana State University, New Orleans.

Lawrence, R. C. and R. W. Bailey. 1970. Biochim. Biophys. Acta 208:77.

Lebeault, J. M., B. Roche, Z. Duvnjak, and E. Azoulay. 1970. Arch. Mikrobiol. 72:140.

Lebeault, J. M., E. T. Lode, and M. J. Coon. 1971. Biochem. Biophys. Res. Commun. 42:413.

Lederer, E. 1964. Biochem. J. 93:449.

Lederer, E. 1969. Quart. Rev. (Lond.) 23:453.

Leegwater, D. C., C. G. Young, J. F. T. Spencer, and B. M. Craig. 1962. Can. J. Biochem. Physiol. 40:847.

Lemieux, R. V. 1951. Can. J. Chem. 29:415.

Lemieux, R. V., J. A. Thorn, and H. F. Bauer. 1953. Can. J. Chem. 31:1054.

Lemieux, R. V. and R. Charanduck. 1951. Can. J. Chem. 29:759.

Lenfant, M., E. Zissman, and E. Lederer. 1967. Tetrahedron Lett. 12:1049.

LenFant, M., G. Farrugia, and E. Lederer. 1969. C. R. Acad. Sci. Paris. Ser. D 268:1986.

Lennarz, W. J. 1970. Annu. Rev. Biochem. 39:359.

Leroux, P. and M. Gredt. 1978a. C. R. Acad. Sci. Paris. t. 286, series D, p. 427.

Leroux, P. and M. Gredt. 1978b. Ann. Phytopathol. 10:45.

Letters, R. 1966. Biochim. Biophys. Acta 116:489.

Letters, R. 1968. Bull. Soc. Chim. Biol. 50:1385.

Letters, R. and P. K. Snell. 1963. J. Chem. Soc. 82:5127.

Lewin, L. M. 1965. J. Gen. Microbiol. 41:215.

Light, R. J., W. J. Lennary, and K. Bloch. 1962. J. Biol. Chem. 237:1793.

Lilly, V. G., H. L. Barnett, and R. F. Krause. 1960. W. Virginia Agric. Exp. Stn. Bull. 441.

Lin, H. K. and H. W. Knoche. 1974. Phytochemistry 13:1795.

Lin, H. K., R. J. Langenbach, and H. W. Knoche. 1972. Phytochemistry 11:2319.

Lindberg, M., F. Gautshi, and K. Bloch. 1963. J. Biol. Chem. 238:1661.

Lindenmayer, A. and L. Smith. 1965. Biochim. Biophys. Acta 93:445.

Linn, T. C. 1967. J. Biol. Chem. 242:984.

Lippel, K. and J. F. Mead. 1968. Biochim. Biophys. Acta 152:669.

Lloyd, G. I., E. O. Morris, and J. E. Smith. 1971. J. Gen. Microbiol. 63:141.

Lloyd, G. I., J. G. Anderson, J. E. Smith, and E. O. Morris. 1972. Trans. British Mycol. Soc. 59:63.

Loesel, D. M. 1967. Ann. Botany 31:417.

Lombardi, B., P. Pani, F. F. Sehlunk, and C. Shi-Hua. 1969. Lipids 4:67.

Longley, R. P., A. H. Rose, and B. A. Knights. 1968. Biochem. J. 108:401.

Lopez, A. and J. Burgos. 1976. Phytochemistry 15:971.

Lowery, G. E., Jr., J. W. Foster, and P. Jurtshuk. 1968. Arch. Mikrobiol. 60:246.

Lundin, H. 1950. J. Inst. Brewing 56:17.

Lust, G. and F. Lynen. 1968. Eur. J. Biochem. 7:68.

Lynen, F. 1959. J. Cell Comp. Physiol. 54:(Suppl.)33.

Lynen, F. 1961. Federation Proc. 20:941.

Lynen, F. 1967. Biochem. J. 102:381.

Lynen, F. 1967. *In* Organizational biosynthesis (H. J. Vogel, J. O. Lanspen, and V. Bryson, eds.). Academic Press, New York.

Lynen, F. 1969. *In* Methods in enzymology (J. M. Lowenstein, ed.). Academic Press, New York, Vol. 14:17.

Lynen, F., B. W. Agranoff, H. Eggerer, U. Henning, and E. M. Möslein. 1959. Angew. Chem. 71:657.

Lynen, F. and E. Reichert. 1951. Angewandte Chemie 63:47.

Lynen, F., E. Reichert, and L. Rueff. 1951. Annalen der Chemie 574:1.

Lynen, F., H. Eggerer, U. Henning, and I. Kessel. 1958. Angew. Chem. 70:738.

Lynen, F., I. Hopper-Kessel, and H. Eggerer. 1964. Biochem. Z. 340:95.

Maas-Forster, M. 1955. Arch. Mikrobiol. 22:115.

Macko, V., J. A. A. Renwick, and J. F. Rissler. 1978. Science 199:442.

MacLean, I. S. and D. Hoffert. 1928. Biochem. J. 17:720.

Madyastha, P. B. and L. W. Parks. 1969. Biochim. Biophys. Acta 176:858.

Maguigan, W. H. and E. Walker. 1940. Biochem. J. 34:804.

Maheshwasi, R. and A. J. Sussman. 1970. Phytopathology 60:1357.

Main, F. A., Z. Fencil, A. Prokop, A. Mohagheghi, and A. Frazeli. 1974. Folia Microbiol (Prague) 19:191.

Maister, H. G., S. P. Rogoirn, F. H. Stodola, and L. J. Wickerham. 1962. Appl. Microbiol. 10:401.

Manet, R. 1972. Arch. Mikrobiol. 81:68.

Mangnall, D. and G. S. Getz. 1973. *In* Lipids and biomembranes of eukaryotic microorganisms (J. A. Erwin, ed.). Academic Press, p. 145.

Manocha, M. S. and C. D. Campbell. 1978. Can. J. Microbiol. 24:670.

Mantle, P. G., L. J. Morris, and S. W. Hall. 1969. Trans. Brit. Mycol. Soc. 53:441.

Marks, D. B., B. J. Keller, and A. J. Guarino. 1969. Biochim. Biophys. Acta 183:58.

Martin, C. E., K. Hiramitsu, Y. Kitajima, Y. Nozawa, L. Skriver, and G. A. Thompson. 1976. Biochemistry 15:5218.

Martin, R. O. and P. K. Stumpf. 1959. J. Biol. Chem. 234:2548.

Mase, Y., W. J. Rabourn, and F. W. Quackenbush. 1957. Arch. Biochem. Biophys. 68:150.

Masschelein, C. A. 1959. Rev. Ferment Ind. Ailment. 14:58.

Maxwell, J. R., A. G. Douglas, G. Eglinton, and A. McCormick. 1968. Phytochemistry 7:2157.

McCarthy, E. D., J. Han, and M. Calvin. 1970. Anal. Chem. 40:1475.

McCorkindale, N. J., S. A. Hutchinson, B. A. Pursey, W. T. Scott, and R. Wheeler. 1969. Phytochemistry 8:861.

McDowell, L. L. and A. A. DeHertogh. 1968. Can. J. Botany 46:449.

McElroy, F. A. and H. B. Stewart. 1967. Can. J. Biochem. 45:171.

McKenna, E. J. and M. J. Coon. 1970. J. Biol. Chem. 245: 3882.

McKenna, E. J. and R. E. Kallio. 1965. Annu. Rev. Microbiol. 19:183.

McMurrough, I. and A. H. Rose. 1971. J. Bacteriol. 107:753.

McNitt, R. 1974. Cytobiologie 9:290.

Meissner, G. and M. Delbruck. 1968. Plant Physiol. 43:1279.

Mendoza, C. G. and J. R. Villanveva. 1967. Biochim. Biophys. Acta 135:189.

Mercer, E. I. and M. W. Johnson. 1969. Phytochemistry 8: 2329.

Merdinger, E. and E. M. Deirne, Jr. 1965. J. Bacteriol. 89: 1488.

Merdinger, E., P. Kohn, and L. C. McClain. 1968. Can. J. Microbiol. 14:1021.

Meyer, K. H. and E. Schweizer. 1972. Biochem. Biophys. Res. Commun. 46:1674.

Meyer, K. H. and E. Schweizer. 1976. Eur. J. Biochem. 65: 317.

Meyer, F. and K. Bloch. 1963. Biochim. Biophys. Acta 77: 671.

Milazzo, F. H. 1965. Can. J. Botany 43:1347.

Miller, W. L. and J. L. Gaylor. 1970a. J. Biol. Chem. 245: 5369.

Miller, W. L. and J. L. Gaylor. 1970b. J. Biol. Chem. 245: 5375.

Mills, G. L. and E. C. Cantino. 1974. J. Bacteriol. 118:192.

Mills, G. L. and E. C. Cantino. 1978. Exp. Mycol. 2:99.

Mills, G. L., R. B. Myers, and E. C. Cantino. 1974b. Phyto-chemistry 13:2653.

Mishina, M., T. Kamiryo, A. Tanaka, S. Fukui, and S. Numa. 1976a. Eur. J. Biochem. 71:295.

Mishina, M., T. Kamiryo, A. Tanaka, S. Kukui, and S. Numa. 1976b. Eur. J. Biochem. 71:301.

Mitchell, R. and N. Subar. 1966. Can. J. Microbiol. 12:471.

Mitropoulos, K. A., G. F. Gibbons, C. M. Cornell, and R. A. Woods. 1976. Biochem. Biophys. Res. Commun. 71:892.

Mizuno, M., Y. Schimojima, T. Iguchi, I. Takeda, and S. Saburo. 1966. Agr. Biol. Chem. 30:506.

Molitoris, H. P. 1963. Arch. Mikrobiol. 47:104.

Montant, C. and M. Someholle. 1969. C. R. Acad. Sci. Paris 269:886.

Mooney, L. A. and E. J. Barron. 1970. Biochemistry 9:2138.

Moore, J. T. and J. L. Gaylor. 1968. Arch. Biochem. Biophys. 124:167.

Moore, Jr., J. T., and J. L. Gaylor. 1969. J. Biol. Chem. 244:6334.

Moore, J. T. and J. L. Gaylor. 1970. J. Biol. Chem. 245:4684.

Moreau, J. P., P. J. Ramm, and E. Caspi. 1975. Eur. J. Biochem. 56:393.

Morell, P. and P. Braun. 1972. J. Lipid Res. 13:293.

Morikawa, M. and S. Yamashita. 1978. Eur. J. Biochem. 84:61.

Morimoto, H., I. Imada, T. Murata, and N. Matsumoto. 1967. Justus Liebigs Ann. Chem. 708:230.

Morris, D. C., S. Safe, and R. E. Subden. 1974. Biochemistry Genet. 12:459.

Morris, L. J. 1967. Lipids 3:260.

Morris, L. J. 1967. Biochem. Biophys. Res. Commun. 29:311.

Morris, L. J. 1970. Biochem. J. 118:681.

Morris, L. J., R. V. Harris, W. Kelly, and A. T. James. 1967. Biochem. Biophys. Res. Commun. 28:904.

Morris, L. J. and S. W. Hall. 1966. Lipids 1:188.

Morris, L. J., S. W. Hall, and A. T. James. 1966. Biochem. J. 100:29.

Motai, H., E. Ichishima, and F. Yoshida. 1966. Nature 210: 308.

Mudd, J. B. and J. Saltzgaber-Müller. 1978. Arch. Biochem. Biophys. 186:359.

Mudgal, R. K., T. T. Tchen, and K. Bloch. 1958. J. Am. Chem. Soc. 80:2589.

Mulheirn, L. J. and E. Caspi. 1971. J. Biol. Chem. 246: 3948.

Mumma, R. O., C. L. Fergus, and R. D. Sekura. 1970. Lipids 5:100.

Mumma, R. O., R. D. Sekura, and C. L. Fergus. 1971. Lipids 6:589.

Murillo, F. J., I. L. Calderon, I. Lopez-Diaz, and E. Cerda-Olmedo. 1978. Appl. and Environ. Microbiol. 36:639.

Murray, S., M. Woodbine, and T. K. Walker. 1953. J. Exp. Botany 4:251.

Naegeli, C. and O. Coew. 1878. Ann. Liebus 193:322.

Nagai, J. and K. Bloch. 1965. J. Biol. Chem. 240:3702.

Nagai, J. and K. Bloch. 1966. J. Biol. Chem. 241:1925.

Nagai, J. and K. Bloch. 1968. J. Biol. Chem. 243:4626.

Nakajima, S. and S. W. Tanenbaum. 1968. Arch. Biochem. Biophys. 127:150.

Naquib, K. 1959. Can. J. Botany 37:353.

Naquib, K. and K. Saddik. 1960. Can. J. Botany 38:613.

Neal, W. D. and L. W. Parks. 1977. J. Bacteriol. 129:1375.

Nelson, R. R. D., Hui Singh, and R. K. Webster. 1967. Phytopathology 57:1081

Nencki, M. 1878. J. fur praktische Chemie 17:105.

Nes, W. R. 1974. Lipids 9:596.

Nes, W. R. 1977. Adv. in Lipid Res. 15:233.

Neujahr, H. Y. and L. Björk. 1970. Acta Chem. Scand. 24: 2361.

Nielson, H. and N. G. Nilsson. 1950. Arch. Biochem. Biophys. 25:316.

Nishi, A. 1961. J. Bacteriol. 81:10.

Noble, A. C. and C. L. Duitschaever. 1973. Lipids 8:655.

Northcote, D. H. and R. W. Thorne. 1952. Biochem. J. 51: 232.

Nowak, R., W. K. Kim, and R. Rohringer. 1972. Can. J. Bot. 50:185.

Nurminen, T. and H. Suomalainen. 1970. Biochem. J. 188:759.

Nurminen, T. and H. Soumalainen. 1971. Biochem. J. 125:963.

Nyns, E. J., N. Chiang, and A. L. Wiaux. 1968. Ant. van Leeuwenhoek J. Microbiol. Serol. 34:197.

Oda, T. 1952. Pharm. Soc. Japan 72:136.

Oda, T. and H. Kamiya. 1958. Chem. Pharm. Bull. 6:682.

Ogiso, T. and M. Sugiura. 1971. Chem. Pharm. Bull. 19:2457.

Olsen, J. A. 1966. Annu. Rev. Biochem. 35:559.

Olsen, Jr., J. A., M. Lindberg, and K. Bloch. 1957. J. Biol. Chem. 226:941.

Ondrusek, V. and M. Prostenik. 1978. Exp. Mycol. 2:156.

Orcutt, D. M. and G. W. Patterson. 1975. Comp. Biochem. Physiol. 50B:579.

Orme, T. W., J. McIntyre, F. Lynen, L. Kuhn, and E. Schweizer. 1972. Eur. J. Biochem. 24:407.

Oró, J., J. L. Laseter, and D. J. Weber. 1966. Science 154: 399.

Osman, H. G., M. Abdel-Akher, A. M. H. El-Refai, and M. A. Nashat. 1969. Z. Allgem. Mikrobiol. 9:283.

Ostrow, D. 1971. Fed. Proc. Fed. Amer. Soc. Exp. Biol. 30: 1226.

O'Sullivan, J. and D. M. Losel. 1971. Arch. Midrobiol. 80: 277.

Ota, Y. and K. Yamada. 1966a. Agr. Biol. Chem. 30:351.

Ota, Y. and K. Yamada. 1966b. Agr. Biol. Chem. 30:1030.

Ota, Y. and K. Yamada. 1967. Agr. Biol. Chem. 31:809.

Ota, Y., T. Nakamiya, and K. Yamada. 1970. Agr. Biol. Chem. 34:1368.

Ottke, R. C., E. L. Tatum, I. Zabin, and C. Bloch. 1950.
    Federation Proc. 9:212.

Ottke, R. C., E. L. Tatum, I. Zabin, and K. Bloch. 1951.
    J. Biol. Chem. 189:429.

Owens, R. G. 1955a. Contrib. Boyce Thompson Inst. 18:125.

Owens, R. G. 1955b. Contrib. Boyce Thompson Inst. 18:145.

Paliokas, A. M. and G. J. Schroepfer. 1967. Biochem. Bio-
    phys. Res. Commun. 26:736.

Paliokas, A. M. and G. J. Schroepfer. 1968. J. Biol. Chem.
    243:453.

Paltauf, F. and G. Schatz. 1969. Biochemistry 8:335.

Paltauf, F. and J. M. Johnston. 1970. Biochim. Biophys.
    Acta 218:424.

Parks, L. W. 1958. J. Am. Chem. Soc. 80:2042.

Parks, L. W., C. McLean-Bowen, F. R. Taylor, and S. Hough.
    1978. Lipids 13:730.

Parks, L. W., F. T. Bond, E. D. Thompson, and P. R. Starr.
    1972. J. Lipid Res. 13:311.

Parks, L. W. and V. K. Stromberg. 1978. Lipids 13:29.

Patrick, M. A., V. K. Sangar, and P. R. Dugan. 1973.
    Mycologia 65:122.

Patterson, G. W. 1974. Comp. Biochem. Physiol. 47B:453.

Patterson, G. W., P. J. Doyle, and J. T. Chan. 1974. Lipids
    9:567.

Paulus, H. and E. P. Kennedy. 1960. J. Biol. Chem. 235:1303.

Pearson, L. K. and H. S. Raper. 1927. Biochem. J. 21:875.

Peck, R. J. 1947. *In* Biology of pathogenic fungi (W. J.
    Nickerson, ed.). Ronald Press, New York.

Pelz, B. F. and H. J. Rehm. 1973. Arch. Mikrobiol. 92:153.

Penman, C. S. and J. H. Duffus. 1976. Zeit. für allmeine
    Mikrobiol. 16:483.

Peters, I. I. and F. E. Nelson. 1948a. J. Bacteriol. 55:581.

Peters, I. I. and F. E. Nelson. 1948b. J. Bacteriol. 55:593.

Peterson, J. A. 1970. J. Bacteriol. 103:714.

Peterson, J. A., E. J. McKenna, R. W. Estabrooks, and M. J.
    Coon. 1969. Arch. Biochem. Biophys. 131:245.

Pettit, G. R. and J. C. Knight. 1962. J. Org. Chem. 27:2696.

Petzoldt, K., M. Kuhne, E. Blanke, K. Kieslick, and E. Kaspar.
    1967. Justus Liebigs Ann. Chem. 709:203.

Phillip, S. E. and T. K. Walker. 1958. J. Sci. Food Agr.
    9:223.

Pierce, A. M., A. M. Unrau, A. C. Oehlschlager, and R. A.
    Woods. 1979. Can. J. Biochem. 57:201.

Pierce, H. D., A. M. Pierce, R. Srinvasan, A. M. Unrau, and
    A. C. Oehlschlager. 1978. Biochim. Biophys. Acta 529:
    429.

Pillai, C. G. P. and J. D. Weete. 1975. Phytochemistry 14: 2347.

Pirson, W. and F. Lynen. 1971. Hoppe-Seylers Z. Physiol. Chem. 352:797.

Ponsinet, G. and G. Durisson. 1965. Soc. Chim. 5^e serie (memories) 3882.

Pontefract, R. D. and J. J. Miller. 1962. Can. J. Microbiol. 8:573.

Popjak, G., G. Schroepfer, and J. W. Cornforth. 1962. Biochem. Biophys. Res. Commun. 6:438.

Popjak, G. and J. W. Cornforth. 1960. *In* Advances in enzymology, Vol. 22 (F. F. Nord, ed.). Interscience, p. 281.

Popjak, G. and J. W. Cornforth. 1966. Biochem. J. 101:553.

Porter, J. K., C. W. Bacon, J. D. Robbins, and H. C. Higma. 1975. J. Agric. Food Chem. 23:771.

Preuss, L. M., E. C. Eichinger, and W. H. Peterson. 1934. Ztschr. Bakt. II 89:370.

Preuss, L. M., H. J. Gorcia, H. C. Green, and W. H. Peterson. 1932. Biochem. Z. 246:401.

Preuss, L. M., W. H. Peterson, H. Steenbock, and E. B. Fred. 1931. J. Biol. Chem. 90:369.

Price, M. J. and G. K. Worth. 1974. Aust. J. Chem. 27:2505.

Prill, E. A., P. R. Wenck, and W. H. Peterson. 1935. Biochem. J. 29:21.

Prostenik, M. 1974. *In* Fundamentals of lipid chemistry. B.I.-Science Publications Div. Webster Groves, Missouri. pp. 307-346.

Prostenik, M. and C. Cosovic. 1974. Chem. Phys. Lipids 13: 117.

Prostenik, M., I. Burcar, A. Castek, C. Coscovic, J. Golem, Z. Jandric, K. Kijaic, and V. Ondrusek. 1978. Chem. and Phys. Lipids 22:97.

Prostenik, M. and N. Z. Stanacev. 1958. Chem. Ber. 91:961.

Prottey, C., M. M. Seidman, and C. E. Ballou. 1970. Lipids 5:463.

Pryce, R. J. 1971. Phytochemistry 10:1303.

Pugh, E. L. and M. Kates. 1973. Biochim. Biophys. Acta 316:305.

Pugh, E. L. and M. Kates. 1975. Biochim. Biophys. Acta 380:442.

Qureshi, A. A., E. Beytia, and J. W. Porter. 1972. Biochem. Biophys. Res. Commun. 48:1123.

Qureshi, A. A., E. D. Beytia, and J. W. Porter. 1973. J. Biol. Chem. 248:1848.

Qureshi, A. A., E. J. Barnes, E. J. Semmler, and J. W. Porter. 1973. J. Biol. Chem. 248:2755.

Qureshi, N., R. E. Dugan, S. Nimmannit, W. H. Wu, and J. W.
Porter. 1976. Biochemistry 15:4185.
Qureshi, N., R. E. Dugan, W. W. Cleland, and J. W. Porter.
1976. Biochemistry 15:4191.
Raab, K. H., N. J. DeSouza, and W. R. Nes. 1968. Biochim.
Biophys. Acta 152:742.
Ragsdale, N. N. 1975. Biochim. Biophys. Acta 380:81.
Rahimtula, A. D. and J. L. Gaylor. 1972. J. Biol. Chem.
247:9.
Rahman, R., K. B. Sharpless, T. A. Spencer, and R. B. Clay-
ton. 1970. J. Biol. Chem. 245:2667.
Rambo, G. W. and G. A. Bean. 1974. Phytochemistry 13:195.
Ramsay, A. M. and L. J. Douglas. 1979. J. Gen. Microbiol.
110:185.
Rasmussen, R. K. and H. P. Klein. 1967. Biochem. Biophys.
Res. Commun. 28:415.
Ratledge, C. and R. K. Saxton. 1968. Anal. Biochem. 26:
288.
Rattray, J. B. M., A. Schibeci, and D. K. Kidby. 1975.
Bacteriol. Rev. 39:197.
Rees, H. H., L. J. Goad, and T. W. Goodwin. 1968. Biochem.
J. 107:417.
Reichert, R. 1945. Helv. Chim. Acta 28:484.
Reindel, F. 1930. Justus Liebigs Ann. Chem. 480:76.
Reindel, F., A. Weickmann, S. Picard, K. Luber, and P. Turula.
1940. Justus Liebigs Ann. Chem. 544:116.
Reisener, H. J. 1976. The fungi spore. J. W. Wiley, New
York. p. 165.
Reisener, H. J., W. B. McConnel, and G. A. Ledingham. 1961.
Can. J. Microbiol. 7:865.
Reitz, R. C. and J. G. Hamilton. 1968. Comp. Biochem.
Physiol. 25:401.
Renaud, R. L., R. E. Subden, A. M. Pierce, and A. C.
Oehlschlager. 1978. Lipids 13:56.
Renauld, R., S. Safe, and R. Subden. 1976. Phytochemistry
15:977.
Retey, J., E. von Stetten, U. Coy, and F. Lynen. 1970. Eur.
J. Biochem. 15:72.
Reuvers, T., E. Tacoronte, C. G. Mendoza, and N. M. Ledieu.
1969. Can. J. Microbiol. 15:989.
Richards, J. H. and J. B. Hendrickson. 1964. The biosynthe-
sis of steroids, terpenes, and acetogenens. W. A. Benja-
min, Inc., New York. p. 135.
Richards, R. L. and F. W. Quackenbush. 1974. Arch. Biochem.
Biophys. 165:780.
Rilling, H. C. 1966. J. Biol. Chem. 241:3233.

Rilling, H. C. and K. Bloch. 1959. J. Biol. Chem. 234:1424.

Rosenthal, R. 1922. Monatsh. Chem. 43:237.

Ruinen, J. and M. H. Deinema. 1964. Antonie van Leeuwenhoek 60:377.

Ruiz-Herrera, J. 1967. Arch. Biochem. 122:118.

Ruzicka, L. 1953. Experimentia. 9:357.

Ruzicka, L. 1959. Proc. Chem. Soc. 341.

Ruzicka, L., R. Denss, and O. Jeger. 1945. Helv. Chim. Acta 28:759.

Ruzicka, L., R. Denss, and O. Jeger. 1946. Helv. Chim. Acta 29:204.

Safe, S. 1973. Biochim. Biophys. Acta 326:471.

Saito, K. and K. Sato. 1968. Biochim. Biophys. Acta 151: 708.

Saladin, T. A. and E. A. Napier. 1967. J. Lipid Res. 8:342.

Salway, J. G., J. L. Harwood, M. Kai, G. White, and J. N. Hawthorne. 1968. J. Neurochem. 15:221.

Sampugna, J. and R. G. Jensen. 1968. Lipids 3:519.

Sargent, J. R., R. F. Lee, and J. C. Vevenzel. 1976. *In* Chemistry and biochemistry of natural waxes (P. E. Kolattukudy, ed.). Elsevier, Amsterdam. p. 49.

Sastry, P. S. and M. Kates. 1964. Biochemistry 3:1271.

Sastry, P. S. and M. Kates. 1965. Can. J. Biochem. 43:1445.

Satina, G. and A. F. Blakeslee. 1928. Biology 14:308.

Sawicki, E. H. and M. A. Pisano. 1977. Lipids 12:125.

Scallen, T. J. and M. W. Schuster. 1968. Steroids 12:683.

Scarborough, G. A. and J. F. Nye. 1967a. J. Biol. Chem. 242:238.

Scarborough, G. A. and J. F. Nye. 1967b. Biochim. Biophys. Acta 146:111.

Schlosser, E., P. D. Shaw, and D. Gottlieb. 1969. Arch. Mikrobiol. 66:147.

Schlossman, D. M. and R. M. Beel. 1977. Arch. Biochem. Biophys. 182:732.

Schlossman, D. M. and R. M. Beel. 1978. J. Bacteriol. 133: 1368.

Schroepfer, G. J., B. N. Lutsky, J. A. Martin, S. Himtoon, B. Fourcans, W. H. Lee, and J. Vermilion. 1972. Proc. Roy. Soc. (Lond.) B 180:125.

Schroepfer, G. J. and K. Bloch. 1965. J. Biol. Chem. 240:54.

Schubert, K., G. Rose, H. Wachtel, C. Horhold, and N. Ikekawa. 1968. Eur. J. Biochem. 5:246.

Schulte, K. E., G. Rucker, and H. Fachmann. 1968. Tetrahedron Lett. 46:4763.

Schultz, J. and F. Lynen. 1971. Eur. J. Biochem. 21:48.

Schweizer, E., B. Kniep, H. Castorph, and V. Holzner. 1973.
    Eur. J. Biochem. 39:353.

Schweizer, E. and H. Bolling. 1970. Proc. Nat. Acad. Sci.
    (USA) 67:660.

Schweizer, E., I. Lerch, L. Droeplin-Rueff, and F. Lynen.
    1970. Eur. J. Biochem. 15:472.

Schweizer, E., K. Werkmeister, and M. K. Jain. 1978. Mole.
    Cell. Biochem. 21:95.

Schweizer, E., K. Willecke, W. Winnewisser, and F. Lynen.
    1970. Vitamins & Hormones 28:329.

Schweizer, E., L. Kuhn, and H. Castorph. 1971. Hoppe-
    Seyler's Z. Physiol. Chem. 352:377.

Schwenk, E. and G. J. Alexander. 1958. Arch. Biochem. Bio-
    phys. 76:65.

Schwenk, E., G. J. Alexander, C. A. Fish, and T. H. Stoudt.
    1955. Federation Proc. 14:752.

Schwinn, F. J. 1969. Phytopathology 2:376.

Seitz, L. M. and J. V. Paukstelis. 1977. J. Agric. Food
    Chem. 25:838.

Semeriva, M., G. Benzonana, and P. Desnuelle. 1967. Bull.
    Soc. Chim. Biol. 49:71.

Semeriva, M., G. Benzonana, and P. Desnuelle. 1967. Bio-
    chim. Biophys. Acata 144:703.

Semeriva, M., G. Benzonana, and P. Desnuelle. 1969. Bio-
    chim. Biophys. Acta 191:598.

SentheShanmuganathan, S. and W. J. Nickerson. 1962. J. Gen.
    Microbiol. 27:451.

Shafai, T. and L. M. Levin. 1968. Biochim. Biophys. Acta
    152:787.

Shafai, T. and L. M. Levin. 1968. Biochim. Biophys. Acta
    218:431.

Shah, V. K. and S. G. Knights. 1968. Arch. Biochem. Bio-
    phys. 127:229.

Shapiro, B. 1967. Annu. Rev. Biochem. 36:247.

Sharpless, K. B., T. E. Snyder, T. A. Spencer, K. K. Make-
    shwari, G. Guhn, and R. B. Clayton. 1968. J. Am. Chem.
    Soc. 90:6874.

Sharpless, K. B., T. E. Snyder, T. A. Spencer, K. K. Make-
    shwari, J. A. Nelson, and R. B. Clayton. 1969. J. Am.
    Chem. Soc. 91:3394.

Shatkin, A. J. and E. L. Tatum. 1961. Am. J. Botany 48:760.

Shaw, R. 1965. Biochim. Biophys. Acta 98:230.

Shaw, R. 1966. Adv. Lipid Research 4:107.

Shaw, R. 1966. Comp. Biochem. Physiol. 18:325.

Shaw, R. 1967. Nature 213:86.

Shechter, I., F. W. Sweat, and K. Bloch. 1970. Biochim. Biophys. Acta 220:463.

Shechter, I. and K. Bloch. 1971. J. Biol. Chem. 246:7690.

Shepherd, C. J. 1957. J. Gen. Microbiol. 26:775.

Sherald, J. L. and H. D. Sisler. 1975. Pest. Biochem. Physiol. 5:477.

Sherr, S. and C. Byk. 1971. Biochim. Biophys. Acta 239:243.

Shih, C-N. and E. H. Marth. 1974. Biochim. Biophys. Acta 338:286.

Shimakata, T., K. Mihara, and R. Sato. 1972. J. Biochem. 72:1163.

Shimizu, I., J. Nagai, H. Hatanaka, E. Saito, and H. Katsuki. 1971. J. Biochem. (Tokyo) 70:175.

Shoppee, C. W. 1964. *In* Chemistry of the steroids. Butterworths, London.

Shu, P., A. C. Neish, and G. A. Ledingham. 1956. Can. J. Microbiol. 2:559.

Shu, P., K. Tamner, and G. A. Ledingham. 1960. Can. J. Botany 32:16.

Sietsma, J. H., D. E. Eveleigh, and R. H. Haskins. 1969. Biochim. Biophys. Acta 184:306.

Sietsma, J. H. and J. T. M. Woutern. 1971. Arch. Mikrobiol. 79:263.

Singh, F. and R. Singh. 1957. Res. Bull. Punjab Univ. 117: 363.

Singh, J. and T. K. Walker. 1956. Biochem. J. 62:286.

Singh, J. and T. K. Walker. 1956. Res. Bull. Punjab Univ. 92:135.

Singh, J., T. K. Walker, and M. L. Meara. 1955. Biochem. J. 61:85.

Skorepa, J., P. Hrabak, P. Mares, and A. Linnarson. 1968. Biochem. J. 107:318.

Slotboom, A. J., G. H. deHaas, P. P. M. Bonsen, G. J. Burbach-Westerhuis, and L. L. M. van Deenen. 1970. Chem. Phys. Lipids 4:5.

Smedley-MacLean, I. 1936. Engeb. Enzymforsch 5:258.

Smith, A. R. H., L. J. Goad, and T. W. Goodwin. 1968. Chem. Commun. p. 926.

Smith, S. W. and R. L. Lester. 1974. J. Biol. Chem. 249: 3395.

Smits, B. L. and W. J. Paterson. 1942. Science 96:210.

Snell, E. E., S. J. DiMari, and R. N. Brady. 1970. Chem. Phys. Lipids 5:116.

Somkuti, G. A. and F. J. Babel. 1968. Appl. Microbiol. 16: 617.

Somkuti, G. A., F. J. Babel, and A. C. Somkuti. 1969.
Appl. Microbiol. 17:606.

Sonderhoff, R. and H. Thomas. 1937. Annalen der Chemie
530:195.

Souza, N. J. and W. R. Nes. 1968. Science 162:363.

Sprecher, E. and K. H. Kubeczka. 1970. Arch. Mikrobiol.
73:337.

Sprinson, D. B. and A. Coulon. 1954. J. Biol. Chem. 207:
585.

Sproston, T. 1971. Photochem. Photobiol. 14:571.

Sproston, T. and R. B. Setlow. 1968. Mycologia 60:104.

Stanacev, N. Z. and M. Kates. 1963. Can. J. Biochem.
Physiol. 41:1330.

Staples, R. C. and W. K. Wynn. 1965. Botan. Rev. 31:537.

Starkey, R. C. 1946. J. Bacteriol. 51:33.

Starratt, A. N. 1976. Phytochemistry 15:2002.

Steele, S. D. and J. J. Miller. 1974. Can. J. Microbiol.
20:929.

Steinberg, M. P. and Z. J. Ordal. 1954. J. Agr. Food Chem.
2:873.

Steiner, M. R. and R. L. Lester. 1970. Biochemistry 9:63.

Steiner, M. R. and R. L. Lester. 1972. Biochim. Biophys.
Acta 260:222.

Steiner, S. and R. L. Lester. 1972. Biochim. Biophys.
Acta 260:82.

Steiner, S., S. Smith, C. J. Waechter, and R. L. Lester.
1969. Proc. Nat. Acad. Sci. (USA) 64:1042.

Stodola, F. H. and L. J. Wickerham. 1960. J. Biol. Chem.
235:2584.

Stodola, F. H., L. J. Wickerham, C. R. Scholfield, and H. J.
Dutton. 1962. Arch. Biochem. Biophys. 98:176.

Stodola, F. H., M. H. Deinema, and J. F. T. Spencer. 1967.
Bacteriol. Rev. 31:194.

Stodola, F. H., R. F. Vesonder, and L. J. Wickerham. 1965.
Biochemistry 4:1390.

Stoffel, W. 1971. Annu. Rev. Biochem. 40:57.

Stoffel, W. 1974. In Fundamentals of lipid chemistry.
BI-Science Publications Div., Webster Groves, MO. p. 339.

Stoffel, W., D. LeKim, and G. Heyn. 1970. Hoppe-Seylers
Z. Physiol. Chem. 351:875.

Stoffel, W. and G. Assmann. 1970. Z. Physiol. Chem. 351:
1041.

Stoffel, W. G. Assmann, and E. Binczek. 1970. Z. Physiol.
Chem. 351:635.

Stoffel, W., G. Sticht, and D. LeKim. 1968. Hoppe-Seylers
Z. Physiol. Chem. 349:1149.

Stoffel, W., G. Sticht, and D. LeKim. 1968. Hoppe-Seylers
Z. Physiol. Chem. 349:1745.

Stone, K. J. and F. W. Hemming. 1965. Biochem. J. 113:727.

Stone, K. J., P. H. W. Butterworth, and F. W. Hemming. 1967.
Biochem. J. 102:443.

Stowe, B. B. 1958. Science 128:421.

Stowe, B. B. 1960. Plant Physiol. 35:262.

Strigina, L. T., Y. N. Elkin, G. B. Elyakov. 1971. Phytochemistry 10:2361.

Strittmatter, P., L. Spatz, D. Corcoran, M. J. Rogers, B.
Setlow, and R. Redline. 1974. Proc. Nat. Acad. Sci.
(USA) 71:4565.

Stumpf, P. K. 1969. Annu. Rev. Biochem. 38:159.

Suberkopp, K. F. and E. C. Cantino. 1973. Arch. Mikrobiol.
89:205.

Sumdell, A. C. and J. L. Gaylor. 1968. J. Biol. Chem. 243:
5546.

Sumner, J. L. 1970. Can. J. Microbiol. 16:1161.

Sumner, J. L. 1973. New Zealand J. of Bot. 11:435.

Sumner, J. L. and E. D. Morgan. 1969. J. Gen. Microbiol.
59:215.

Sumner, J. L., E. D. Morgan, and H. C. Evans. 1969. Can.
J. Microbiol. 15:515.

Sumner, J. L. and N. Colotelo. 1970. Can. J. Microbiol.
16:1171.

Sumper, M. and C. Riepertinger. 1972. Eur. J. Biochem. 29:
237.

Sumpter, M., D. Dasterhelt, C. Riepertinger, and F. Lynen.
1969. Eur. J. Biochem. 10:377.

Suomalainen, H. 1969. Ant. van Leeuwenhoek, J. Microbiol.
Serol. (Suppl.) 35:83.

Suomalainen, H. and A. J. A. Keranen. 1968. Chem. Phys.
Lipids 2:296.

Suomalainen, H. and A. J. H. Keranen. 1963. Biochim. Biophys. Acta 70:493.

Suomalainen, H. and T. Nurmenin. 1970. Chem. Phys. Lipids
4:247.

Sussman, A. S. 1954. Mycologia 46:143.

Sussman, A. S. and B. T. Lingappa. 1959. Science 130:1343.

Svennerholm, L. 1964. J. Lipid Res. 5:145.

Sweeley, C. C. 1959. Biochim. Biophys. Acta 36:268.

Sweeley, C. C. and E. A. Moscatelli. 1959. J. Lipid Res.
1:40.

Tajima, M., N. Okada, K. Suzuki, and S. Yoshikawa. 1976.
Nippon Shokuhen Kogys Gakkaishi 23:562.

Talamo, B., N. Chang, and K. Bloch. 1973. J. Biol. Chem. 248:2738.

Tamura, Y., Y. Yoshida, R. Sato, and H. Kumaska. 1976. Arch. Biochem. Biophys. 175:284.

Tanahashi, Y. and T. Takahashi. 1966. Bull. Chem. Soc. (Japan) 39:848.

Tanner, W. 1968. Arch. Mikrobiol. 64:158.

Tanret, C. 1889. C. R. Acad. Sci. Paris 108:98.

Tanret, C. 1908. Compt. Rend. Acad. Sci. Paris Ser. D 147:75.

Tauro, P., U. Holzner, H. Castorph, F. Hill, and E. Schweizer. 1974. Molec. Gen. Genet. 129:131.

Tchen, T. T. 1958. J. Biol. Chem. 233:1100.

Terroine, E. F. and R. Bonnet. 1927. Bulletin de la Societé de chime biologique 9:588.

Thompson, E. D., B. A. Knights, and L. W. Parks. 1973. Biochim. Biophys. Acta 304:132.

Thompson, E. D. and L. W. Parks. 1974. J. Bacteriol. 120: 779.

Thompson, E. D., R. B. Bailey, and L. W. Parks. 1974. Biochim. Biophys. Acta 334:116.

Thompson, W. and R. M. C. Dawson. 1964. Biochem. J. 91: 237.

Thorpe, R. F. and C. Ratledge. 1972. J. Gen. Microbiol. 72:151.

Tillman, R. W. and G. A. Bean. 1970. Mycologia 62:428.

Topham, R. W. and J. L. Gaylor. 1967. Biochem. Biophys. Res. Commun. 27:644.

Topham, R. W. and J. L. Gaylor. 1970. J. Biol. Chem. 245: 2319.

Topham, R. W. and J. L. Gaylor. 1972. Biochem. Biophys. Res. Commun. 47:180.

Tornabene, T. C., E. Gelpi, and J. Oró. 1967. J. Bacteriol. 94:333.

Tornabene, T. G. and J. Oró. 1967. J. Bacteriol. 94:349.

Tornabene, T. G. and S. P. Markey. 1971. Lipids 6:190.

Trevelyan, W. E. 1966. J. Inst. Brewing 72:184.

Trevelyan, W. E. 1966. J. Lipid Res. 7:445.

Trione, E. J. and T. M. Ching. 1971. Phytochemistry 10: 227.

Trocha, P. J., S. J. Jasne, and D. B. Sprinson. 1974. Biochem. Biophys. Res. Commun. 59:666.

Truesdell, L. C. and E. C. Cantino. 1970. Arch. Mikrobiol. 70:378.

Tseng, T. C. and D. F. Bateman. 1969. Phytopathology 59: 359.

Tulloch, A. P. 1963. Can. J. Biochem. Physiol. 41:1115.

Tulloch, A. P. and G. A. Ledingham. 1960. Can. J. Microbiol. 6:425.

Tulloch, A. P. and G. H. Ledingham. 1962. Can. J. Microbiol. 8:379.

Tulloch, A. P. and J. F. T. Spencer. 1964. Can. J. Chem.
42:830.

Tulloch, A. P., J. F. T. Spencer, and M. H. Dienema. 1968.
Can. J. Chem. 46:345.

Tulloch, A. P., J. F. T. Spencer, and P. A. J. Gorin. 1962.
Can. J. Chem. 40:1326.

Turner, J. R. and L. W. Parks. 1965. Biochim. Biophys. Acta
98:394.

Turner, W. B. 1971. Fungal metabolites. Academic Press,
New York. p. 66.

Tyoringa, K., T. Nurminen, and H. Suomalainen. 1974. Biochem. J. 141:133.

Tyrrell, D. 1967. Can. J. Microbiol. 13:755.

Tyrrell, D. 1968. Lipids 3:568.

Tyrrell, D. 1971. Can. J. Microbiol. 17:1115.

Uen, W. N. and K. H. Ling. 1969. J. Formosa Med. Assoc.
68:244.

Valadon, L. R. G. 1964. J. Exp. Bot. 15:219.

Valadon, L. R. G. 1966. Advancing Frontiers of Plant
Science 15:183.

Valadon, L. R. G. 1976. Trans. Br. Mycol. Soc. 67:1.

Valadon, L. R. G. and R. S. Mummery. 1966. J. Gen. Microbiol. 45:531.

Valadon, L. R. G. and R. S. Mummery. 1968. Nature, Lond.
217:1066.

Valadon, L. R. G. and R. S. Mummery. 1969. Ann. Bot. 33:
879.

Valadon, L. R. G. and R. S. Mummery. 1975. Trans. Br.
Mycol. Soc. 65:485.

van Deenen, L. L. M. and G. H. de Haas. 1966. Annu. Rev.
Biochem. 35:157.

van den Bosch, H., A. J. Aarsman, A. J. Slotboom, and L. L. M.
van Deenen. 1968. Biochim. Biophys. Acta 164:215.

van den Bosch, H., H. A. Bonte, and L. L. M. van Deenen.
1965. Biochim. Biophys. Acta 98:648.

van den Bosch, H., H. M. van der Elzen, and L. L. M. van
Deenen. 1967. Lipids 2:279.

Van Den Bossche, H., G. Willemsens, W. Cools, W. F. J.
Lauwers, and L. LeJeune. 1978a. Current Chemotherapy
1:228.

Van Den Bossche, H., G. Willemsens, W. Cools, W. F. J. Lauwers, and L. LeJeune. 1978b. Chem.-Biol. Interactions 21:59.

Van der Lindon, A. C. and G. J. E. Thiisse. 1965. Adv. Enzymol. 27:469.

Van Eijk, G. W. 1972. Ant. van Leeuwenhoek 38:163.

Van Etten, J. L. and D. Gottlieb. 1965. J. Bact. 89:409.

Vanghelovici, M. and F. Serban. 1940. Acad. Romana Mem. Sect. Stiint. 22:287.

Vanghelovici, M. and F. Serban. 1941. Acad. Romana Mem. Sect. Stiint. 23:436.

van Tamelin, E. E., J. D. Willet, R. B. Clayton, and K. E. Lord. 1966. J. Am. Chem. Soc. 88:4752.

Vareune, J., J. Polonsky, N. Cagnoli-Bellavista, and P. Ceccherelli. Biochimie 53:261.

Vaver, V. A., N. V. Prokozova, A. N. Ushakov, L. S. Golovkina, and L. D. Bergel'son. 1967. Biokhimya 32:310.

Vaver, V. A., V. A. Shckennikov, and L. D. Bergel'son. 1967. Biokhimya 32:1027.

Vignais, P. M., J. Nachbaur, J. Huet, and P. V. Vignais. 1970. Biochem. J. 116:42p.

Villanuera, V. R. 1971. Phytochemistry 10:427.

Villanueva, V. R., M. Barbier, and E. Ledener. 1967. Bull. Soc. Chim. Biol. 49:389.

Volpe, J. J. and P. R. Vagelos. 1973. Annu. Rev. Biochem. 42:21.

Vrkoc, J., M. Budesinsky, and L. Dolejs. 1976. Phyto-chemistry 15:1782.

Waechter, C. J., M. R. Steiner, and R. L. Lester. 1969. J. Biol. Chem. 244:3419.

Waechter, C. J. and R. L. Lester. 1971. J. Bacteriol. 105:837.

Wagner, A. F. and K. Folkers. 1961. Adv. Enzymol. 23:471.

Wagner, F., W. Zahn, and U. Bühring. 1967. Angew. Chem. 79:314.

Wagner, H. and W. Zofcsik. 1966a. Biochem. Z. 346:333.

Wagner, H. and W. Zofcsik. 1966b. Biochem. Z. 346:343.

Wakil, S. J. 1958. J. Amer. Chem. Soc. 80:6465.

Wakil, S., E. L. Pugh, and F. Sauer. 1964. Proc. Nat. Acad. Sci. (USA) 52:106.

Wakil, S. J. and J. Ganguly. 1959. J. Amer. Chem. Soc. 81:2597.

Ward, G. E., L. B. Lockwood, O. E. May, and H. T. Herrick. 1935. Ind. and Eng. Chem. 27:318.

Wardle, K. S. and L. C. Schisler. 1969. Mycologia 61:305.

Wassef, M. K. 1977. Adv. Lipid Res. 15:159.

Wassef, M. K. and J. W. Hendrix. 1977. Biochim. Biophys. Acta 486:172.

Weber, D. and E. Trione. 1980. Can. J. Botany (in press).

Weete, J. 1972. Phytochemistry 11:1201.

Weete, J. D. 1973. Phytochemistry 12:1843.

Weete, J. D. 1974. Fungal lipid biochemistry. Plenum Publ. Corp., New York. p. 392.

Weete, J. D. 1976. *In* Chemistry and biochemistry of natural waxes (P. E. Kolattukudy, ed.). Elsevier, Amsterdam. p. 349.

Weete, J. D., D. J. Weber, and D. J. Letourneau. 1970. Arch. Mikrobiol. 75:59.

Weete, J. D., D. J. Weber, and J. L. Laseter. 1970. J. Bacteriol. 103:536.

Weete, J. D., G. C. Lawler, and J. L. Laseter. 1973. Arch. Biochem. Biophys. 155:411.

Weete, J. D. and J. L. Laseter. 1974. Lipids 9:575.

Weete, J. D., J. L. Laseter, D. J. Weber, W. M. Hess, and D. L. Stocks. 1969. Phytopathology 59:545.

Weete, J. D., S. Venketeswaran, and J. L. Laseter. 1971. Phytochemistry 10:939.

Weete, J. D., W. Huang, and J. L. Laseter. 1979. Soil, air and water pollution (in press).

Weete, J. D. and W. D. Kelley. 1977. Lipids 12:398.

Weete, J. D., W. D. Kelley, and C. Hollis. 1979. Can. J. Microbiol. 25:1481.

Weiss, B. and R. L. Stiller. 1967. J. Biol. Chem. 242:2903.

Weiss, B. and R. L. Stiller. 1972. Biochemistry 11:4522.

Weiss, B., R. L. Stiller, and R. C. M. Jack. 1973. Lipids 8:25.

Wenck, P. R., W. H. Peterson, and H. C. Green. 1935a. Zentr. Bakteriol. Parasitenk. Abt. II 92:324.

Wenck, P. R., W. H. Peterson, and E. B. Fred. 1935b. Zentr. Bakteriol. II 92:330.

Whereat, A. F., M. W. Orishimo, J. Nelson, and S. J. Phillips. 1969. J. Biol. Chem. 244:6498.

White, A. G. C. and C. H. Werkman. 1948. Arch. Biochem. Biophys. 17:475.

White, G. L. and J. N. Hawthorne. 1970. Biochem. J. 117:203.

White, H. B. and S. S. Powell. 1966. Biochim. Biophys. Acta 116:391.

White, J. D. and S. I. Taylor. 1970. J. Am. Chem. Soc. 92:5811.

White, J. D., W. D. Perkins, and S. I. Taylor. 1973.
   Biorg. Chem. 2:163.

Wickerham, L. J. and F. H. Stodola. 1960. J. Bacteriology
   80:484.

Wieland, F., M. Sturzer, and F. Lynen. 1977. Abstracts of
   the 11th FEBS Meeting, A5-1-756-1/2.

Wieland, H., F. Rath, and H. Hesse. 1941. Justus Liebigs
   Ann. Chem. 548:34.

Wieland, H., F. Rath, and W. Benend. 1941. Justus Liebigs
   Ann. Chem. 548:19.

Wieland, H. and G. Coutelle. 1941. Justus Liebigs Ann.
   Chem. 548:275.

Wieland, H., H. Pasedach, and A. Ballauf. 1937. Ann. Chem.
   529:68.

Wieland, H. and M. Asano. 1929. Justus Liebigs Ann. Chem.
   473:300.

Wieland, H. and W. Benend. 1943. Justus Liebigs Ann. Chem.
   554:1.

Wieland, H. and W. M. Stanley. 1931. Justus Liebigs Ann.
   Chem. 489:31.

Wieland, O. and M. Suyter. 1957. Biochem. 329:320.

Willecke, K., E. Ritter, and F. Lynen. 1969. Eur. J. Biochem. 8:503.

Willecke, K. and F. Lynen. 1969. Prog. Biochem. Pharmacol.
   5:91.

Wilton, D. C., K. A. Munday, S. J. M. Skinner, and M.
   Akhtar. 1968. Biochem. J. 106:803.

Winters, K., P. L. Parker, and C. Van Baalen. 1969. Science
   163:467.

Wirth, J. C., S. R. Annand, and Z. L. Kish. 1964. Can. J.
   Microbiol. 10:811.

Wolstenholme, G. E. W. and M. O'Connor (eds.). 1959. Ciba
   Foundation symposium on the biosynthesis of terpenes and
   sterols, Little, Brown, and Co., Boston.

Wood, H. G. and R. E. Barden. 1977. Annu. Rev. Biochem.
   46:385.

Wood, S. G. and D. Gottlieb. 1978a. Biochem. J. 170:343.

Wood, S. G. and D. Gottlieb. 1978b. Biochem. J. 170:355.

Woodbine, M. 1959. Progress in Industrial Microbiology 1:
   181 (Heywood, London).

Woodbine, M., M. E. Gregory, and T. K. Walker. 1951. J.
   Exp. Bot. 25:206.

Woodward, R. B. and K. Bloch. 1953. J. Am. Chem. Soc. 75:
   2023.

Yamada, K. and H. Machida. 1962. J. Agric. Chem. Soc.,
   Japan 36:858.

Yamada, K., H. Okuyama, Y. Endo, and H. Ikezawa. 1977.
  Arch. Biochem. Biophys. 183:281.
Yamada, M. and P. K. Stumpf. 1964. Biochem. Biophys. Res.
  Commun. 14:165.
Yamashita, S., N. Nakaya, Y. Miki, and S. Numa. 1975. Proc.
  Natl. Acad. Sci. (USA) 72:600.
Yanagita, T. and F. Rogane. 1963. J. Gen. Appl. Microbiol.
  9:179.
Yoshida, F., H. Motai, and E. Ichishima. 1968. Appl.
  Microbiol. 16:845.
Yoshida, Y. and H. Kumaoka. 1969. Biochim. Biophys. Acta
  189:461.
Yoshida, Y. H. Kumaoka, and R. Sato. 1974a. J. Biochem.
  75:1201.
Yoshida, Y., H. Kumaoka, and R. Sato. 1974b. J. Biochem.
  75:1211.
Youngblood, W. W., M. Blumer, R. R. L. Guillard, and F.
  Fione. 1971. Marine Biol. 8:190.
Yuan, C. and K. Bloch. 1961. J. Biol. Chem. 236:1277.
Zabin, I. 1957. J. Am. Chem. Soc. 79:5834.
Zabin, I. and J. F. Mead. 1953. J. Biol. Chem. 205:271.
Zabin, I. and J. F. Mead. 1954. J. Biol. Chem. 211:87.
Zalokar, M. 1957. Arch. Biochem. Biophys. 70:568.
Zellner, J. 1911. Monatsh. Chem. 32:1054.
Zonneveld, B. J. M. 1971. Biochim. Biophys. Acta 249:506.
Zurzycka, A. 1963. Acta Soc. Bot. Pol. 32:715.